A MANUAL OF SAMPLING TECHNIQUES

A Manual of Sampling Techniques

[Ranjan Kumar Som,] D. PHIL.

Chief, Population Programme Centre
United Nations Economic Commission for Africa
Formerly Professor and Head, National Sample Survey Population Division
Indian Statistical Institute

Distributed in the United States by
CRANE, RUSSAK & COMPANY, INC.
347 Madison Avenue
New York, New York 10017

Heinemann Educational Books Ltd
LONDON EDINBURGH MELBOURNE AUCKLAND TORONTO
HONG KONG SINGAPORE KUALALUMPUR
IBADAN NAIROBI JOHANNESBURG NEW DELHI

ISBN 0 435 53865 9

First published 1973

Published by
Heinemann Educational Books Ltd
48 Charles Street, London W1X 8AH

Text set in 11/12 pt. IBM Press Roman, printed by photolithography,
and bound in Great Britain at The Pitman Press, Bath

Foreword

The rôle of sampling methods is now generally recognized in all countries as permitting the study of characteristics and relationships by the selection, by rigorous methods, of part of a 'population' to be representative of the whole. Nowhere is this method more important than in the developing countries and, indeed in other institutions, where the limitation of resources compels the use of the most economical procedures for the collection and analysis of statistics.

Deeply concerned with the ways and means to assist governments to obtain the required data, the Statistical Office of the United Nations, under the guidance of the United Nations Statistical Commission, has been promoting programmes in this field. These include the guidelines by the Sub-Commission on Statistical Sampling, the provision on request of sampling experts to a large number of countries, the preparation of manuals, and the support of national and regional seminars and training courses on sampling.

The United Nations Secretariat is also fortunate in having on its staff experts such as Dr R. K. Som, who in addition to their normal responsibilities have undertaken to write scholarly books.

The book by Dr Som, I believe, meets an important need for a textbook at an intermediate level. To his task of preparing the book, Dr Som brings not only his knowledge of sampling theory but also, what is perhaps more important for the type of readers in view, his rich experience of large-scale nation-wide sampling in practice: in India, he was the technical officer in charge of one of the largest scale demographic sample surveys ever undertaken; at the secretariat of the United Nations Economic Commission for Africa, as Chief of its Population Programme Centre, he is also responsible for the regional component, including co-ordination, of an African Population Census Programme that has a strong sampling content and has been evolved to assist the African countries to meet the requirements of demographic data collection and analysis.

I am glad to note that this book is neither too elementary nor too academic but deals with real-life problems and provides guidelines for the design and analysis of sample surveys. The welcome features of the book are a minimum of theory, emphasis on practical uses, introduction to methods such as inverse sampling and the use of Bayes's method as well as the provision of a case-study of a large-scale national sample survey.

I am sure that this book will meet the needs of both the graduate students and others studying sampling as well as those of professional workers from other fields who need in one volume practical guidelines for designing, conducting, and analysing sample surveys.

1973

P. J. Loftus
Director, Statistical Office
United Nations

To
KANIKA, SUJATO,
and BISHAKH

Preface

Using only simple mathematics and statistics, the book aims to present the principles of scientific sampling and to show in a practical manner how to design a sample survey and analyse the resulting data. It seeks a *via media* between on the one hand, the monographs on sampling and special chapters in books in related fields that set forth the basic ideas of sampling, and on the other hand, the advanced sampling texts with detailed proofs and advanced theory. It is more in the nature of a manual or a handbook which attempts at systematizing ideas and practices and could also be used as a textbook for a one- or two-term course. For an understanding of the book, the basic requirement is a knowledge of elementary statistics (including the use of multiple subscripts and summation and product notations), and algebra, but not of calculus.

The approach is essentially practical and the book is primarily aimed at the non-specialist interested in sampling, namely, graduate students of statistics and students and professional and research workers from other fields such as sociology, demography, agriculture, economics, administration, public health, psychology, public opinion, and market research. For these people, the book presents what is believed to be the basic knowledge required to permit, as a minimum, intelligent consultation of professional sampling statisticians, and at best, the design, implementation, and analysis of sampling inquiries when such consultative services are not available. The latter applies particularly in most developing countries where many a professional or researcher must work in his own subject field without the ready benefit of consultation with sampling experts. It is hoped that survey practitioners would find the book useful.

With the above objectives, the book is entitled 'a manual': however, it is not an elementary book.

The requirements of students have been considered by including proofs of some theorems of sampling and by following pedagogical methods: for the latter purpose, some repetitions, the avoidance of which would have made the treatment too terse, have been allowed. For both types of reader, the book tries to be as complete as possible to permit self-study.

After an introduction to 'with and without replacement' schemes for simple random sampling, attention has been confined mainly to sampling with replacement, which results in a considerable simplicity in the presentation and understanding of principles. The theoretical results have generally been approached from simple principles and derived from illustration, intuitive discussion and some fundamental theorems. Proofs have, however, been supplied only for those theorems which do not require a knowledge of calculus or advanced probability, and even these have been placed in an appendix in order not to interrupt the discussion.

While remaining reasonably rigorous at a low mathematical level, the main emphasis has been not so much on providing proofs of all the theorems, as on imparting an understanding of why these theorems are true, what makes them work and how they are applied. An effort has been made to avoid misleading generalizations, and the theories have been tested by their usefulness and relevance in application. References to advanced theoretical texts have been given for those who would pursue the subject further, and the basis is provided for at least a nodding acquaintance with topics such as inverse sampling. At the same time, an attempt has been made to shield the beginner from an over-exposure to terminologies and concepts: theories have not been identified with authors, except when it simplifies references. Inevitably some results are more or less pulled out of the hat.

After an introductory chapter, the book is divided into five parts dealing successively with: single-stage sampling; stratified single-stage sampling; multi-stage sampling, stratified multi-stage sampling; and miscellaneous topics including errors in data and estimates, and planning, execution and analysis of surveys. The arrangement of the topics within the first four parts is such that all the aspects related to those parts are covered. For example, Part I, 'Single-stage Sampling', covers simple random sampling (with and without replacement), use of ancillary variables with simple random sampling (ratio and regression estimators), systematic sampling, unequal probability sampling (sampling with probability proportional to size), choice of sampling unit (cluster sampling), size of sample (cost and error considerations), etc. For each type of sampling, estimating procedures are shown not only for the total and the mean of a study variable, but also for the ratio of the totals of two study variables and the correlation coefficient between them. It is intended that the first four parts of the book should be read in logical sequence. Although the entire book should be read for an intelligent application of the principles and choosing the sample design, at a later stage the reader who is concerned, for example, only with a stratified single-stage design need refer only to Parts I and II for estimation purposes.

Appendix I gives the list of notations and symbols, Appendix II elements of probability and proofs of some theorems in sampling, Appendix III some mathematical and statistical tables, Appendix IV a hypothetical universe from which samples are drawn, and Appendix V a case study of the Indian National Sample Survey, 1964–5. References to publications cited are also supplied.

Drawing from different fields, numerous examples have been provided, as well as exercises with hints for solution. Working out non-trivial exercises should be as important as following the textual material. The specially constructed hypothetical universe has been used for selecting samples, which have been followed through, in order to illustrate different sampling designs. Live examples from nation-wide inquiries have also been provided.

The book has evolved from the author's association with the Indian National Sample Survey (from its inception until 1962), and from lectures delivered at national and international statistical and demographic courses in India and a number of African countries.

Acknowledgements

My grateful thanks are due to Dr W. E. Deming for permission to quote an extract from his book *Some Theory of Sampling*; to John Wiley and Dr W. E. Deming for permission to quote a table from his book *Sample Design in Business Research*; to John Wiley and Sons and Professor W. G. Cochran for permission to quote some examples and exercises from his book *Sampling Techniques*; to the Econometric Society and Mrs Rani Manalanobis for permission to reproduce a figure from an article by P. C. Mahalanobis (1960) that appeared in *Econometrica*; to the Central Statistical Office, Imperial Government of Ethiopia for permission to use the data of Example 15.1; to the Food and Agriculture Organization of the United Nations for permission to reproduce (in an adapted form) a figure from *Quality of Statistical Data* by S. S. Zarkovich; to the Department of Statistics, Government of India for the permission to quote two tables from *Preliminary Estimates of Birth and Death Rates and of the Rate of Growth of Population* by Som *et al.*; to Charles Griffin for the permission to quote an extract and some examples from *Sampling Methods for Censuses and Surveys* by F. Yates; to Dr J. Hájek, Dr V. Dupač and Academia, Publishing House of the Czechoslovak Academy of Sciences for the permission to quote an exercise from their book *Probability in Science and Engineering*; to Dr C. R. Rao for permission to adapt an exercise from his *Advanced Statistical Methods in Biometric Research*; to Dr Walter A. Hendricks for permission to quote an example and a table from his book *The Mathematical Theory of Sampling*; to the Iowa State University Press for permission to quote an example from *Statistical Methods* by G. W. Snedecor and W. G. Cochran; to Professor Leslie Kish for permission to quote extracts from an article by him; to the Literary Executor of the late Professor R. A. Fisher, Dr F. Yates, and Oliver and Boyd for permission to reprint two tables from their *Statistical Tables for Biological, Agricultural and Medical Research*; to Dr M. R. Sampford for permission to reproduce an example from his book *An Introduction to Sampling Theory*; to Col M. J. Slonim for permission to quote an extract from his book *Sampling*; to the Statistical Publishing Society for permission to reproduce some exercises from *Sampling Theory and Methods* by M. N. Murthy; to Dr P. V. Sukhatme and Dr B. V. Sukhatme for permission to quote an extract and some examples from their book *Sampling Theory of Surveys with Applications*; to the United Nations Statistical Office for permission to reproduce (in adapted forms) some examples from *A Short Manual on Sampling*, Volume I, and extensive extracts from *Preparation of Sample Survey Reports, Provisional Issue*; to the U.S. Bureau of the Census for permission to quote a table from their *Technical Paper No. 2*; to Dr A. Weber for permission to quote two examples from his mimeographed lecture notes *Les Méthodes de Sondage*; to John Wiley and Sons for permission to quote some exercises from *Handbook on Methods of Applied Statistics, Vol. II* by Chakravarti *et al.*

Some chapters of the book were prepared in a draft form in September 1969 and I have benefitted greatly from the comments, suggestions, and information received in particular from: B. C. Brookes (Reader in Information Studies, University of London); Dr W. A. Ericson (Professor and Chairman, Department of Statistics, the University of Michigan); Mr Benjamin Gura (Deputy Chief, International Statistical Programs Center), Mr Leo Solomon (Chief, Research & Analysis Branch), Dr B. J. Tepping (Chief, Center for Research for Measurement Methods), Dr Peter Rober Ohs (Mathematical Statistician), and Mr David Bateman (Sampling Statistician) at the U.S. Bureau of the Census; Mr Jack Harewood (Professor of Statistics, University of Trinidad and Tobago);

Dr D. N. Lal (Director, United Nations East African Statistical Training Centre); Mr C. M. H. Morojele (Regional Statistician for Africa, Food and Agriculture Organization of the United Nations); Professor C. A. Moser (Director, Central Statistical Office, U.K.); Dr M. N. Murthy (UN Lecturer in Statistical Methods, Asian Statistical Institute); Mr M. D. Palekar (Chief, Special Projects Section, Statistical Office, United Nations); Dr Julius S. Prince (Principal Adviser, Africa Bureau), Mr Milton D. Lieberman (Chief, Manpower Division, Office of Population), and Mr John C. Rumford (Demographic Surveys Adviser to Liberia), at the U.S. Agency for International Development; Mr K. Purakayastha (Senior Deputy Director, State Statistical Bureau, West Bengal, India); Mr Jacques Royer (Director, Projections and Programming Division, United Nations Economic Commission for Europe); Dr C. S. O'd Scott (Statistician, United Nations Educational, Cultural and Scientific Organization); Dr H. Schubnell (Director, Statistical Office, Federal Republic of Germany); Dr S. R. Sen (Executive Director for India, International Bank for Reconstruction and Development); Dr Egon Szabady (President, Demographic Committee of the Hungarian Academy of Sciences); Dr R. Zasep (Professor, Central School of Planning and Statistics, Warsaw); and my thanks are due to all of them. Mr A. S. Roy (Statistician, Sampling Division, National Sample Survey Department, Indian Statistical Institute) read the final draft and offered many constructive and extremely valuable suggestions, and I am most grateful to him. The responsibility for the form and content of the book is of course entirely mine.

I am thankful to Mrs Tanawork Tadesse for her patient and careful typing of the manuscript.

I have great pleasure in expressing my gratitude to Mr Alan Hill, Chairman and Managing Director, Mr Hamish MacGibbon, Scientific Director, and Miss Marian Miller, Science Editor, Heinemann Educational Books Ltd., who have throughout been most helpful.

Finally, my thanks are due to Dr Robert K. A. Gardiner, Executive Secretary of the United Nations Economic Commission for Africa, and for his interest in this study. The book was, however, prepared outside official duties and does not necessarily reflect the views of the United Nations secretariat.

A textbook, it has been said, should be dedicated to one's children. With my sons Sujato and Bishakh, my wife Kanika also suffered from my neglect when I was preparing this book outside office hours.

1973 R. K. S.

Contents

PART III: MULTI-STAGE SAMPLING

PART IV: STRATIFIED MULTI-STAGE SAMPLING

Guide to Readers

Topics which may be considered too advanced or too detailed may be deferred at the first reading. These have been indicated by distinguishing marks • at the beginning and the end.

Other topics which may be of interest to some readers only, such as agricultural surveys and opinion and marketing research, may be skipped over by other readers.

In classroom teaching, the choice of topics will of course depend on the views of the teacher and the discipline under which sampling is taught; topics such as the extension to two-stage sampling to three-stage sampling may be indicated briefly and the miscellaneous topics (Part V) either omitted or condensed in an introductory course.

1 Basic Concepts of Sampling

1.1 Introduction

Sampling is the process by which inference is made to the whole by examining only a part. It is woven into the fabric of our personal and public lives. In some cultures, a couple enters into a marriage partnership on the basis of a short courtship; with a single grain of rice, an Asian village housewife tests if all the rice in the pot has boiled; from a cup of tea, a tea-taster determines the quality of the brand of tea; medicine dosages are set on the basis of sampling investigations; and a sample of moon rocks provides scientists with information on the origin of the moon.

As a part of the information-collection and decision-making process, sample surveys are conducted on different aspects of life, culture, and science. The purpose of sampling is to provide various types of statistical information of a quantitative or qualitative nature about the whole by examining a few selected units: sampling method is the scientific procedure of selecting those sampling units which would provide the required estimates with associated margins of uncertainty, arising from examining only a part and not the whole.

An indication of the scope of sampling may be obtained from the following partial listing from the United Nations report, *Sample Surveys of Current Interest* (Eleventh Report), prepared in 1967: national sample surveys in Ethiopia and India; demographic sample surveys in Australia, Hungary, and Saudi Arabia; housing surveys in the Federal Republic of Germany, Japan, and Taiwan; labour force sample surveys in Barbados, Iran, and Panama; a survey of the handicapped in Denmark; family budget surveys in Burma, Ceylon, Denmark, Hong Kong, Japan, Philippines, Tanzania, and Egypt; a survey on poverty in the U.S.A.; a survey of pig population in Belgium; a survey of the age structure of cows in Czechoslovakia; surveys on rice yield in Ceylon, Malaysia, and Taiwan; agricultural surveys in Pakistan and Sierra Leone; a survey on men's, youths' and boys' outerwear in Australia; radio listening surveys in Israel and Malaysia; and a survey on use of leisure time in Denmark.

1.2 Complete enumeration or sample

Leaving aside the rôle of sampling in personal lives and the design of experiments under controlled laboratory or field conditions, there arises in every country a demand for statistical information in two ways: first, when the data obtained from routine administrative and other sources have to be analysed speedily, and second, and more frequently, when accurate data are not available from the conventional sources. Consider, for example, the problem of how to conduct an inquiry into family budgets by employing field interviewers. Three alternatives seem possible for collecting accurate data:

1. a complete enumeration of all the units of observation (commonly known as the census), such as all the families in the area of the inquiry, using the best available enumerators,

2. a complete enumeration of all the units by relatively less competent enumerators,
3. a sample of a selected number of units by the best available enumerators.

In reality, especially for large parts of the world for even a limited number of inquiries, and for all the countries of the world for an inquiry into every facet of community life, an adequate number of competent staff are simply not available. In practice, therefore, to enumerate fully all the units and still make use of the best available staff, the enumeration has to be spread over a long period. But the situation may change with time: we might then be measuring different things.

In general, therefore, only two alternatives remain, namely, a complete enumeration of all the units by relatively less efficient enumerators and a sampling of units with more efficient staff. The advantages of sampling (see section 1.3) are such that the latter method is the one usually chosen, although errors can arise as a result of generalizing about the whole from the surveyed part.

There are situations where a complete enumeration would be essential, and if situations do not permit this, efforts should be made to establish it as soon as possible. Examples are the counting of population for census purposes, income tax returns and a voters' list.

Conversely, sampling is the only recourse in destructive tests such as testing the life of electric bulbs, the effectiveness of explosives, and haematological testing.

1.3 Advantages of sampling

Sample surveys have these potential advantages over a complete enumeration – greater economy, shorter time-lag, greater scope, higher quality of work, and actual appraisal of reliability.

A sample requires relatively better resources for designing and executing it adequately and so the cost per unit of observation is higher in a sample than in a complete enumeration. But the total cost of a sample will be much less than that of a complete enumeration covering the same items of inquiry. One airline found, for example, that inter-company settlements for the preparation of tickets due to other air carriers could be estimated quite accurately, within 0.07 per cent, from a 10 per cent sample of the total of about 100 000 documents per month and at a considerable saving of funds (Slonim, Chapter XVII).

With a smaller number of observations, it is also possible to provide results much faster than in a complete enumeration.

Sampling has a greater scope than a complete enumeration regarding the variety of information by virtue of its flexibility and adaptability and the possibility of studying the interrelations of various factors.

Data obtained from a complete enumeration or a sample are subject to different types of errors and biases, the magnitude of which depends on the particular survey procedure: this aspect is considered in greater detail in Chapter 25. However, if the same survey procedure is followed in both a sample and a complete enumeration, the accuracy of a single observation will be the same; but with a comparatively small scale of operations, a sample survey makes it possible to adopt a superior survey procedure by exercising better control over the collection and processing of data, by employing better staff and providing them with intensive training and better equipment, and in interview surveys by employing in-depth interviews. Under suitable conditions – and this is one of the main objectives of sampling – more accurate data can be provided by a sample than by a complete enumeration.

For example, the Current Population Survey, conducted by the U.S. Bureau of the Census, produced, in April 1950, a more accurate count of the labour force in the U.S.A. than did the population census of April 1950. Comparison made through a case-by-case matched study with the sample returns, and individual comparisons made with the special sample interviews taken by the sample enumerators soon after the date of the census enumeration, revealed significant net differences in the labour force participation rates and the number unemployed between the census and the sample. The sample indicated about 2½ million more persons in the labour force than the census and half a million more unemployed. These differences appeared to occur because the marginal classifications were more adequately identified in the sample than in the census, mainly because of the type of enumerators, and their training and supervision; the census enumerators, with necessarily limited training on labour force problems, were apparently less effective in identifying marginal labour force groups (Hansen and Pritzker, 1956). Further studies made at the U.S. Bureau of the Census have shown that because of the response biases of a substantial order in difficult items such as occupation, industry, work status, income, and education, the amount lost by collecting such items for a sample of 25 per cent instead of a complete census was far less than one might assume, even for very small areas. These considerations, plus those of economies and timeliness of results, led to the adoption in the 1960 census of a 25 per cent sample of households as the basic procedure for such information (Hansen and Tepping, 1969).

To be useful, a sample must be scientifically designed and conducted. A classic example of a biased sample is the ill-fitted *Literary Digest* poll of the 1936 U.S. presidential election, where a mammoth sample of ten million individuals was taken. But the sample was obtained from automobile registration lists, telephone directories and similar sources and only 20 per cent of the mail ballots were returned; both these factors resulted in a final sample biased heavily in favour of more literate and affluent individuals, with political affiliations generally different from the rest: the poll predicted the defeat of Roosevelt by 20 per cent of the votes cast but he won by a 20 per cent majority. The failure of the 1948 polls for the U.S. presidential election, in which Truman was elected, is mainly explained by changes in the preference of the people polled between the time they were interviewed and the time they had voted, but could be partly attributed to the quota system of sampling that was deficient in organized labour. In the 1970 British General Election also, pollsters performed very poorly when they predicted the return of the Labour Party.

1.4 Limitations of sampling

When basic information is required for every unit, obviously a complete enumeration has to be undertaken. Errors due to sampling also tend to be high for small administrative areas (but on this point see the preceding section) or for some cross-tabulations where the number of sample observations falling in a certain cell may be very small.

1.5 Relationship between a complete enumeration and a sample

Almost invariably a complete enumeration of population, housing, and farms provides the frame (such as a set of enumeration areas) from which the sample can be drawn in the first instance. In addition, the information obtained about the units from a complete enumeration, even if out of date, can be used to provide supplementary information with a view to improving the efficiency of sample designs.

Sampling is also used as an integral part of a complete enumeration, e.g. of population, housing, and farms, in the following operations: tests of census procedures; enumeration of items in addition to those for which universal coverage is required; evaluation through post-enumerative field checks; quality control of data processing; advance tabulation of selected topics; extending the scope of analysis, such as that of the interrelations of various factors, for which results are required only for large areas and for a country as a whole; and for enumerating difficult items which may require the use of special techniques that are costly or time-consuming. For such purposes sampling is often used in connection with the census of population such as in the U.K. and the U.S.A., as well as in the recent censuses in a number of east African countries, in India, in the U.S.S.R., and elsewhere. Sampling is also often used for updating census results.

Sampling and complete enumeration are thus complementary and, in general, not competitive. For example, in areas where routine systems of data collection and analysis do not exist or are defective and early establishment and operation of such systems are not feasible, the required data and analysis for population programming could be obtained from integrated sample surveys where the interrelations between the demographic, social and economic sectors are covered. For such programming at the local level, rough estimates could often be adequate; but at the stage of implementation, detailed data at these local levels (obtained only through a complete enumeration) would be required and would necessarily be secured.

1.6 Probability *versus* purposive samples

A sample is called a *probability sample* (and also sometimes, a *random sample*) when the method of its selection is based on the theory of probability. In order to provide a valid estimate of an unknown value along with a measure of its reliability, a probability sample must be used. The advantages of sampling referred to in the preceding section accrue only from a probability sample.

Only a probability sample, in addition to providing valid estimates, can provide measures of the reliability of these estimates by indicating the extent of the error due to sampling a part of the whole being surveyed, and can also set limits within which the unknown value that is being estimated from the sample data is expected to lie with a given probability, were all the units to be completely enumerated using the same survey procedure. With the required information, it is also possible to design a sample that would yield data of maximum reliability at a given cost or *vice versa.*

In addition, a probability sample, especially designed and using specific techniques, can provide estimates of errors and biases other than those due to sampling, that affect both a sample and a complete enumeration, such as the response or observational errors, the differential bias of enumerators, errors arising from incomplete samples, processing errors etc.

A sample selected by a non-random method is known as a *non-random sample:* with such a sample it is not possible to measure the degree of reliability of the sample results. Examples are *purposive* samples (selected by some purposive method and therefore subject to biases of personal selection), and *quota samples* (usually of human beings, in which each enumerator is instructed to collect information from the 'quota' of the assigned number of individuals of either sex, and belonging to a certain age group, social class etc., but the individuals to be selected are left to his choice); in quota sampling, errors of frame and non-response are not recognized. There still remains, especially in opinion polls and marketing research, the vestige of purposive and quota sampling, but their use is decreasing and is to be discouraged. This is not to say that non-random samples cannot be used for some tests of procedure, such as the pre-testing of questionnaires, but wherever quantitative measures are required that can also provide a scientific and continuing basis for planning inquiries of a related nature, a probability sample must be used.

In what follows, we consider only probability samples.

1.7 Terms and definitions

An *elementary unit*, or simply a unit, is an element or group of elements on which information is required. Thus persons, families, households, farms, etc. are examples of units. A *recording unit* is an element or group of elements for which information is recorded. In an inquiry, if information is required on sex and age of individuals then the individuals constitute the recording unit; but if information is required on family size, the family is the recording unit. A *unit of analysis* is the unit for which analysis is made. In a family budget inquiry, the units of analysis might be the persons in the family or all the families in a certain expenditure group.

The *universe* (also called *population,* but we shall not use this term in order to avoid a possible confusion with the population of human beings) is the collection of all the units of a specified type defined over a given space and time. The universe may be sub-divided into a number of *sub-universes.* Thus in a country-wide inquiry into family budgets, all the families in the country, defined in a certain manner, would constitute the universe. The sub-universes might consist of the totality of families living in rural and urban areas or of families with incomes below and above a certain level.

The universe may be *finite,* that is, it may consist of a finite number of elements. Almost all sample surveys deal with finite universes, but there are some advantages in considering a sample as having been obtained from an *infinite* universe. This is so when the universe is very large in relation to the sample size, or when the sampling plan is such that a finite universe is turned into an infinite one: the latter procedure is known as *sampling with replacement* and will be explained later.

A characteristic of the units which may take any of the values of a specified set with a specific relative frequency or probability for the different units is called a *random variable* (or a *variate*). If such a characteristic is to be studied on the basis of information obtained through an inquiry, the variable is termed a *study variable.* Of course, in an inquiry there will be more than one study variable. In a family budget survey, the study variables are per family and *per capita* expenditure on food and other items. An *ancillary variable* is a variable which, although not a subject of study of a particular inquiry, is one on which information can be collected which may help in providing, for the study variable, estimators with better approximations to the unknown, true value. Thus in a demographic inquiry, the ancillary information might be the previous census population for the sample areas.

1.8 Sampling frames

For drawing a sample from the universe, a frame of all the units in the universe with proper identification should be available. This frame may consist either of a list of the units or of a map of areas, in case a sample of areas is being taken. The frame should be accurate, free from omissions and duplications, adequate and up to date, and the units should be identified without ambiguity. Supplementary information available for the field covered by the frame may also be of value in improving the sample design.

Two obvious examples of frames are lists of households (and persons) enumerated in a population census, and a map of areas of a country showing boundaries of area units.

The first, however attractive, suffers from two limitations, namely, of possible under- or over-enumeration of households (and population) in censuses and of change in the population due to births, deaths, and migration, unless the sample is conducted simultaneously with the census. There are remedial methods for these difficulties. One, for example, consists of selecting a sample of area units, such as villages and urban blocks, from the frame available from a population census and listing the households residing in these sample areas at the time of the survey-enumeration; these households may then be completely enumerated with regard to the study variable or a sub-sample might be drawn for the inquiry.

1.9 Types of probability sampling

If, from the universe, a sample is taken in such a manner that the sample selected has the same *a priori* chance of selection as any other sample of the same size, it is known as *simple random sampling.* This is the most direct method of sampling but has most often to be modified in order to make better approximations to the unknown universe values, or to obviate the difficulty of lack of accurate and up-to-date frames, or to simplify the selection procedure.

One variant is *systematic sampling.* If we are to draw, say, a five per cent sample from a universe, then a random number is first chosen between 1 and 20; suppose that the selected random number is 12. Then the 12th, 32nd, 52nd etc. units would constitute the systematic sample. This method is often used because of its simplicity.

Another method is to use the available information on an ancillary variable in drawing the units. If no information is available on the units except their identification, there will be no reason to prefer one unit to another for inclusion in the sample, i.e. all the units would have the same chance of selection; and this is what happens with simple random and systematic samples. However, if information is available on an ancillary variable that is known to have a high degree of positive correlation with the study variable, the information on the ancillary variable may be used to draw the units in such a manner that different units get different probabilities of selection with a view to obtaining more reliable results. This is known as *sampling with varying probability* (of which *sampling with probability proportional to size* is the most common case). In a survey of establishments, for example, if, while preparing the list of all the establishments, information is also collected on their sizes (either number of persons employed or the value of production), then a sample of establishments could be drawn such that each establishment has an unequal, but known, *a priori* probability of selection which is proportional to its size.

Before drawing a sample, the units of the universe could be sub-divided into different sub-universes or strata and a sample drawn from each stratum according to one or more of the three procedures. This is known as *stratified sampling.* Thus in a family budget inquiry, the families constituting the universe could be sub-divided into a number of strata depending on residence in different areas, or into a number of strata according to the size of the family if such information is available prior to sampling; different systems of stratification might be combined in a sampling plan. Stratified sampling is used when separate estimates are required for the different strata or to obtain more reliable results.

So far we have considered *single-stage sampling.* The sample might, however, be drawn in a number of stages. Thus in a household inquiry where a list of the existing households is not available, a sample of geographical areas, such as villages and urban-blocks (first-stage units) may first be chosen, then a current list made of the households residing in these sample areas and the required information collected from a sample of households (second-stage units) drawn from this list; and this is an example of a *two-stage sample design.* In practice *multi-stage samples* are frequently used because a current and accurate frame is not often available.

The above types of sampling may be used singly or in combination. For example, there may be unstratified single-stage sampling, stratified single-stage sampling, unstratified multi-stage sampling and, the most common of all, stratified multi-stage sampling. Other types of sampling, such as multi-phase sampling, will be discussed later.

We shall use the term 'sampling plan' to mean the set of rules or specifications for selecting the sample; the term 'sample design' to cover in addition the method of estimation; and 'survey design' to cover other aspects of the survey also, e.g. choice and training of enumerators, tabulation plans, etc.

1.10 Universe parameters

Let the finite universe comprise N units, and let us denote by Y_i the value of a study variable for the ith unit ($i = 1, 2, \ldots, N$). For example, in a family budget inquiry, where the unit is the family, one study variable may be the total family size, when Y_i would denote the size of the ith family; another study variable may be the total family income in a particular reference period (e.g. the thirty days preceding the date of inquiry), when Y_i would denote the value of the income of the ith family; and so on.

A *universe parameter* (or simply a *parameter*) is a function of the frequency values of all the N units. Some important parameters are described below.

In general we are interested in the *universe total* and the *universe mean* of the values of the study variable. The total of the values Y_i is denoted by Y,

$$Y_1 + Y_2 + \ldots + Y_N = \sum_{i=1}^{N} Y_i = Y \tag{1.1}$$

and the universe mean is the universe total divided by the number of universe units and denoted by $\overline{Y}$,

$$\overline{Y} = Y/N \tag{1.2}$$

In addition, interest centres also on the variability of the units in the universe, for the variability of any measure computed from the sample data depends on it. This variability of the values of the study variable in the universe is measured by the mean of the squared deviations of the values from the mean, and is called the *universe variance per unit*; it is denoted by σ_Y^2,

$$\sigma_Y^2 = \frac{1}{N} \sum_{i=1}^{N} (Y_i - \overline{Y})^2 \tag{1.3}$$

The positive square root of σ_Y^2 is termed the *universe standard deviation per unit.*

Note: In the notation for the universe variance per unit, σ_Y^2, the subscript Y is included to denote that the universe variance refers to a particular study variable; in later sections, the subscripts are used in a somewhat different way.

Note that both the universe mean and the universe standard deviation are in the same unit of measurement; for example, in a family budget inquiry, if the study variable is the total family income, then both these measures are expressed in the same currency, be it pounds, dollars, roubles, or rupees.

To obtain a measure of the universe variability independent of the unit of measurement, the universe standard deviation is divided by the universe mean; it is called the *universe coefficient of variation* and denoted by CV,

$$\text{CV}_Y = \sigma_Y / \overline{Y} \tag{1.4}$$

It is often expressed as a percentage. The square of the CV is called the *relative variance.* With the CV it becomes possible to compare the variability of different items, e.g. the variability of family consumption in different countries and areas.

Another universe measure of interest is the ratio of the totals or means of the values of two study variables, e.g. the proportion of income that is spent on food in a family budget inquiry; it is denoted by R,

$$R = \frac{Y}{X} = \frac{\overline{Y}}{\overline{X}} \tag{1.5}$$

where X is the universe total and $\overline{X}$ the universe mean of another study variable, similarly defined as for Y and $\overline{Y}$. Let the corresponding universe standard deviation per unit be denoted by σ_X.

The *universe covariance* between two study variables is obtained on taking the mean of the products of deviations from their respective means, and is denoted by σ_{YX},

$$\sigma_{YX} = \frac{1}{N} \sum_{i=1}^{N} (Y_i - \overline{Y})(X_i - \overline{X}) \tag{1.6}$$

The *universe linear (product-moment) correlation coefficient* between two study variables is obtained on dividing the product of the two respective standard deviations into the covariance and is denoted by ρ_{YX},

$$\rho_{YX} = \frac{\sigma_{YX}}{\sigma_Y \cdot \sigma_X} \tag{1.7}$$

Note that the correlation coefficient is a *pure number.* It varies from -1 (perfect negative correlation) through zero (no linear correlation) to $+1$ (perfect positive correlation).

If N' of the total N units possess a certain attribute or belong to a certain category (such as the number of houses possessing piped water supply), the universe proportion P of the number of such units is

$$P = N'/N \tag{1.8}$$

In addition to the parameters already defined, there are other types of parameters. The values of the universe parameters are generally unknown and the primary objective of a sampling inquiry is to obtain estimates of these parameters along with measures of the reliability of the estimates from the data of a sample.

1.11 Sample estimators

An *estimator* is a rule or method of estimating a universe parameter. Usually expressed as a function of the sample values, it is called a *sample estimator* (or simply, an *estimator*). Note that there can be more than one estimator for the same universe parameter. The particular value yielded by the sample estimator for a given sample is called a *sample estimate* (or simply, an *estimate*).

Let a simple random sample of n units be selected from the universe of N units, the value of a study variable for the ith sample unit being denoted by y_i ($i = 1, 2, \ldots, n$). The ith sample unit may be any of the N universe units. The sample mean $\bar{y}$, defined by

$$\bar{y} = \frac{1}{n}(y_1 + y_2 + \ldots + y_n) = \frac{1}{n}\sum_{i=1}^{n} y_i \tag{1.9}$$

is an estimator of the universe mean $\bar{Y}$.

A sample estimator of the universe total Y, defined in equation (1.1), is

$$y = N\bar{y} \tag{1.10}$$

In general, the estimates obtained from different samples of the same size and taken from the same universe will vary among themselves, and will only by accident coincide with the universe value being estimated, even if the same survey procedure is followed in both the sample and the complete enumeration. This is simply because a part, and not the whole, of the universe is covered in a sample. This variability due to sampling is measured by the sampling variance of the estimator. For example, the sampling variance of the sample mean in sampling with replacement is

$$\sigma_{\bar{y}}^2 = \frac{\sigma_Y^2}{n} \tag{1.11}$$

In general, of course, σ_Y^2 will not be known, and the sampling variance of the sample estimator has to be estimated from the sample data. For a simple random sample of n units, the sample estimator of the universe variance per unit is

$$s_y^2 = \frac{\sum_{i=1}^{n} (y_i - \bar{y})^2}{n - 1} \tag{1.12}$$

If sampling is made with replacement, so that the same universe unit can occur more than once in the sample, or if the universe is considered infinite, then an estimator of the variance of the sample mean $\bar{y}$ in simple random sampling is

$$s_{\bar{y}}^2 = s_y^2/n \tag{1.13}$$

The sample estimators for other universe parameters can also be defined. The sample estimator of the universe covariance of two study variables obtained from a simple random sample of n units is

$$s_{yx} = \frac{\sum_{i=1}^{n} (y_i - \bar{y})(x_i - \bar{x})}{n - 1} \tag{1.14}$$

An estimator of the universe ratio R of the totals two study variables, defined in equation (1.5), is the ratio of the estimators of the respective totals, namely

$$r = \frac{y}{x} = \frac{\bar{y}}{\bar{x}} \tag{1.15}$$

and an estimator of the correlation coefficient ρ_{YX} is the sample correlation coefficient

$$\hat{\rho}_{yx} = \frac{s_{yx}}{s_y s_x} \tag{1.16}$$

1.12 Criteria of estimators

Several criteria have been established for the sample estimators. One is unbiasedness. A sample estimator is said to be *unbiased* if the average values of the sample estimates for all possible samples of the same size is mathematically identical with the value of the universe parameter; this average over all possible samples of the same size is also known as the *mathematical expectation* of, or the (mathematically) *expected survey value.* In the example above, the sample estimators y and $\bar{y}$ are unbiased estimators of the respective parameters Y and $\bar{Y}$; this will be shown in Chapter 2. That a sample estimator is unbiased does not, of course, mean that the estimate obtained from a particular sample will give the exact universe value.

In general, sample estimators which are obtained by taking a ratio of two sample estimators (which themselves might be unbiased) are not unbiased. Thus the sample ratio of the totals of two study variables, defined in equation (1.15), is generally not unbiased.

The second criterion is that of *consistency.* An estimator is said to be *consistent* if it tends to the universe value with increasing sample size; an alternative definition for a finite universe is whether the universe value is obtained, if the universe frequencies are put in the formula for the sample estimator. The sample mean, sample total, sample variance of mean (whether or not the sample is drawn with replacement), the sample ratio, the sample correlation coefficient etc., defined above, are all consistent estimators. An example of an inconsistent estimator is the following: suppose that from the data of a simple random sample, the ratios of the values of two study variables are computed, namely,

$$r_i = y_i/x_i \tag{1.17}$$

and a mean is taken of these ratios, namely,

$$\bar{r} = \frac{1}{n} \sum_{i=1}^{n} r_i \tag{1.18}$$

The mean of the ratios, $\bar{r}$, is not a consistent estimator of the universe ratio R, for in sampling without replacement, if the sample is ultimately enlarged to cover the whole universe, the sample estimator $\bar{r}$ will take the value

$$\frac{1}{N}\sum_{i=1}^{N}(Y_i/X_i)$$

which will not be the same as the universe value, namely,

$$R=\frac{Y}{X}=\frac{\sum_{i=1}^{N}Y_i}{\sum_{i=1}^{N}X_i} \tag{1.5}$$

except in the trival case of all the ratios Y_i/X_i being the same.

An unbiased estimator is not necessarily consistent or *vice versa*; but in the general cases that we shall consider, an unbiased estimator can be taken to be consistent. Also, if a consistent estimator is biased, its bias will decrease with increasing sample size.

The third criterion is that of *precision* or *efficiency*. Of two estimators for the same universe parameter, one is said to be more efficient than the other when its sampling variance is smaller than the other's. The *precision* of the estimator t relative to that of t' is defined as

$$\text{precision}\ (t,\ t')=\frac{\sigma_{t'}^2}{\sigma_t^2} \tag{1.19}$$

The *efficiency* of the estimator t relative to that of t' is defined as

$$\text{efficiency}\ (t,\ t')=\frac{\text{MSE}_{t'}}{\text{MSE}_t} \tag{1.20}$$

the MSE_t denoting the mean square error defined below. For unbiased estimators, precision and efficiency are equivalent.

Another concept is that of *cost efficiency*. If C and C' are the respective costs of surveys for obtaining the estimators t and t', the cost efficiency of t relative to that of t' is defined as

$$\text{cost efficiency}\ (t,\ t')=\frac{C'.\,\text{MSE}_{t'}}{C.\,\text{MSE}_t} \tag{1.21}$$

When, as is common, the universe values of the variances are not available in the above expressions, the sample estimators are substituted.

It can be shown that in large samples (simple random), the variance of the median as an estimator of the universe mean $\overline{Y}$ is $\frac{1}{2}\pi\sigma_Y^2/n$, (where π is the ratio of the circumference of a circle to its diameter = 3.1416 approximately), so that the efficiency of the mean relative to the median is, from equations (1.11) and (1.20),

$$(\tfrac{1}{2}\pi\sigma_Y^2/n)/(\sigma_Y^2/n)=\tfrac{1}{2}\pi=1.57$$

approximately, i.e. the sample mean is 57 per cent more efficient than the sample median for simple random samples of the same size.

A related concept is that of a *minimum variance estimator.* A well-known inequality in mathematical statistics (*Cramér-Rao inequality*) states that the variance of an estimator of $\overline{Y}$ cannot be smaller than σ_Y^2/n. But as the sample mean itself has sampling variance σ_Y^2/n in simple random sample (with replacement), we can thus say that under this sampling plan, the sample mean is an unbiased, consistent, minimum variance estimator of the universe mean.

1.13 Mean square error

If the sample estimator t has a mathematical expectation T', different from the universe parameter T which it seeks to measure, the sampling variability of the estimator t around its mathematical expectation T' is given by the sampling variance σ_t^2; but the variability of the estimator t around the true value T is given by the mean square error (defined in section 25.3.4). These are connected by the relation

$$\text{mean square error} = \text{sampling variance} + (\text{bias})^2$$

$$\text{MSE}_t = \sigma_t^2 + B_t^2 \tag{1.22}$$

where $B_t = T' - T$ is the bias in the estimator t.

1.14 Principal steps in a sample survey

The principal steps that should be taken in planning and executing a sample survey and in analysing the results relate to the following topics:

1. preparatory work – objectives; legal basis; publicity and co-operation of respondents; budget and cost control; survey calendar; administrative organization; co-ordination with other inquiries; cartographic work and pre-listing; tabulation programme; methods of data collection; questionnaire preparation; survey design; pilot inquiries and pre-tests; and staff recruitment and training,
2. collection of data and supervision,
3. data processing,
4. evaluation and analysis of results,
5. preparation of reports.

This book is on sample designs and does not cover the whole field of survey designs. General considerations to some aspects of survey designing are, however, given in Chapter 26, which includes reference to some specific subject fields.

1.15 Symbols and notations

Capital (and sometimes small) letters are used to denote the universe values of the study and ancillary variables (such as $Y_1, Y_2, \ldots, Y_N; X_1, X_2, \ldots, X_N$), and small letters those in the sample (such as $y_1, y_2, \ldots, y_n$). N is the size (i.e. the total number of units) of the universe, and n the sample size.

Universe parameters are denoted by capital or Greek letters (such as $\overline{Y}$, the universe mean; σ^2, the universe variance per unit; ρ, the universe correlation coefficient of two study variables) and the sample estimators by small letters or with the circumflex (ˆ) on the corresponding symbols for the universe parameters (such as $\overline{y}$, the sample mean as estimator of $\overline{Y}$; s^2, the sample estimator for σ^2; and $\hat{\rho}$, the sample estimator for ρ).

Σ stands for summation (either for the universe or for the sample), and the number of terms to be summed is indicated by the letters at the bottom and the top thus: $Y_1 + Y_2 + \ldots + Y_N = \sum_{i=1}^{N} Y_i$ or simply $\sum_{1}^{N} Y_i$ or $\sum^{N} Y_i$ or $\sum_{i} Y_i$ or, when there is no risk of ambiguity, $\Sigma\, Y_i$.

Double and triple summation notations may be required. Thus, consider N (universe) first-stage units, the ith first-stage unit ($i = 1, 2, \ldots, N$) containing M_i second-stage units. Let the value of the study variable for the jth second-stage unit ($j = 1, 2, \ldots, M_i$) in the ith first-stage unit be denoted by Y_{ij}. Then the sum of the values of the study variable for any first-stage unit and for all first-stage units can be represented as follows:

First-stage units	*Sum of the values of the study variable of the second-stage units*
1	$Y_{11} + Y_{12} + \ldots + Y_{1i} + \ldots + Y_{1M_1} = \sum_{j=1}^{M_1} Y_{1j}$
2	$Y_{21} + Y_{22} + \ldots + Y_{2j} + \ldots + Y_{2M_2} = \sum_{j=1}^{M_2} Y_{2j}$
.	.
.	.
.	.
i	$Y_{i1} + Y_{i2} + \ldots + Y_{ij} + \ldots + Y_{iM_i} = \sum_{j=1}^{M_i} Y_{ij}$
.	.
.	.
.	.
N	$Y_{N1} + Y_{N2} + \ldots + Y_{NJ} + \ldots + Y_{NM_N} = \sum_{j=1}^{M_N} Y_{Nj}$

Grand total $Y_{11} + \ldots Y_{1M_1} + Y_{21} + \ldots + Y_{2M_2} + \ldots + Y_{N1} + \ldots + Y_{NM_N} = \sum_{i=1}^{N} \sum_{j=1}^{M_i} Y_{ij}$

In addition, the following notations will often be used for sums of squares and products. $\sum^{n} y_i^2$ will be called the '*raw*' or '*crude*' *sum of squares* of the y_i values and the sum of squares of deviations from the mean

$$\mathrm{SS}y_i = \sum^{n} (y_i - \bar{y})^2 = \sum^{n} y_i^2 - \Big(\sum^{n} y_i\Big)^2/n = \sum^{n} y_i^2 - n\bar{y}^2 = \sum^{n} y_i^2 - \bar{y}\Big(\sum^{n} y_i\Big) \quad (1.23)$$

will be termed the *corrected sum of squares.* The choice of the particular expressions for $\mathrm{SS}y_i$ in (1.23) will depend on the computational convenience.

Similarly, $\sum^{n} y_i x_i$ will be called the 'raw' or 'crude' sum of products of the y_i and x_i values and the sum of products of deviations from the respective means

$$\mathrm{SP}y_i x_i = \sum^{n} (y_i - \bar{y})(x_i - \bar{x}) = \sum^{n} y_i x_i - \Big(\sum^{n} y_i\Big)\Big(\sum^{n} x_i\Big)/n = \sum^{n} y_i x_i - n\bar{y}\bar{x}$$
$$= \sum^{n} y_i x_i - \bar{y}\Big(\sum^{n} x_i\Big) = \sum^{n} y_i x_i - \bar{x}\Big(\sum^{n} y_i\Big) \quad (1.24)$$

the corrected sum of products. Thus, the sample variance is given by

$$s_y^2 = \frac{\sum^{n} (y_i - \bar{y})^2}{n-1} = \frac{SSy_i}{n-1} \qquad (1.25)$$

and the sample covariance is given by

$$s_{yx} = \frac{\sum^{n} (y_i - \bar{y})(x_i - \bar{x})}{n-1} = \frac{SPy_ix_i}{n-1} \qquad (1.26)$$

Π (capital pi) stands for product. Thus $Y_1 \times Y_2 \times \ldots \times Y_N = \prod_{i=1}^{N} Y_i$ or simply $\prod^{N} Y_i$.

Sub- and super-scripts are added, as necessary. A list of notations and symbols is given in Appendix I.

Further reading

References have been made only to the names of the authors; complete references are given at the end of the book.

For an introduction to the ideas of scientific sampling, see the book by Slonim (a popular account), Stuart (a monograph that includes the principal theory and methods), or Conway, and special chapters on sampling in the book by Moser and Kalton and in textbooks on statistics (with generally the same mathematical level as this book) such as those by Brookes and Dick; Hájek and Dupač; Johnson; Snedecor and Cochran; Tippet; and Yule and Kendall.

A short Manual on Sampling by the United Nations and *Sampling Lectures* by the U.S. Bureau of the Census would also be valuable introductory reading.

For definitions of statistical terms, see the dictionary by Kendall and Buckland.

As a follow-up to this book, a text on sampling can be chosen, depending on the special interest of the reader and the level of mathematical sophistication desired, such as the books by Cochran; Deming (1960); Hansen; Hurwitz and Madow; Kish; Murthy; Raj; Sukhatme and Sukhatme (first edition available in Spanish), and Yates. Deming's book is oriented towards business research, and the books by Yates and the Sukhatmes to agriculture. The first volume by Hansen *et al.* and the books by Kish and Yates are on sampling methods and techniques; the second volume of Hansen *et al.* and Raj's book are mainly on the mathematical theory of sampling, and Hendricks, Yamane, and, at an advanced level, Kendall and Stuart provide introduction to the latter. The books by Cochran, Murthy, and the Sukhatmes cover both sampling theory and methods.

For readers in French, the books by Thionet, Desabie and Yates are recommended as further reading.

References to the application of sampling methods in some subject fields are given under 'Further reading' in Chapter 26.

Note: In following the literature on sampling one difficulty stems from the differences in notations and concepts used by different authors. It is suggested that the reader acquaints himself with the notations and concepts used in this book, and then notes the differences in notations and concepts used by others. For example, some authors (including this one) use the y-notation for the (first) study variable while others use the x-notation. In dealing with single-stage sampling, all the authors cited use N and n to denote the number of units in the universe and the sample respectively; for two-stage sampling some authors (including this one) employ M_i and m_i to denote respectively the universe and the sample number of second-stage units in the ith (universe or sample) first-stage unit, but some others denote by M and m the universe and sample numbers of first-stage units and by N_i and n_i the universe and sample numbers of second-stage units in the ith first-stage unit. Again in two-stage sampling, some authors use the term 'size' of a first-stage unit to mean the number of second-stage unit it (the first-stage unit) contains, but this author and others have used the term 'size' in a more general sense of the value of an ancillary variable for selection with probability proportional to size: a special case of the 'size' is, of course, the number of second-stage units. Note also that some authors introduce multi-stage sampling under cluster sampling.

As further reading to Chapter 1, the following are recommended: Cochran, Chapter 1; Deming (1966), Chapter 1; Hansen *et al.*, Vols. I and II, Chapters 1–3; Kish, Chapters 1 and 2; Murthy, Chapters 1 and 2; Raj, Chapters 1 and 2; Sukhatme and Sukhatme, Chapter 1; Yates, Chapters 1–4.

Exercises

1. In which of the following situations would you prefer a sample survey to a complete enumeration?

(a) Determination of the average number of matchsticks in a box, and the proportion that actually light, from a carton containing a gross of matchboxes.

(b) A survey of tuberculosis prevalence in the islands of Seychelles (population about 50 000).

(c) A longitudinal fertility inquiry in England (In a *longitudinal survey*, used especially in sociological and medical investigations, a sample of individuals or other units is observed at intervals over a period of time, so as to study their development as individuals. It may be contrasted with a *cross-sectional* survey, in which a sample of individuals in various stages of development, e.g. children of different ages, is observed at one time.)

(d) Preparation of electoral registers.

(e) Output of steel in a country.

(f) Handloom cloth production in India.

(g) Production of coffee in Ethiopia. (*Note:* coffee grows in a wild state in some parts of Ethiopia.)

(h) A survey of preferences of television viewers.

2. In family living and budget surveys, it is possible to use the questionnaire and interview method or the account book (or 'diary') method, or a combination of the two. What are the main advantages and disadvantages of each method?

(*Hint:* The 'diary' method is not feasible in a pre-literate society. The very fact that records are being kept may also change the existing habits and patterns.)

3. A survey is required to determine the sex ratio of children (under fifteen years). Four procedures are suggested:

(a) each boy in a sample of boys is asked to state the number of brothers and sisters he has;
(b) each girl in a sample of girls is asked to state the number of sisters and brothers she has;
(c) a sample of children is taken, and the number of boys and girls in the sample counted; and
(d) a sample of families is asked to give the number of boys and girls in the family.

Why would procedures (a) and (b) give biased results – over-representation of boys in one, and over-representation of girls in the other? (Rao, Misc. Problem 20, adapted.)

(*Hint:* In procedures (a) and (b), a single child and siblings consisting only of the opposite sex would not be represented at all. Of the two procedures (c) and (d), the former is to be preferred, for reasons given in Chapter 6. Note that children in institutions are to be sampled too. Account has also to be taken of multiple reporting by siblings.)

4. Comment on the following extract from the proceedings of the Seminar on African Demography, Paris, 1959 (International Union for the Scientific Study of Population, 1960).

> '... suggested that under some conditions the principles of probability sampling might be sacrificed in the interest of obtaining properly qualified and reliable information in the villages to be included in the programme. These villages would then provide merely a "purpose sample"; but the extent to which they are representative of a large population could be measured, at least partially, by comparing their characteristics in other respects on which information can be obtained with those of the whole population of the region.'

(*Hints:* 'The suggestion ... is a hazardous one ... Gini and Galvani (1929) did precisely what ... suggested and they published their results to tell statisticians that the suggestion is a hazard. They selected and rejected counties in Italy until they found a sample that matched seven characteristics for the whole country, only to find that the sample broke down completely on all other characteristics': personal communication from Dr. W. E. Deming. Note also the observations by Yates (1960, section 3.13): 'Gini and Galvani (1929) selected a sample from the Italian census data of 1921 which consisted of all the returns of 29 out of the 214 *circondari* into which the country is divided, using seven control characters. Agreement of the average values of other characters in the sample and population was poor, and that of the frequency-distribution of such characters was even worse. The real weakness here is the use of excessively large units, though even with smaller units the use of purposive selection without rigorous rules of selection is always liable to give unsatisfactory results. There is, moreover, no means of judging its reliability. For these reasons purposive selection has ceased to be extensively used and in modern sampling work it has largely been replaced by more thorough application of the principles of stratification, etc.).

Part I:

Single-stage Sampling

2 Simple Random Sampling

2.1 Introduction

In this chapter we shall consider simple random sampling without stratification of the universe and without the selection of the sample units in different stages; unstratified single-stage simple random sampling is sometimes known as *unrestricted simple random sampling.* In the next chapter will be shown the use of ancillary information in improving the efficiency of estimators obtained from a simple random sample. We shall often use the abbreviation 'srs' for 'simple random sample' or 'simple random sampling'.

We shall first consider the problem of sampling the universe of six households a, b, c, d, e, and f, with respective sizes (number of persons) 8, 6, 3, 5, 4, and 4. This is a hypothetical example, for, with such a small universe, it would be normal to enumerate all the units completely rather than take a sample. However, we shall use it to illustrate the principles.

2.2 Characteristics of the universe

We note first some characteristics of the universe. The total number of units, $N = 6$; if by Y_i we denote the size of the ith household ($i = 1, 2, \ldots, N$), the total number of persons in the universe is

$$Y = \sum_{i=1}^{N} Y_i \qquad (2.1)$$

$$= 8 + 6 + 3 + 5 + 4 + 4$$

$$= 30$$

The average household size (number of persons per household) is

$$\overline{Y} = Y/N$$

$$= \sum^{N} Y_i/N \qquad (2.2)$$

$$= 30/6 = 5$$

The variance per unit of the universe is defined as

$$\sigma^2 = \sum^{N} (Y_i - \overline{Y})^2/N$$

$$= 16/6 = 2.\dot{6} \qquad (2.3)$$

The computations are shown in Table 2.1. The standard deviation per unit of the universe is

$$\sigma = \sqrt{2.6} = 1.6330$$

Table 2.1 Sizes (numbers of persons) of a hypothetical universe of six households

Household i	*Size* Y_i	$Y_i - \overline{Y}$	$(Y_i - \overline{Y})^2$
a	8	3	9
b	6	1	1
c	3	−2	4
d	5	0	0
e	4	−1	1
f	4	−1	1
Total	30	0	16
Mean	$\overline{Y} = 5$	0	$\sigma^2 = 2.6$

The coefficient of variation of the universe is defined as

$$\begin{aligned} \text{CV} &= \sigma/\overline{Y} \\ &= 1.6330/5 \\ &= 0.3266 \quad \text{or} \quad 32.66 \text{ per cent} \end{aligned} \tag{2.4}$$

Note: The characteristics of the universe will not be known in general, and the main objective in sample surveys is to estimate these characteristics.

2.3 Simple random sampling

If the sampling frame consists of an identifiable list of all the units in the universe, but without any information on the value (or magnitude) of the variable under study (the 'study variable') or of any ancillary variable, and if we are sampling only one unit from the universe, then *a priori* there would be no reason to choose one unit over another, i.e. all the units should have the same chance of being selected. Similarly, if we are selecting n (>1) units from the universe of N units, then each combination of the n units should have the same chance of selection as every other combination: such a sample is termed a simple random sample of n units.

Note: In simple random sampling, the probability that a universe unit will be selected at any given draw is the same as that at the first draw, namely, $1/N$, and the probability that the specific unit is included in the sample of n units is n/N; see Appendix II, section A2.3.2, notes 1 and 2.

If we draw a simple random sample such that no unit occurs more than once in the sample, it is said to be *without replacement*; if a unit can occur more than once in the sample, it is said to be *with replacement*. We shall illustrate these with the hypothetical universe of six households given in Table 2.1, the objective being to estimate the average size of household from samples of different sizes.

2.4 Sampling without replacement

2.4.1 Mean and variance for the hypothetical universe

If we sample only one unit from the universe of six households without replacement, the six possible samples of size 1 are the six households in the universe. The estimator of the household size of the universe from each sample will be the household size of the sample unit itself; the average of the size estimators is 5, the same value as the universe household size. But we cannot compute the sampling variance from one sample in this case, as the sample consists of one unit only.

If we draw samples of two units each from the universe, such that no unit occurs more than once in the sample, and if we disregard the order of drawing so

Table 2.2 All possible samples of size 2 drawn without replacement from the universe of Table 2.1

Sample no.	*Sample units*	*Total size* $y_1 + y_2$ *	*Mean size* $(\bar{y})$ †	s_y^2 ‡	$s_{\bar{y}}^2$ §
1	ab	14	7.0	2.0	0.6667
2	ac	11	5.5	12.5	4.1667
3	ad	13	6.5	4.5	1.5000
4	ae	12	6.0	8.0	2.6667
5	af	12	6.0	8.0	2.6667
6	bc	9	4.5	4.5	1.5000
7	bd	11	5.5	0.5	0.1667
8	be	10	5.0	2.0	0.6667
9	bf	10	5.0	2.0	0.6667
10	cd	8	4.0	2.0	0.6667
11	ce	7	3.5	0.5	0.1667
12	cf	7	3.5	0.5	0.1667
13	de	9	4.5	0.5	0.1667
14	df	9	4.5	0.5	0.1667
15	ef	8	4.0	0	0
Total	—	—	75.0	48.0	16.0000
Mean	—	—	5.0	3.2	1.06

* y_1 and y_2 are the values of the two units included in the sample.
† $\bar{y} = \frac{1}{2}(y_1 + y_2)$ from equation (2.5).
‡ $s_y^2 = \frac{1}{2}(y_1 - y_2)^2$ from equation (2.8).
§ $s_{\bar{y}}^2 = (1 - f)s_y^2/n = \frac{1}{3}s_y^2$ from equation (2.7), as $n = 2$, $N = 6$, and $n/N = \frac{1}{3}$.

that the same combination of two units such as (ab, ba) is not repeated, we shall have ${}^6C_2 = 15$ combinations: these are written down in Table 2.2. For each of the 15 combinations (samples) we compute the following measures, as estimators of the corresponding universe values:

Estimator of the universe mean: sample mean $$\bar{y} = \sum_{i=1}^{n} y_i/n \qquad (2.5)$$

Estimator of variance per unit: $$s_y^2 = \sum^{n} (y_i - \bar{y})^2/(n-1) \qquad (2.6)$$

Estimator of variance of sample mean: $$s_{\bar{y}}^2 = s_y^2(1 - f)/n \qquad (2.7)$$

where n is the sample size,

y_i $(i = 1, 2, \ldots, n)$ is the value of the ith unit of the study variable included in the sample,

$f = n/N$ is the sampling fraction.

Note the following points:

1. For $n = 2$, the formula for s_y^2 is simplified to

$$s_y^2 = \tfrac{1}{2}(y_1 - y_2)^2 \qquad (2.8)$$

where y_1 and y_2 are the values of the two sample units included in the sample.

2. In the general case, a workable formula for s_y^2 may be chosen from among the following, depending on the convenience of computation:

$$\begin{aligned} s_y^2 &= \frac{SSy_i}{n-1} \\ &= \frac{\sum^n (y_i - \bar{y})^2}{n-1} \\ &= \frac{\sum^n y_i^2 - (\sum^n y_i)^2/n}{n-1} \\ &= \frac{\sum^n y_i^2 - n\bar{y}^2}{n-1} \\ &= \frac{\sum^n y_i^2 - \bar{y}(\sum^n y_i)}{n-1} \end{aligned} \qquad (2.9)$$

Table 2.3 All possible samples of size 3 drawn without replacement from the universe of Table 2.1

Sample no.	*Sample units*	*Total sizes*	*Mean size* $\bar{y}$ *	s_y^2 †	$s_{\bar{y}}^2$ ‡
1	abc	17	5.67	6.3333	1.0556
2	abd	19	6.33	2.3333	0.3889
3	abe	18	6.00	4.0000	0.6667
4	abf	18	6.00	4.0000	0.6667
5	acd	16	5.33	6.3333	1.0556
6	ace	15	5.00	7.0000	1.1667
7	acf	15	5.00	7.0000	1.1667
8	ade	17	5.67	4.3333	0.7222
9	adf	17	5.67	4.3333	0.7222
10	aef	16	5.33	5.3333	0.8889
11	bcd	14	4.67	2.3333	0.3889
12	bcd	13	4.33	2.3333	0.3889
13	bcf	13	4.33	2.3333	0.3889
14	bde	15	5.00	1.0000	0.1667
15	bdf	15	5.00	1.0000	0.1667
16	bef	14	4.67	1.3333	0.2222
17	cde	12	4.00	1.0000	0.1667
18	cdf	12	4.00	1.0000	0.1667
19	cef	11	3.67	0.3333	0.0556
20	def	13	4.33	0.3333	0.0556
Total	–	–	100.00	64.0000	10.6667
Mean	–	–	5.00	3.2000	0.5333

* $\bar{y} = \frac{1}{3}\sum_1^3 \bar{y}_i$

† $s_y^2 = SSy_i/n(n-1) = \frac{1}{6}SSy_i$;

‡ $s_{\bar{y}}^2 = s_y^2(1-f)/n = \frac{1}{6}s_y^2$, as $n = 3$, $N = 6$ and $f = n/N = \frac{1}{2}$.

The results of all possible samples of size 3 drawn without replacement from the same universe are given in Table 2.3. The computations of the results of all possible samples of sizes 4, 5, and 6 drawn without replacement are not shown separately.

2.4.2 Distribution of sample mean

The distribution of the sample mean in all possible samples of different sizes are given in Table 2.4. Two points may be noted:

1. The average of the estimates of all possible samples for any sample size (this is known as the *mathematical expectation*) is the true universe value, namely 5. If the mathematical expectation of a sample estimator is the universe parameter, the estimator is said to be *unbiased* for the universe parameter; if not, the estimator is *biased.* We have seen that the sample mean, obtained from a simple random sample without replacement, is an unbiased estimator of the universe mean (see also section 1.12).
2. As the size of the sample increases, the sample estimates concentrate around the universe value; thus 73 per cent of the sample estimates of size two, 90 per cent of size three, and all from four and larger sizes fall within the range of household size 4–6 (see Figure 2.1). This characteristic of a sample estimator is known as *consistency* (see also section 1.12).

Table 2.4 Estimates of average household size from all possible samples of different sizes drawn without replacement from the universe of Table 2.1

Average size of household	*Number of samples with estimated average size of household for samples of size n*					
	n = 1	*n = 2*	*n = 3*	*n = 4*	*n = 5*	*n = 6*
3.00–3.24	1					
3.25–3.49						
3.50–3.74		2	1			
3.75–3.99						
4.00–4.24	2	2	2	1		
4.25–4.49			3	1	1	
4.50–4.74		3	2	2		
4.75–4.99				2	1	
5.00–5.24	1	2	4	2	3	1
5.25–5.49			2	3	1	
5.50–5.74		2	3	2		
5.75–5.99				2		
6.00–6.24	1	2	2			
6.25–6.49			1			
6.50–6.74		1				
6.75–6.99						
7.00–7.24		1				
7.25–7.49						
7.50–7.74						
7.75–7.99						
8.00–8.24	1					
Number of samples	6	15	20	15	6	1
Mean	5	5	5	5	5	5
Variance of mean	2.6667	1.0667	0.5333	0.2667	0.1067	0

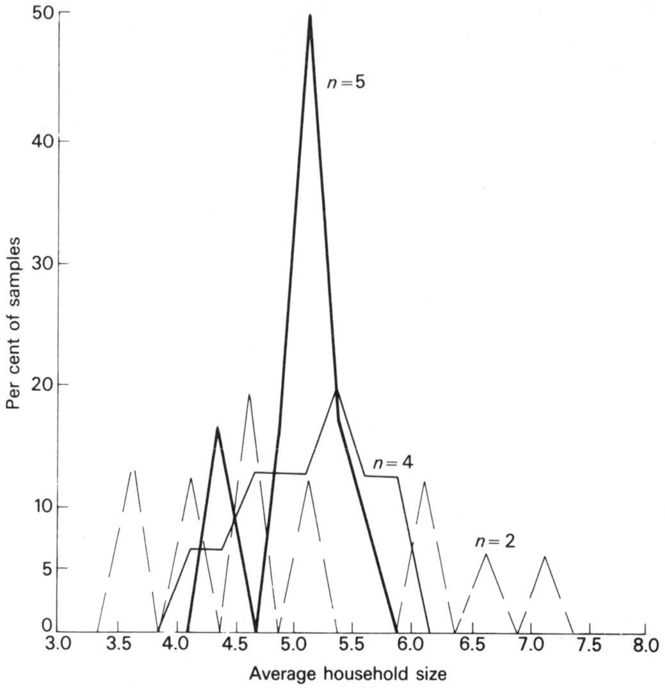

Figure 2.1 Distribution of estimates of average household size in simple random sampling without replacement from the universe of Table 2.1 (Source: Table 2.4).

2.4.3 Variance estimators

A measure of the degree by which the estimate $\overline{y}$ obtained from a sample differs from the true value $\overline{Y}$ is given by $s_{\overline{y}}^2$. The values of s_y^2 for the different sample sizes are given in Table 2.5. The following points may be noted from the data:

1. The average of s_y^2 for all the sample sizes is a constant and is 3.2, whereas the universe variance is $\sigma^2 = 2.\dot{6}$ (Table 2.1). Thus $s_{\overline{y}}^2$ is not an unbiased estimator of the universe variance σ^2. The expected value of s_y^2 is

$$E(s_y^2) = \frac{N}{N-1}\,\sigma^2$$

$$= \frac{\sum^N (Y_i - \overline{Y})^2}{N-1} \tag{2.10}$$

Table 2.5 Results of all possible samples of different sizes drawn without replacement from the universe of Table 2.1

Sample size n	*Number of possible samples*	*Mean of sample means*	*Mean of* s_y^2	*Mean of* $s_{\overline{y}}^2$
1	6	5	3.2	2.6667
2	15	5	3.2	1.0667
3	20	5	3.2	0.5333
4	15	5	3.2	0.2667
5	6	5	3.2	0.1067
6	1	5	3.2	0

In some text-books, the expression on the right-hand side is denoted by S^2 and in others the universe variance is defined by this expression. In our example,

$$S^2 = \frac{\sum^{N} (Y_i - \bar{Y})^2}{N-1} = \frac{16}{5} = 3.2$$

which is the average of s_y^2 for all the sample sizes.

2. The universe variance of the sample mean for srs without replacement is

$$\sigma_{\bar{y}}^2 = \frac{\sigma^2}{n} \cdot \frac{(N-n)}{(N-1)}$$
$$= \frac{S^2(1-f)}{n} \qquad (2.11)$$

Note that if $n = 1$, $\sigma_{\bar{y}}^2 = \sigma^2$, the universe variance per unit, as it should be.

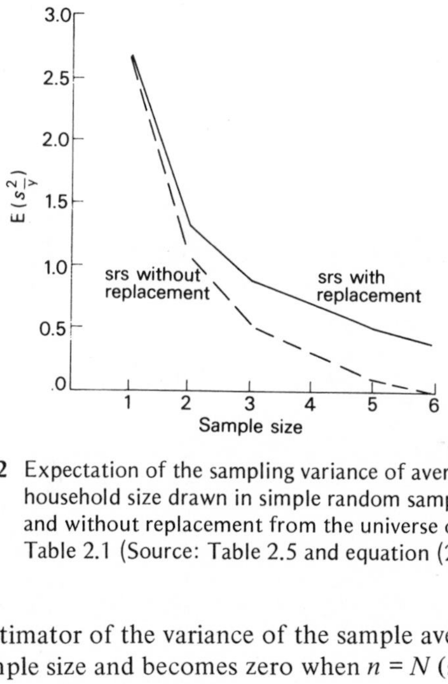

Figure 2.2 Expectation of the sampling variance of average household size drawn in simple random sampling with and without replacement from the universe of Table 2.1 (Source: Table 2.5 and equation (2.20)).

3. The value of the estimator of the variance of the sample average $s_{\bar{y}}^2$ decreases with increasing sample size and becomes zero when $n = N$ (see Figure 2.2).

4. Tables 2.4 and 2.5 give the results obtained from the theoretical probabilities of the different samples. The same results are approximated when actual samples are drawn a large number of times from a universe.

2.5 Sampling with replacement

2.5.1 Estimators of mean and variance

We have seen that in simple random sampling without replacement of n units out of a total of N units, there may be ${}^NC_n$ combinations (samples) if no unit is repeated in the sample and the order of selection is not taken into account; the probability of the occurrence of any combination is $1/{}^NC_n$, the total of all the probabilities for the ${}^NC_n$ combinations being 1. In practice, however, the above procedure of writing out all the ${}^NC_n$ combinations and then selecting one at

Table 2.6 All possible samples of size 2 drawn with replacement from the universe of Table 2.1

Sample no.	*Sample units*	*Total size* $y_1 + y_2$ *	*Average size* $\bar{y}$ †	s_y^2 ‡	$s_{\bar{y}}^2$ §
1	aa	16	8.0	0	0
2	ab	14	7.0	2.0	1.00
3	ac	11	5.5	12.5	6.25
4	ad	13	6.5	4.5	2.25
5	ae	12	6.0	8.0	4.00
6	af	12	6.0	8.0	4.00
7	ba	14	7.0	2.0	1.00
8	bb	12	6.0	0	0
9	bc	9	4.5	4.5	2.25
10	bd	11	5.5	0.5	0.25
11	bd	10	5.0	2.0	1.00
12	bf	10	5.0	2.0	1.00
13	ca	11	5.5	12.5	6.25
14	cb	9	4.5	4.5	2.25
15	cc	6	3.0	0	0
16	cd	8	4.0	2.0	1.00
17	ce	7	3.5	0.5	0.25
18	cf	7	3.5	0.5	0.25
19	da	13	6.5	4.5	2.25
20	db	11	5.5	0.5	0.25
21	dc	8	4.0	2.0	1.00
22	dd	10	5.0	0	0
23	de	9	4.5	0.5	0.25
24	df	9	4.5	0.5	0.25
25	ea	12	6.0	8.0	4.00
26	eb	10	5.0	2.0	1.00
27	ec	7	3.5	0.5	0.25
28	ed	9	4.5	0.5	0.25
29	ee	8	4.0	0	0
30	ef	8	4.0	0	0
31	fa	12	6.0	8.0	4.00
32	fb	10	5.0	2.0	1.00
33	fc	7	3.5	0.5	0.25
34	fd	9	4.5	0.5	0.25
35	fe	8	4.0	0	0
36	ff	8	4.0	0	0
Total	—	—	180.00	96.00	48.00
Mean	—	—	5.00	2.6	1.3

* y_1 and y_2 are the values of the two units included in the sample.
† $\bar{y} = \frac{1}{2}(y_1 + y_2)$ from equation (2.5).
‡ $s_y^2 = \frac{1}{2}(y_1 - y_2)^2$ from equations (2.13) and (2.8).
§ $s_{\bar{y}}^2 = \frac{1}{2} s_y^2$ from equation (2.14).

random is not followed as this would be a needless, tedious procedure. Actually, the n units of a sample are selected one by one with the help of random numbers (or in some other random way), and if a unit is once selected in the sample, it is not repeated.

In simple random sampling with replacement, on the other hand, a unit can appear more than once in the sample. In Table 2.6 are shown the 36 possible samples of size 2 drawn with replacement from the hypothetical universe of Table 2.1: the probability of occurrence of any sample is $\frac{1}{36}$. Note the repetition of units in the sample such as aa, bb etc. and ab, ba etc. For each of the 36 combinations (samples) we compute the following measures, as estimators of the universe parameters:

Estimator of the universe mean: sample mean $\bar{y} = \sum^{n} y_i/n$ (2.12)

Estimator of variance per unit: $s_y^2 = \sum^{n} (y_i - \bar{y})^2/(n-1)$ (2.13)

Estimator of variance of sample mean: $s_{\bar{y}}^2 = s_y^2/n$ (2.14)

Note that the first two estimators are the same as those for srs without replacement (equations (2.5) and (2.6) respectively), but the third estimator is different.

The results of Table 2.6 show the following:

1. The mean of the estimates of all possible samples is the universe value, namely 5. Thus the sample mean is an unbiased estimator of the universe mean.
2. The mean of all the possible sample estimates of variance s_y^2 is 2.6, the universe variance, showing that s_y^2 is an unbiased estimator of σ^2.

2.5.2 Consistency of mean and sampling variance

In general, in srs with replacement of n units out of a total of N units, there will be N^n combinations, and the probability of selection of any combination is $1/N^n$. The results noted for srs with replacement of 2 units from the universe of 6 units will also hold for samples of other sizes. In addition, the following may be noted:

1. As the size of the sample increases, the sample estimates concentrate around the universe value. The sample mean is again seen to be a consistent estimator of the universe average.
2. The sampling variance of the sample mean for srs with replacement is

$$\sigma_{\bar{y}}^2 = \sigma^2/n \tag{2.15}$$

and its unbiased estimator is

$$s_{\bar{y}}^2 = s_y^2/n \tag{2.16}$$

This decreases with increasing sample size, but does not reach zero (Figure 2.2).

2.6 Important results

2.6.1 For simple random sampling

We shall summarize the important results for simple random sampling. We shall often refer to srs with replacement as 'srswr' and srs without replacement as 'srswor'.

1. The sample mean $\bar{y}$ is an unbiased estimator of the universe mean $\bar{Y}$.
2. The variance of the sample mean is

$$\sigma_{\bar{y}}^2 = \frac{\sigma^2}{n} \quad \text{in srswr} \tag{2.17}$$

$$= \frac{\sigma^2}{n}\frac{(N-n)}{(N-1)}$$

$$= \frac{S^2}{n}(1-f) \quad \text{in srswor} \tag{2.18}$$

where

$$S^2 = \frac{N}{N-1}\sigma^2 \tag{2.19}$$

and $f = n/N$ is the sampling fraction.

3. The sample estimator of variance per unit

$$s^2 = \frac{\sum^n (y_i - \bar{y})^2}{n-1}$$

is an unbiased estimator of σ^2 in srswr and of S^2 in srswor.

4. Unbiased variance estimators of the sample mean $\bar{y}$ are

$$s_{\bar{y}}^2 = \frac{s^2}{n} \quad \text{in srswr} \tag{2.20}$$

$$s_{\bar{y}}^2 = \frac{s^2}{n}(1-f) \quad \text{in srswor} \tag{2.21}$$

From the numerical examples, these results have been seen to hold for the universe of six households. Theoretical proofs are given in Appendix II that they hold for a universe of any size (sections A2.3.2–A2.3.4).

2.6.2 Variance of the ratio of random variables

Another theorem of wide generality relates to the ratio of two random variables. If y and x are two random variables with expectations Y and X, variances σ_Y^2 and σ_X^2, and covariance $\sigma_{YX} = \rho\sigma_Y\sigma_X$ (where ρ is the correlation coefficient), then the variance of the ratio of two random variables $r = y/x$ is given approximately by

$$\sigma_r^2 = \frac{1}{X^2}(\sigma_Y^2 + R^2\sigma_X^2 - 2R\sigma_{YX}) \tag{2.22}$$

$$= R^2(\text{CV}_Y^2 + \text{CV}_X^2 - 2\rho\text{CV}_Y\text{CV}_X) \tag{2.23}$$

where $R = Y/X$. Also,

$$\text{CV}_r^2 = \text{CV}_Y^2 + \text{CV}_X^2 - 2\rho\text{CV}_Y\text{CV}_X \tag{2.24}$$

Theoretical proofs for simple random sampling are given in Appendix II, section A2.3.6. Estimators of σ_r^2 and CV_r are obtained on substituting the sample estimators for the universe parameters in the respective expressions.

Notes

1. In effect, sampling with replacement is equivalent to drawing samples from an infinitely large universe, and in the estimating formulae for the variance of the sample mean (and other related measures), the factor

$$1 - f = 1 - \frac{n}{N}$$

known variously as the *finite sampling correction, finite multiplier,* or *finite population correction*, does not enter.

2. As the finite sampling correction $(1-f)$ is less than 1, the estimator of the sample mean (and other related measures) will, on an average, have less variance (i.e. be more efficient) for sampling without replacement than with replacement. However, sampling with replacement greatly simplifies the computations required for estimating universe parameters; for

this reason, even if sampling is made without replacement, the sample may be treated as if it was obtained with replacement if the sample size is small relative to that of the universe. A useful rule of thumb for the latter is that the sampling fraction $f = n/N$ should be less than 10 per cent, and preferably less than 5 per cent.

3. *Behaviour of sampling variance with increasing sampling size:* In sampling without replacement, the expression for the estimator of the variance of the sample mean, $s_y^2(1-f)/n$, becomes zero when the universe is completely enumerated, for then $n = N$ and $(1-f) = 0$. In sampling with replacement, however, the expression for the estimator of the variance of the sample mean, namely s_y^2/n, will not necessarily become zero even when the sample size n is made equal to the universe size N. This is because in sampling with replacement, a unit can be selected more than once, and hence a sample of N units may not necessarily include all the distinct N universe units. We have already noted that sampling with replacement is equivalent to sampling an infinite universe, which can never be exhausted by sampling, as the units are replaced after each draw. The coefficient of variation of the sample mean is

$$\text{CV}(\bar{y}) = \sigma_{\bar{y}}/\bar{Y} \tag{2.25}$$

$$= \sigma/\bar{Y}\sqrt{n}$$

$$= \text{universe CV per unit}/\sqrt{n} \text{ in srswr} \tag{2.26}$$

$$= \frac{\sigma}{\bar{Y}\sqrt{n}}\sqrt{\frac{N-n}{N-1}}$$

$$= \frac{\text{universe CV per unit}}{\sqrt{n}}\sqrt{\frac{N-n}{N-1}} \text{ in srswor} \tag{2.27}$$

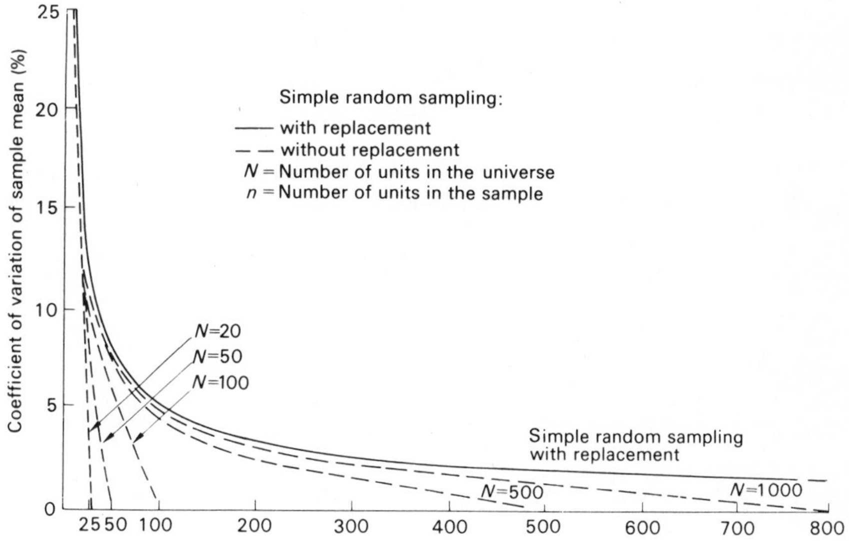

Figure 2.3 Behaviour of the coefficient of variation of sample mean with increasing sample size in simple random sampling with and without replacement when the universe coefficient of variation per unit is 50%.

The behaviour of the coefficient of variation of the sample mean with increasing sample size in simple random sampling both with and without replacement is illustrated in Figure 2.3 for selected values of N and n, assuming the universe CV per unit to be 50 per cent.

Figure 2.3 illustrates the fact that when the sampling fraction is relatively large, sampling without replacement has substantial superiority over sampling with replacement; with a small sampling fraction, however, the gain is marginal.

In sampling with replacement, and approximately also in sampling without replacement when the sampling fraction is relatively small, the sampling variance of the sample mean is inversely related to the sample size, and does not depend on the size of the universe.

4. Although the variance estimators per unit in the universe s_y^2 and of the sample average $s_{\bar{y}}^2$ are unbiased estimators of the corresponding universe values, the sample estimators of the standard deviations, obtained on taking the positive square root of the variance estimators, are not unbiased estimators of the corresponding universe standard deviations: the reader may verify this for the numerical example of section 2.4. The bias, however, is negligible for large samples.
5. The sample CV is also not an unbiased estimator of the universe CV.

2.7 Fundamental theorems

We now introduce some theorems of fundamental importance in sampling.

1. If t_i $(i = 1, 2, \ldots, n)$ are n independent unbiased estimators of the universe parameter T, then a combined unbiased estimator of T is the arithmetic mean

$$\bar{t} = \frac{1}{n} \sum^{n} t_i \tag{2.28}$$

and an unbiased estimator of the variance of $\bar{t}$ is

$$s_{\bar{t}}^2 = \frac{\sum^{n} (t_i - \bar{t})^2}{n(n-1)} = \frac{\mathrm{SS}t_i}{n(n-1)} \tag{2.29}$$

If u_i $(i = 1, 2, \ldots, n)$ are n independent unbiased estimators of another universe parameter U, then a combined unbiased estimator of U is similarly

$$\bar{u} = \frac{1}{n} \sum^{n} u_i \tag{2.30}$$

and an unbiased estimator of the variance of $\bar{u}$ is similarly

$$s_{\bar{u}}^2 = \frac{\sum^{n} (u_i - \bar{u})^2}{n(n-1)} = \frac{\mathrm{SS}u_i}{n(n-1)} \tag{2.31}$$

2. An unbiased estimator of the covariance of $\bar{t}$ and $\bar{u}$ is the estimator

$$s_{\bar{t}\bar{u}} = \frac{\sum^{n} (t_i - \bar{t})(u_i - \bar{u})}{n(n-1)} = \frac{\mathrm{SP}t_i u_i}{n(n-1)} \tag{2.32}$$

3. A consistent, but generally biased, estimator of the ratio of two universe parameters

$$R = T/U$$

is the ratio of the sample estimators $\bar{t}$ and $\bar{u}$

$$r = \bar{t}/\bar{u} \tag{2.33}$$

and an estimator of the variance of r is

$$s_r^2 = (s_{\bar{t}}^2 + r^2 s_{\bar{u}}^2 - 2r s_{\bar{t}\bar{u}})/\bar{u}^2 \tag{2.34}$$

4. A consistent, but generally biased, estimator of the universe correlation coefficient ρ between the two study variables is the sample correlation coefficient

$$\rho_{\bar{t}\bar{u}} = s_{\bar{t}\bar{u}} / s_{\bar{t}} s_{\bar{u}} \tag{2.35}$$

Notes

1. The above theorems have a very wide range of application in sampling. These can be generalized to stratified and multi-stage designs, with sample units being selected with equal or varying probabilities. The only requirement is that for the universe parameters, a number of independent and unbiased estimators be available from the sample. The independence of the sample estimators is ensured by selecting the sample units with replacement (at least at the first stage in a multi-stage sample design and in each stratum separately in a stratified design); in practice, when sampling is carried out without replacement this is approximated by making the (first-stage) sampling fraction fairly small (10 per cent or less, preferably 5 per cent or less: see note 2, section 2.6).
2. The estimators $\bar{t}$ and $\bar{u}$ generally refer to the totals of the characteristics; the advantage of this will be seen later.
3. The estimator r is unbiased if the linear regression of $\bar{t}$ on $\bar{u}$ passes through the origin (0, 0). An estimator of the bias (this estimator itself is generally biased) is

$$\frac{r s_{\bar{t}}^2 - s_{\bar{t}\bar{u}}}{\bar{u}^2} \tag{2.36}$$

4. Whenever a universe ratio or its sample estimator is considered, it is assumed that the denominator is not zero.

2.8 Standard errors of estimators and setting of probability limits

The *standard error* of an estimator is given by the positive square root of the sampling variance of the estimator. Expressed as a ratio of the expected value of the estimator, it is called the *relative standard error* or the *Coefficient of Variation* (CV) and is conveniently expressed as a percentage measure. The square of the CV is sometimes called the *relative variance* of the estimator. When the universe values are not available, sample estimators are substituted in computing the CV.

The standard error of the estimator helps in setting probability limits to the estimator by assigning limits which are expected to include the (unknown) universe value with a certain probability: these limits are known as *confidence limits,* and the interval set by the limits, the *confidence interval.*

If $\bar{t}$ is an unbiased estimator of a universe parameter T (using the notations of section 2.7), and its sampling distribution moderately normal (which will happen when the sample size is not too small), and $s_{\bar{t}}$, the standard error of $\bar{t}$, is estimated from the sample data, then the statistic

$$t' = \frac{\bar{t} - T}{s_{\bar{t}}} = \frac{(\bar{t} - T)\sqrt{n}}{s_t} \tag{2.37}$$

is distributed as *Student's t-distribution* with $(n - 1)$ *degrees of freedom* on which $s_{\bar{t}}^2$, the sample estimator of the variance of t, is based; here s_t^2 is defined as

$$s_t^2 = \frac{\sum^n (t_i - \bar{t})^2}{n - 1} \tag{2.38}$$

Denoting by $t'_{\alpha, n-1}$ the 100α percentage point of the t-distribution corresponding to $(n - 1)$ degrees of freedom, we see that the inequality

$$|\bar{t} - T|\sqrt{n}/s_t \leqslant t'_{\alpha, n-1} \tag{2.39}$$

is expected to occur on an average with probability $(1 - \alpha)$. The chances are, therefore, $100(1 - \alpha)$ per cent that the universe value T will be contained by the limits

$$\bar{t} - (t'_{\alpha, n-1}\, s_t)/\sqrt{n}, \qquad \bar{t} + (t'_{\alpha, n-1}\, s_t)/\sqrt{n} \tag{2.40}$$

Some illustrative values of the t-distribution are given in Appendix III, Table 2. For example, for $n = 10$, and $\alpha = 0.05$, $t'_{\alpha, n-1} = 2.26$. For large n, the normal distribution may be used, where for $\alpha = 0.05$, $t'_{\alpha, \infty} = 1.96$ or 2 approximately.

The following points should be noted:

1. When the universe standard error of $\bar{t}$, namely $\sigma_{\bar{t}} = \sigma/\sqrt{n}$, is known, then the statistic

$$(\bar{t} - T)/\sigma_{\bar{t}} \tag{2.41}$$

follows the normal distribution.

2. When the sample size is small, the t-distribution may still be used in setting probability limits to the universe values, but with the additional assumption that the sample itself comes from a normal distribution.

2.9 Estimation of totals, means, ratios and their variances

2.9.1 Simple random sampling with replacement

We have seen in sections 2.5 and 2.6 that if a simple random sample of n units is selected with replacement from the universe of N units, the sample mean $\bar{y}$ is an unbiased estimator of the universe mean $\bar{Y}$, and an unbiased estimator of the variance of $\bar{y}$ is

$$s_{\bar{y}}^2 = \frac{s_y^2}{n} = \frac{\sum^n (y_i - \bar{y})^2}{n(n-1)} \tag{2.16}$$

Let us see, however, how the above and other results follow from the fundamental theorems of section 2.7.

With a simple random sample of n units selected with replacement, let the sample values of the study variable be $y_1, y_2, \ldots, y_n$. Each of these y_i values $(i = 1, 2, \ldots, n)$, when multiplied by N, the total number of units in the universe, gives an independent, unbiased estimator of the universe total Y

$$y_i^* = Ny_i \tag{2.42}$$

The combined unbiased estimator of Y is (from equation (2.28))

$$y_0^* = \frac{1}{n}\sum^n y_i^*$$

$$= \frac{N}{n}\sum^n y_i = N\bar{y} \tag{2.43}$$

An unbiased estimator of the variance of the estimator y_0^* is, from equation (2.29),

$$s_{y_0^*}^2 = \frac{\sum^n (y_i^* - y_0^*)^2}{n(n-1)}$$

$$= \frac{\text{SS}y_i^*}{n(n-1)}$$

$$= \frac{N^2 \sum^n (y_i - \bar{y})^2}{n(n-1)}$$

$$= \frac{N^2 s_y^2}{n}$$

$$= \frac{N^2 \text{SS}y_i}{n(n-1)} \qquad (2.44)$$

An unbiased estimator of the universe mean $\bar{Y}$ is obtained on dividing the unbiased estimator of the universe total by the number of universe units N, namely,

$$y_0^*/N = \bar{y} \qquad (2.45)$$

i.e. the sample mean, and an unbiased estimator of the variance of $\bar{y}$ on dividing the unbiased estimator of variance of the estimator of the universe total by N^2, namely,

$$s_{y_0^*}^2/N^2 = s_y^2/n = s_{\bar{y}}^2 \qquad (2.46)$$

From the same sample, unbiased estimators of the total and the mean of another study variable may be computed from estimation formulae of the above types. An unbiased estimator of the covariance between the estimators y_0^* and x_0^* of the two universe totals Y and X is, from equation (2.32),

$$s_{y_0^* x_0^*} = \frac{\sum^n (y_i^* - y_0^*)(x_i^* - x_0^*)}{n(n-1)}$$

$$= \frac{\text{SP}y_i^* x_i^*}{n(n-1)}$$

$$= \frac{N^2 \sum^n (y_i - \bar{y})(x_i - \bar{x})}{n(n-1)}$$

$$= \frac{N^2 s_{yx}}{n}$$

$$= \frac{N^2 \text{SP}y_i x_i}{n(n-1)} \qquad (2.47)$$

A consistent, but generally biased, estimator of the universe ratio $R = Y/X$ of two universe totals Y and X is the ratio of the sample estimators from equation (2.33), namely,

$$r = y_0^*/x_0^* \qquad (2.48)$$

and an estimator of the variance of r is (from equation (2.34))

$$s_r^2 = \frac{s_{y_0^*}^2 + r^2 s_{x_0^*}^2 - 2r s_{y_0^* x_0^*}}{x_0^{*2}}$$

$$= \frac{s_{\bar{y}}^2 + r^2 s_{\bar{x}}^2 - 2r s_{yx}}{n\bar{x}^2}$$

$$= \frac{\sum^n (y_i - \bar{y})^2 + r^2 \sum^n (x_i - \bar{x})^2 - 2r \sum^n (y_i - \bar{y})(x_i - \bar{x})}{n(n-1)\bar{x}^2}$$

$$= \frac{\text{SS}y_i + r^2\,\text{SS}x_i - 2r\,.\,\text{SP}y_i x_i}{n(n-1)\bar{x}^2}$$

$$= \frac{\sum^n y_i^2 + r^2 \sum^n x_i^2 - 2r \sum^n y_i x_i}{n(n-1)\bar{x}^2} \qquad (2.49)$$

2.9.2 Simple random sampling without replacement

Here the unbiased estimators of universe totals and means, and the estimator of the universe ratio of two universe totals (or means) remain the same as for srs with replacement, but the finite sampling correction, $(1-f) = (1-n/N)$, is applied to the variance and covariance estimators. Thus the unbiased variance estimator of $\bar{y}$ is

$$s_{\bar{y}}^2 = (1-f)\frac{s_y^2}{n} \qquad (2.21) \text{ or}$$

$$= (1-f)\frac{\text{SS}y_i}{n(n-1)} \qquad (2.50)$$

An unbiased variance estimator of y_0^* is

$$s_{y_0^*}^2 = N^2 s_{\bar{y}}^2 = (1-f) N^2 \frac{s_y^2}{n}$$

$$= (1-f)\frac{N^2\,\text{SS}y_i}{n(n-1)} \qquad (2.51)$$

An unbiased estimator of covariance of y_0^* and x_0^* is

$$s_{y_0^* x_0^*} = (1-f) N^2 \frac{s_{yx}}{n}$$

$$= (1-f)\frac{N^2\,\text{SP}y_i x_i}{n(n-1)} \qquad (2.52)$$

A variance estimator of r is

$$s_r^2 = \frac{1}{\bar{x}^2}(s_{\bar{y}}^2 + r^2 s_{\bar{x}}^2 - 2r s_{\bar{y}\bar{x}})$$

$$= \frac{(1-f)}{n\bar{x}^2}(s_y^2 + r^2 s_x^2 - 2r s_{yx})$$

$$= \frac{(1-f)(\text{SS}y_i + r^2\,\text{SS}x_i - 2r\,.\,\text{SP}y_i x_i)}{n(n-1)\bar{x}^2} \qquad (2.53)$$

2.10 Use of random sample numbers

The first two plates of random sample numbers from *Statistical Tables for Biological, Agricultural and Medical Research* by R. A. Fisher and F. Yates are reproduced in Appendix III, Table 1. The use of these tables in selecting probability samples will be illustrated with examples. Random number tables are also provided by Kendall and Smith and by the Rand Corporation.

Notes

1. When a number of plates of random numbers is available, the student should select one plate by any manner and follow through the columns of random numbers from one exercise to the next until the end of the plates and then start again from the first plate.
2. When drawing a sample from a small number of units, such as selecting one adult from the total number of adults in a household, the 'lottery method' may be followed by selecting one out of a number of plastic cards after proper shuffling.

2.11 Examples

In Appendix IV are given for thirty villages located in three zones, the data on the current total population and also the size of the households and the population obtaining during a census conducted five years ago. For some sample villages additional information on items such as the monthly income of the households will also be given. These will constitute our universe, from which samples will be drawn and the principles of sample selection and procedures of estimation of universe values illustrated.

Some universe values follow; normally, of course, not all these values will be known and will have to be estimated from a probability sample:

total number of zones = 3
total number of villages = 30
total number of households = 600
total current number of persons = 3042
previous census number of persons = 2815
total area = 270 km^2
average number of households per village = 20
average number of persons per village = 101.42
average number of persons per household = 5.07

This is, of course, an artificial, hypothetical universe, as can be seen from the small number of households in a village. Examples from live data will, however, be given later.

Example 2.1

From village number 8 in zone I (Appendix IV), select a simple random sample of 5 households without replacement, and on the basis of the data on size and the monthly income of these 5 sample households, estimate for the 24 households in the village the total number of persons, total monthly income, average household size, and average monthly income per household and per person, together with their standard errors.

Of the two plates of random numbers given in Appendix III (Table 1), we choose one plate at random; suppose the second plate is chosen. To obtain a simple random sample of 5 households without replacement from the present universe of 24 households in the village, we start from the top left-hand corner of the plate, and read down the first two columns (as the total number of households is of two digits).

One method of procedure is to select any random number that lies between 01 and 24, and not to consider all numbers between 25 and 99 as also 00; the five numbers thus selected would constitute our required sample of 5 households. (As selection is without replacement, if a random number is repeated in the draws we would reject it, and continue drawing until five different numbers are selected).

However, this procedure will lead to a large number of rejections of the random numbers, on an average 100 − 24 = 76 per cent. A modified procedure is to divide all two-digit random numbers greater than 24 by 24 and take the remainder as the random number selected. Obviously, we should not consider numbers 97 to 99 and also 00 in order to give all the 24 households equal chances of selection.

The random numbers as they appear (and after division by 24 and taking the remainder, if the random number is larger than 24) are: 53 (remainder 5); 97 (rejected); 35 (remainder 11); 63 (remainder 15); 02; 98 (rejected); 64 (remainder 16).

$$SPy_ix_i = \sum^n y_ix_i - \left(\sum^n y_i\right)\left(\sum^n x_i\right)/n$$
$$= 1067 - 21 \times 236/5$$
$$= 1067 - 991.2 = 75.8$$

The unbiased estimate of the average household size is (from equation (2.45)) $\bar{y} = \sum^n y_i/n = 4.2$.

An unbiased variance estimate of $\bar{y}$ is (from equation (2.50)) $s_{\bar{y}}^2 = (1 - f) \times SSy_i/n(n-1) = (0.79167 \times 6.8)/(5 \times 4) = 0.269178$, so that the estimated standard error of $\bar{y}$ is

$$s_{\bar{y}} = \sqrt{(0.269178)} = 0.519$$

and the estimated CV of $\bar{y}$ is

$$s_{\bar{y}}/\bar{y} = 0.519/4.2$$
$$= 0.1236 \quad \text{or} \quad 12.36\%$$

Table 2.7 Size and monthly income of the 5 simple random samples of households selected from 24 households in village no. 8 in zone I

Sample household no.	*Household serial no.*	*Number of persons* y_i	*Total monthly income in* \$ x_i	y_i^2	x_i^2	y_ix_i
1	2	3	33	9	1089	99
2	5	4	40	16	1600	160
3	11	3	34	9	1156	102
4	15	5	68	25	4624	340
5	16	6	61	36	3721	366
Total	–	21	236	95	12 190	1067
Mean	–	4.2 ($\bar{y}$)	47.2 ($\bar{x}$)			

The five households that constitute our sample are then households with serial numbers 5, 11, 15, 2, and 16. The data on the size and monthly income of these 5 sample households are given in Table 2.7, along with the required computations. The household size is denoted by y_i and the monthly household income by x_i.

Here $N = 24$; $n = 5$; $f = n/N = 0.2083$; and $1 - f = 0.7916$. The corrected sums of squares and products are

$$SSy_i = \sum^n y_i^2 - \left(\sum^n y_i\right)^2/n$$
$$= 95 - 21^2/5$$
$$= 95 - 88.2 = 6.8$$

$$SSx_i = \sum^n x_i^2 - \left(\sum^n x_i\right)^2/n$$
$$= 12\,190 - 236^2/5$$
$$= 12\,190 - 11\,139.2 = 1050.8$$

An unbiased estimate of the total number of persons in the village is (from equation (2.43))

$$y_0^* = N\bar{y} = 24 \times 4.2 = 100.8 \quad \text{or} \quad 101$$

with an unbiased variance estimate (from equation (2.51))

$$s_{y_0^*}^2 = N^2 s_{\bar{y}}^2 = 24^2 \times 0.269178$$

so that the estimated standard error of y_0^* is

$$s_{y_0^*} = Ns_{\bar{y}} = 24 \times 0.519 = 12.46$$

An unbiased estimate of the average monthly income per household is $\bar{x} = \$47.20$, with an unbiased variance estimate $s_{\bar{x}}^2 = (1 - f)SSx_i/n(n-1) = 41.5943$; estimated standard error $s_{\bar{x}} = \$6.45$; and estimated CV $s_{\bar{x}}/\bar{x} = \$6.45/\$47.20 = 0.1367$ or 13.67%.

An unbiased estimate of total monthly income of all the 24 households in the village is $x_0^* = N\bar{x} = \$1132.80$ or \$1133, with an unbiased variance estimate $s^2_{x_0^*} = 24^2 \times 41.5943$, so that the estimated standard error of x_0^* is

$$s_{x_0^*} = 24 \times \$6.45 = \$154.80$$

The estimated monthly income per person is, from equation (2.48), the estimated total income divided by the estimated total number of persons, or

$$r = x_0^* / y_0^*$$
$$= \sum^n x_i / \sum^n y_i$$
$$= \$236/21 = \$11.24$$

The estimated variance of r is, from equation (2.53),

$$s_r^2 = (1-f)(\text{SS}x_i + r^2\,\text{SS}y_i - 2r.\,\text{SP}y_ix_i)/n(n-1)\bar{y}^2 = 0.6797$$

Therefore, the estimated standard error of r is $\sqrt{0.6797}$ = \$0.68 and the estimated CV of r is

$$s_r/r = \$0.68/\$11.24$$
$$= 0.0605 \quad \text{or} \quad 6.05\%$$

much smaller than those for the numerator and the denominator, namely, total income and total number of persons.

Note: The standard errors of the totals and the means are rather large: the standard errors could be decreased by increasing the size of the sample and/or by adopting appropriate sampling techniques, detailed in later chapters.

Example 2.2

From the list of 600 households residing in 30 villages (Appendix IV), select a simple random sample of 20 households with replacement, and, on the basis of the data on the size of these 20 sample households, estimate for the whole universe the total number of persons and average household size, together with their standard errors, and the 95 per cent probability limits of the universe values.

Before the required sample can be drawn, we have to obtain a serial list of all the 600 households. The households in the 30 villages need not actually be re-numbered, but the purpose would be served by taking the cumulative totals of the number of households from the 30 villages, given in Table 2.8.

Starting with the 3-digit numbers from where we had left off, and rejecting numbers between 601 and 999 as also 000, we get the following 20 acceptable random numbers:

585 348 39 84 70 18 451 433 504 226
317 366 72 101 551 538 518 359 377 29

Table 2.8 Total and cumulative total number of households in the 30 villages: data of Appendix IV

Zone no.	*Village no.*	*Number of households*	*Cumulative number of households*	*Zone no.*	*Village no.*	*Number of households*	*Cumulative number of households*
I	1	17	17	II	6	19	322
	2	18	35		7	19	341
	3	26	61		8	23	364
	4	18	79		9	25	389
	5	24	103		10	23	412
	6	17	120	III	11	18	430
	7	20	140		1	15	445
	8	24	164		2	21	466
	9	24	188		3	18	484
	10	22	210		4	21	505
II	1	15	225		5	20	525
	2	22	247		6	21	546
	3	17	264		7	21	567
	4	19	283		8	17	584
	5	20	303		9	16	600

Random number 585 refers to (585 − 584 = 1), the first household in the ninth village is zone III; random number 348 refers to (348 − 341 = 7), the seventh household in the eighth village in zone II; and so on. The selected sample of 20 households and the data on household size are given in Table 2.9.

Here $N = 600$ and $n = 20$. The reader should

Table 2.9 Size of the srs of 20 households selected from 600 households in 30 villages

	Zone no. I	Zone no. II	Zone no. III
Village no.	2 2 3 4 4 5 5	2 6 8 8 9 9	1 2 4 5 6 7 9
Household no.	1 12 4 9 11 5 22	1 14 7 18 2 13	3 6 20 13 13 5 1
Size y_i	6 4 5 4 4 4 3	9 5 5 5 4 8	7 3 4 7 8 1 6

verify the following sums, sums of squares and products:

$$\sum y_i = 102 \qquad \sum y_i^2 = 594 \qquad SSy_i = 73.8$$

Therefore,

$$\bar{y} = \frac{1}{n}\sum y_i = \frac{102}{20} = 5.1$$

The estimated total number of persons in the 600 households in the universe is, from equation (2.43),

$$y_0^* = N\bar{y} = 600 \times 5.1 = 3060$$

The estimated variance of y_0^* is, from equation (2.44),

$$s_{y_0^*}^2 = 600^2 \times 73.8/380 = 600^2 \times 0.1942$$

Therefore the estimated standard error y_0^* is

$$s_{y_0^*} = 600 \times 0.4407 = 264.42$$

and the estimated CV of y_0^* is

$$s_{y_0^*}/y_0^* = 264.42/3060 = 0.0863 \text{ or } 8.63\%$$

From Appendix III, Table 2, the value of t for $n - 1 = 19$ degrees of freedom and 0.05 probability is 2.09. Therefore, from equation (2.40), the 95 per cent probability limits of the total number of persons in the 30 villages are $3060 \pm 2.09 \times 264.42$ or 2508 and 3612.

The estimated average household size is $\bar{y} = 5.1$ and the estimated standard error of $\bar{y}$ is

$$s_{\bar{y}} = \sqrt{\frac{SSy_i}{n(n-1)}} = 0.4407$$

The 95 per cent probability limits of the universe average household size are $5.1 \pm 2.09 \times 0.4407$ or 4.2 and 6.0.

Note: The total number and the serial numbering of households in each village will not in general be known beforehand; also, in the above sampling scheme, not all the villages will necessarily be represented. We shall see later how these limitations can be overcome by using stratified multi-stage sample designs.

Example 2.3

From the list of 30 villages (Appendix IV), select a simple random sample of 4 villages with replacement, and from the data on the numbers of households and persons, estimate for the universe the total numbers of persons and households, the average numbers of households and persons per village, and the average household size, together with their standard errors.

With the 30 villages in the 3 zones, it is easy to number these villages serially from 1 to 30. Starting from where we had left off, and using two-digit random numbers, we get the following: 19, 24, 18, 19, 15. In a 'with replacement' sample, we would have to take the first four indicated villages, with village number 19 occurring twice; however, although the sampling fraction 4/30 = 1/7.5 is over 10 per

Table 2.10 Number of households, number of persons, and total monthly income of the srs of 4 sample villages, selected from the 30 villages in Appendix IV

Serial no. i	*Village serial no.*	*No. of households* h_i	*No. of persons* y_i
1	19	25	127
2	24	18	105
3	18	23	114
4	15	20	105
Total	–	86	451
Mean	–	$\bar{h} = 21.5$	$\bar{y} = 112.75$

cent, we shall, for the sake of illustration, reject the repeated random number 19, and select the next non-repetitive random number, 15. The sample then becomes 'without replacement', but we shall treat it as if it were selected with replacement. The required data for these sample villages are given in Table 2.10.

Here $N = 30$, $n = 4$. The reader should verify the following sums of squares and products: $\Sigma h_i^2 = 1878$; $SSh_i = 29$; $\Sigma y_i^2 = 51\,175$; $SSy_i = 324.75$; $\Sigma h_i y_i = 9787$; $SPh_i y_i = 90.5$.

From equations of the type (2.42) to (2.49), the estimated total number of households in the 30 villages is

$$h_0^* = N\bar{h} = 30 \times 21.5 = 645$$

The estimated variance of h_0^* is

$$s_{h_0^*}^2 = 30^2 \times 2.4167$$

so that the estimated standard error of h_0^* is

$$s_{h_0^*} = 30 \times 1.5055 = 45.16$$

and the estimated CV of h_0^* is

$$45.16/645 = 0.0700 \quad \text{or} \quad 7.00\%$$

The estimated total number of persons in the 30 villages is

$$y_0^* = N\bar{y} = 30 \times 112.75 = 3382.5$$

The estimated variance of y_0^* is

$$s_{y_0^*}^2 = 30^2 \times 27.0625$$

so that the estimated standard error of y_0^* is

$$s_{y_0^*} = 30 \times 5.2022 = 156.066$$

and the estimated CV of y_0^* is 156.066/3382.5 = 0.0461 or 4.61 per cent.

The estimated average number of households per village is $\bar{h} = 21.5$, with estimated standard error of $s_{\bar{h}} = 1.5055$. The estimated average number of persons per village is $\bar{y} = 112.75$, with the estimated standard error of $s_{\bar{y}} = 5.2022$.

The estimated average household size is given by the estimated total number of persons divided by the estimated total number of households, or

$$r = y_0^*/h_0^* = 451/86 = 5.24$$

The estimated variance of r is

$$s_r^2 = (SSy_i + r^2 SSh_i - 2r\ SPy_i h_i)/n(n-1)\bar{h}^2$$
$$= 172.5775/5547 = 0.03111,$$

so that the estimated standard error of r is $s_r = 0.1706$ and the estimated CV of r is 0.1706/5.24 = 0.0326 or 3.26 per cent.

2.12 Estimation for sub-universes

Estimates of totals, means, ratios, etc. are often required for the study variables for sub-divisions of the universe or *domains of study,* e.g. for households in different occupation or income groups in household budget surveys, or for different family structures (such as nuclear and extended families) in social investigations. The general estimating equations for a simple random sample apply equally here, the only difference being that the study variable will take the sample value for units belonging to the particular sub-universe, and will be considered to have a zero value otherwise.

Thus, the sub-universe total is given by

$$Y' = \sum^{N} Y_i' \tag{2.54}$$

where $Y_i' = Y_i$ for units belonging to the sub-universe, and zero otherwise.

The sample values are y_i ($i = 1, 2, \ldots, n$) for the study variable, and we define $y_i' = y_i$ if the sample unit belongs to the sub-universe, and $y_i' = 0$ otherwise.

Then if $y_i'^*$ is an unbiased estimator of the total of the sub-universe Y' obtained from the ith unit, a combined unbiased estimator of Y' is the arithmetic mean

$$y_0'^* = \frac{1}{n}\sum^{n} y_i'^* \tag{2.55}$$

and an unbiased estimator of the variance of $y_0'^*$ is

$$s_{y_0'^*}^2 = \frac{\sum^{n} (y_i'^* - y_0'^*)^2}{n(n-1)} \tag{2.56}$$

The covariance and ratio of two sample estimators $y_0'^*$ and $x_0'^*$ are similarly defined, as also is the estimated variance of the ratio $y_0'^*/x_0'^*$.

From equation (2.43), the combined unbiased estimator of Y' from *a simple random sample* (with replacement) is

$$y_0'^* = \frac{N}{n}\sum^n y_i' = N\bar{y}' \tag{2.57}$$

where

$$\bar{y}' = \frac{1}{n}\sum^n y_i' \tag{2.58}$$

is the sample mean of the y_i' values.

An unbiased estimator of the variance of $y_0'^*$ is, from equation (2.44),

$$s^2_{y_0'^*} = \frac{N^2 \sum^n (y_i' - \bar{y}')^2}{n(n-1)} \tag{2.59}$$

If the number of units in the sub-universe N' is known, then an unbiased estimator of the mean of the sub-universe

$$\bar{Y}' = Y'/N' \tag{2.60}$$

is

$$\bar{y}_0'^* = y_0'^*/N' \tag{2.61}$$

An unbiased estimator of the variance of $\bar{y}'^*$ is

$$s^2_{\bar{y}_0'^*} = s^2_{y_0'^*}/N^2 \tag{2.62}$$

If N' is not known, then an estimator of $\bar{Y}'$ is

$$\bar{y}_0'^* = y_0'^*/n_0'^* \tag{2.63}$$

where $n_0'^*$ is the combined unbiased estimator of N'. For a simple random sample, $n_0'^*$ is obtained on putting $y_i' = 1$ for the sample units which belong to the sub-universe, namely,

$$n_0'^* = Nn'/n \tag{2.64}$$

where n' is the number of sample units that belong to the sub-universe. Thus from equation (2.63), the sample average $\bar{y}'$ (equation (2.58)) is an estimator of $\bar{Y}'$.

An estimator of the variance of $\bar{y}'$ is

$$s^2_{\bar{y}'} = \frac{\sum^n (y_i' - \bar{y}')^2}{n'(n'-1)} \tag{2.65}$$

2.13 Estimation of proportion of units

Estimates are often required of the total number or the proportion of the *units* that possess a certain qualitative characteristic or attribute or fall into a certain class. Thus, in a household survey, we might wish to know the proportion that live in houses they own.

If N' is the number of units in the universe possessing the attribute, then the universe proportion of the number with the attribute is

$$P = N'/N \tag{2.66}$$

where N is the total number of units in the universe. If we define our study variable so that it takes the value

$$Y_i = 1, \quad \text{if the unit has the attribute; and}$$
$$Y_i = 0, \quad \text{otherwise,}$$

then the universe number N' with the attribute is

$$N' = \sum_{1}^{N} Y_i$$

and the mean of the Y_i s is

$$\bar{Y} = \sum^{N} Y_i/N = N'/N = P \tag{2.67}$$

Since Y_i can either be 1 or 0, the universe variance of Y_i is

$$\sigma^2 = \sum^{N} (Y_i - \bar{Y})^2/N = \sum^{N} Y_i^2/N - \bar{Y}^2 = P - P^2 = P(1 - P) \tag{2.68}$$

If in the simple random sample of n units (drawn with replacement), r units possess the attribute, then the sample mean

$$\bar{y} = \sum^{n} y_i/n = r/n = p \tag{2.69}$$

is an unbiased estimator of the universe proportion P, where $y_i = 1$ if the sample unit has the attribute; and $y_i = 0$ otherwise.

The sampling variance of p is

$$\sigma_p^2 = \sigma_{\bar{y}}^2 = \frac{\sigma^2}{n} = \frac{P(1 - P)}{n} \tag{2.70}$$

the coefficient of variation per unit is

$$\frac{\sigma}{P} = \sqrt{\frac{P(1 - P)}{P}} = \sqrt{\frac{1 - P}{P}} \tag{2.71}$$

and the CV of p is

$$\frac{\sigma_p}{P} = \frac{\sigma_{\bar{y}}}{P} = \sqrt{\frac{1 - P}{nP}} \tag{2.72}$$

An unbiased estimator of σ^2 is

$$s^2 = \frac{\sum^{n} (y_i - \bar{y})^2}{n - 1} = \frac{(np - np^2)}{n - 1} = \frac{np(1 - p)}{n - 1} \tag{2.73}$$

and an unbiased estimator of the variance of p is

$$s_p^2 = s_{\bar{y}}^2 = \frac{s^2}{n} = \frac{p(1 - p)}{n - 1} \tag{2.74}$$

An estimator of the universe CV is

$$\frac{s}{p} = \sqrt{\left[\frac{(1 - p)}{p} \cdot \frac{n}{(n - 1)}\right]} \tag{2.75}$$

and an estimator of the CV of p is

$$\frac{s_p}{p} = \frac{s_{\bar{y}}}{p} = \sqrt{\left[\frac{1-p}{p(n-1)}\right]} \tag{2.76}$$

An unbiased estimator of N', the number of units in the universe with the attribute, is

$$n'^* = Np \tag{2.77}$$

an unbiased estimator of whose variance is

$$s^2_{n'^*} = N^2 s^2_p = N^2 p(1-p)/(n-1) \tag{2.78}$$

In large samples, the 100 $(1-\alpha)$ per cent probability limits of the universe proportion p are provided by

$$p \pm t'_\alpha s_p \tag{2.79}$$

where t'_α is the α percentage point of the normal distribution.

Note: The results of this section could be derived from those of the preceding section by setting $y'_i = 1$, if the sample unit has the attribute, and zero otherwise. Theoretical proofs are given in Appendix II, section A2.3.5.

Example 2.4

Table 2.11 Size of sample households with piped water supply in the srs of 20 households selected in Example 2.2

	Zone no.				
	I		*II*		*III*
Village no.	2	5	6	8	6
Household no.	1	5	14	18	13
Size y_i	6	4	5	5	8

In the sample of 20 households selected from 600 households given in Example 2.2, the data on the households which were found to have piped water supply are given in Table 2.11. Estimate the proportion and the total number of households having piped water supply and the total number of persons in these households, together with their standard errors.

Here $N = 600$, $n = 20$, and r (the sample number of households possessing the attribute) = 5. From equation (2.69), an unbiased estimate of the universe proportion of households with piped water supply is $p = r/n = 5/20 = 0.25$, and an unbiased estimate of the variance of p is, from equation (2.74),

$$s^2_p = p(1-p)/(n-1)$$
$$= 0.25 \times 0.75/19 = 0.00986842$$

so that the standard error of p is $s_p = 0.09934$.

The unbiased estimate of the number of households with piped water supply is, from equation (2.77),

$$n'^* = Np = 600 \times 0.25 = 150$$

An unbiased estimator of the variance of n'^* is, from equation (2.78)

$$s^2_{n'^*} = N^2 s^2_p = 600^2 \times 0.00976742$$

so that the estimated standard error of n'^* is

$$s_{n'^*} = N s_p = 600 \times 0.09934 = 39.74$$

The estimated number of persons in these households is obtained from equation (2.57) on putting $y'_i = y_i$ for households with piped water supply, and $y'_i = 0$ for other households in the sample. Then

$$y'^*_0 = N \Sigma y'_i/n$$
$$= 600 \times 28/20 = 840$$

An unbiased estimate of the variance of y'^*_0 is obtained similarly from equation (2.59),

$$s^2_{y'^*_0} = 600^2 \times 0.333684$$

so that an estimated standard error of $y_0'^*$ is $600 \times 0.5776 = 346.56$.

Example 2.5

In the sample of 20 households selected from the 600 households (Example 2.2), the numbers of males and females are given in Table 2.12. Estimate the universe proportion of males to total population and its standard error.

The procedure of section 2.13 will be inappropriate in this case, for the sample persons were not selected directly out of the universe of persons. However, we shall illustrate both the appropriate and the inappropriate procedures.

Assuming that the 102 sample persons were selected directly from the universe of persons, we apply the procedure of section 2.13, noting that $n = 102$, and r (the number of males) = 52. Hence from equation (2.69) the estimated proportion of males in the universe is $p = r/n = 52/102 = 0.5098$.

From equation (2.74) the estimated variance of the sample proportion p is $s_p^2 = p(1-p)/(n-1) = 0.00206545$, so that $s_p = 0.04545$.

The above is an inappropriate procedure. The appropriate procedure is that of section 2.9, and the estimated proportion of males in the universe is equal to the estimated total number of males divided by the estimated total number of persons. Note that $N = 600$ households and $n = 20$ households. In Example 2.2, we have already obtained the following for the estimated total number of persons:

$$y_0^* = N\bar{y} = 600 \times 5.1 = 3060$$

with corrected sum of squares

$$SSy_i = 73.8$$

Denoting by x_i the number of males in the ith sample household, we have

$$\bar{x} = \Sigma x_i/n = 52/20 = 2.6$$

$$SSx_i = 24.8 \qquad SPy_ix_i = 37.8.$$

The estimated total number of males is, from equation (2.43),

$$x_0^* = N\bar{x} = 600 \times 2.6 = 1560$$

and the estimated universe proportion of males is, from equation (2.48),

$$r = x_0^*/y_0^* = 1560/3060 = 0.5098$$

as in the inappropriate procedure.

The estimated variance of r is, from equation (2.49),

$$s_r^2 = (SSx_i + r^2 SSy_i - 2r\, SPy_ix_i)/n(n-1)\bar{y}^2$$
$$= 0.0005391082$$

so that $s_r = 0.02322$, considerably different from the value of s_p obtained by the inappropriate procedure.

Table 2.12 Numbers of males and females in the srs of 20 households selected in Example 2.2

																					Total
Male	3	2	3	2	2	2	1	5	3	2	2	2	5	4	2	2	4	3	1	2	52
Female	3	2	2	2	2	2	2	4	2	3	3	2	3	3	1	2	3	5	0	4	50
Total	6	4	5	4	4	4	3	9	5	5	5	4	8	7	3	4	7	8	1	6	102

• 2.14 Method of random groups for estimating variances and covariances

A short-cut method of estimating universe variances and covariances is to group the total of n sample units (selected with replacement) into k groups of m units each ($n = km$), and to compute estimates of the variances and covariances from the group means. The results follow from the fundamental theorems of section 2.7, but will be illustrated for a simple random sample (with replacement).

Let y_{jl} be the value of the study variable for the lth unit ($l = 1, 2, \ldots, m$) in the jth group ($j = 1, 2, \ldots, k$) and the mean for the jth group

$$\bar{y}_j = \frac{1}{m}\sum_{l=1}^{m} y_{jl} \tag{2.80}$$

The mean of all the n sample units is, as before,

$$\bar{y} = \frac{1}{n}\sum_{i=1}^{n} y_i = \frac{1}{km}\sum_{j=1}^{k}\sum_{l=1}^{m} y_{jl} = \frac{1}{k}\sum_{j=1}^{k}\bar{y}_j \tag{2.81}$$

i.e. the mean of the group means.

The overall mean $\bar{y}$ is an unbiased estimator of the universe mean $\bar{Y}$ with variance σ^2/n, so also the group means $\bar{y}_j$ are each an unbiased estimator $\bar{Y}$ but with variance σ^2/m.

An unbiased estimator of $\sigma_{\bar{y}}^2$ is

$$s_{\bar{y}}'^2 = \sum^{k}(\bar{y}_j - \bar{y})^2/k(k-1) \tag{2.82}$$

and an unbiased estimator of the universe variance σ^2 is

$$s'^2 = ns_{\bar{y}}'^2 = m\sum^{k}(\bar{y}_j - \bar{y})^2/k(k-1) \tag{2.83}$$

For two study variables an unbiased estimator of the covariance of $\bar{y}$ and $\bar{x}$ is

$$s_{\bar{y}\bar{x}}' = \sum^{k}(\bar{y}_j - \bar{y})(\bar{x}_j - \bar{x})/k(k-1) \tag{2.84}$$

Note: The variance and covariance estimators $s_{\bar{y}}'^2$ and $s_{\bar{y}\bar{x}}'$ are based on $(k-1)$ degrees of freedom, and are, therefore, less efficient than the variance and covariance estimators, obtained respectively from equations (2.44) and (2.47) based on $(n-1)$ degrees of freedom.

Example 2.6

The simple random sample of 20 households selected (with replacement) from the universe of 600 households in Example 2.2 is arranged in 4 random groups of 5 households each: the random groups were formed in order of selection of the sample households. The results are given in Table 2.13. Obtain estimates of the variance of the sample mean. How would the result differ if there were 2 groups of 10 households each?

Here $n = 20$, $k = 4$, $m = 5$. An unbiased estimate of the variance of $\bar{y}$ is, from equation (2.82), $s_{\bar{y}}'^2 = 0.43$ so that the estimated standard error is $s_{\bar{y}}' = 0.656$.

If the 20 households were formed into 2 groups of 10 households each (taking again the households in the order of selection), then the two group means are

$$\bar{y}_1 = (24 + 29)/10 = 5.3$$
$$\bar{y}_2 = (17 + 32)/10 = 4.9$$

The estimated standard error is, from equation (2.82),

$$s_{\bar{y}}' = \tfrac{1}{2}|\bar{y}_1 - \bar{y}_2| = 0.5 \bullet$$

Further reading

Cochran, Chapters 2 and 3; Deming (1966), Chapter 4; Hansen *et al.*, Vols. I and II, Chapter 4; Kish, Chapter 2; Murthy, Chapter 3; Raj, Chapter 3; Sukhatme and Sukhatme, Chapter II; Yates, sections 6.1–6.4, 6.9, 7.1–7.4, and 7.8.

Table 2.13 Random groups of sample households: data of Example 2.2

Random group no.	*Total number of persons* $\sum_{l=1}^{4} y_{jl}$	*Average size* $\bar{y}_j$	$\bar{y}_j - \bar{y}$	$(\bar{y}_j - \bar{y})^2$
1	24	4.8	−0.3	0.09
2	29	5.8	+0.7	0.47
3	17	3.4	−1.7	2.89
4	32	6.4	+1.3	1.69
All groups	102	5.1	0	5.16

Exercises

1. A simple random sample of 10 agricultural plots drawn without replacement from 100 plots in a village gave the following areas (in acres) under a certain crop:

2.4, 3.2, 2.9, 4.6, 1.9, 2.8, 3.1, 1.8, 3.6, 2.8

Compute unbiased estimates of the mean and total area under the crop for the village, along with their standard errors and coefficients of variation.

2. From a list of 75 308 (= N) farms in a province, a simple random sample of 2072 (= n) farms was drawn. Data for the number of cattle for the sample were:

$$\Sigma y_i = 25\,883; \quad \Sigma y_i^2 = 599\,488.$$

Ignoring the finite sampling correction, estimate from the sample (a) the total number of cattle in the province and (b) the average number of cattle per farm, along with their standard errors and the 95% probability limits. (United Nations *Manual*, Example 1).

3. From village number 8 in zone I (Appendix IV) draw a simple random sample of 5 households without replacement. Obtain estimates of the same characteristics (other than income) for the village as a whole as in Example 2.1, and compare the results.

4. From the list of 30 villages (Appendix IV) draw an srsf of 4 villages with replacement and obtain estimates of the same universe characteristics as in Example 2.3. Compare your results with those in Example 2.3.

5. In a simple random sample of 50 households drawn with replacement from a total of 250 households in a village, only 8 were found to possess transistor radios. These households had respectively 3, 5, 3, 4, 7, 4, 4, and 5 members. Estimate the total number of households that possess transistor radios and the total number of persons in these households, along with their standard errors (Chakravarti, Laha, and Roy, Example 3.1, modified to 'with replacement' scheme).

6. Show that in a simple random sample, unbiased estimators of the universe mean and universe total of a study variable have the same coefficient of variation, and that the CV of the estimated number of units possessing a certain attribute is the same as that of the estimated proportion of such units in the universe.

7. Table 2.14 gives the number of persons belonging to 43 *kraals* which form a random sample of the 325 *kraals* in the Mondora Reserve in Southern Rhodesia, and also the numbers of persons absent from these *kraals.*

(a) Estimate (i) the total number of persons belonging to the reserve, (ii) the number absent from the reserve, and (iii) the proportion of persons absent, and their standard errors.

(b) What would be the estimated standard error of the proportion of persons absent had the sample been a sample of individuals selected directly from all the persons belonging to the reserve? (Yates, Examples 6.9.b and 7.8.b, modified to 'with replacement' sampling for computation of variance; note the differences in notations in Yate's book and this book).

The following corrected sums of squares and products are given:

$$SSy_i = 55\,199.0; \quad SSx_i = 7218.5; \quad SPy_ix_i = 13\,286.5.$$

Table 2.14 Total number of persons (including absentees), y, and number of absentees, x; data from an srs of 43 *kraals*

y	x	y	x	y	x	y	x
95	18	89	7	75	12	159	36
79	14	57	9	69	16	54	26
30	6	132	26	63	9	69	27
45	3	47	7	83	14	61	2
28	5	43	17	124	25	164	69
142	15	116	24	31	3	132	41
125	18	65	16	96	45	82	10
81	9	103	18	42	25	33	8
43	12	52	16	85	35	86	22
53	4	67	27	91	28	51	19
148	31	64	12	73	13		
						3427	799

8. Table 2.15 shows the number of persons ($\bar{y}_i$), the weekly family income (x_i), and the weekly expenditure on food (w_i) in a simple random sample of 33 families selected from 660 families. Estimate from the sample (a) the total number of persons, (b) the total weekly family income, (c) the total weekly expenditure on food, (d) the average number of persons per family, (e) the average weekly expenditure on food per family, (f) the average weekly expenditure on food per person, (g) the average weekly income per person, and (h) the proportion of income that is spent on food, along with their standard errors (adapted from Cochran, 1963, Example, pp. 31–33).

The following raw sums of squares and products are given: $\Sigma\, y_i^2 = 533$; $\Sigma\, x_i^2 = 177\,524$; $\Sigma\, w_i^2 = 28\,224$; $\Sigma\, y_i x_i = 8889$; $\Sigma\, y_i w_i = 3595.3$; $\Sigma\, x_i w_i = 66\,678.0$.

Table 2.15 Size, weekly income, and food cost of the srs of 33 families selected from 660 families

Family number i	*Size* y_i	*Income* x_i	*Food cost* w_i	*Family number* i	*Size* y_i	*Income* x_i	*Food cost* w_i
1	2	$62	$14.3	18	4	$83	$36.0
2	3	62	20.8	19	2	85	20.6
3	3	87	22.7	20	4	73	27.7
4	5	65	30.5	21	2	66	25.9
5	4	58	41.2	22	5	58	23.3
6	7	92	28.2	23	3	77	39.8
7	2	88	24.2	24	4	69	16.8
8	4	79	30.0	25	7	65	37.8
9	2	83	24.2	26	3	77	34.8
10	5	62	44.4	27	3	69	28.7
11	3	63	13.4	28	6	95	63.0
12	6	62	19.8	29	2	77	19.5
13	4	60	29.4	30	2	69	21.6
14	4	75	27.1	31	6	69	18.2
15	2	90	22.2	32	4	67	20.1
16	5	75	37.7	33	2	63	20.7
17	3	69	22.6				
				Total	123	$2394	$907.2

3 Use of Ancillary variables in Simple Random Sampling

3.1 Introduction

If ancillary information is available for each of the units of the universe, it can, under suitable conditions, be used in several ways to improve the efficiency of the estimators of the study variable. One way, to be described in this chapter, is to use the ancillary information, *after sample selection,* in providing what are known as *ratio estimators* and *regression estimators.* The other, described in Chapter 5, is to use the ancillary information *in sample selection* and so also in estimation. The two could be used in combination for two ancillary variables.

3.2 Ratio method of estimation

From a simple random sample (with replacement) of n units out of the universe of N units, an unbiased estimator of the universe total Y has been seen to be

$$y_0^* = \frac{1}{n}\sum^n y_i^* = \frac{N}{n}\sum^n y_i = N\bar{y} \tag{3.1}$$

If, however, the values of an ancillary variable z are available for all the units of the universe, then

$$Z = \sum^N z_i \tag{3.2}$$

the universe total for the ancillary variable is also known, and this information can be utilized to improve the efficiency of the estimators relating to the study variable.

An unbiased estimator of the universe total Z is first obtained *from the sample,* namely,

$$z_0^* = \frac{1}{n}\sum^n z_i^* = \frac{N}{n}\sum^n z_i = N\bar{z} \quad \text{in srs} \tag{3.3}$$

As an estimator of the universe ratio

$$R_{Y/Z} = Y/Z \tag{3.4}$$

the ratio of the estimators of the universe totals of the study and the ancillary variables is taken,

$$r_{y/z} = y_0^*/z_0^* = \sum^n y_i / \sum^n z_i = \bar{y}/\bar{z} \quad \text{in srs} \tag{3.5}$$

As $Y = Z.\ Y/Z = Z.\ R_{Y/Z}$, the ratio estimator of Y is defined as

$$y_R^* = Z.\ r_{y/z} = Z.\ y_0^*/z_0^* = Z.\ \bar{y}/\bar{z} \quad \text{in srs} \tag{3.6}$$

(the subscript 'R' in y_R^* standing for 'ratio').

The sampling variance of $r_{y/z}$ is, from equation (2.22),

$$\sigma_{r_{y/z}}^2 = (\sigma_{y_0^*}^2 + R_{Y/Z}^2\,\sigma_{z_0^*}^2 - 2R_{Y/Z}\,\sigma_{y_0^* z_0^*})/Z^2$$

A variance estimator of $r_{y/z}$ is, from equation (2.49),

$$s^2_{r_{y/z}} = (s^2_{y_0^*} + r^2_{y/z}\, s^2_{z_0^*} - 2r_{y/z}\, s_{y_0^* z_0^*})/Z^2 \qquad (3.7)$$

$$= N^2(\mathrm{SS}y_i + r^2_{y/z}\,\mathrm{SS}z_i - 2r\,\mathrm{SP}y_i z_i)/n(n-1)Z^2 \quad \text{in srswr} \qquad (3.8)$$

as the value of Z is known.

The variance of y_R^* is

$$\sigma^2_{y_R^*} = Z^2\sigma^2_{r_{y/z}} \qquad (3.9)$$

an estimator of which is

$$s^2_{y_R^*} = Z^2 s^2_{r_{y/z}}$$

$$= (s^2_{y_0^*} + r^2_{y/z} s^2_{z_0^*} - 2r_{y/z} s_{y_0^* z_0^*}) \qquad (3.10)$$

$$= N^2(\mathrm{SS}y_i + r^2_{y/z}\,\mathrm{SS}z_i - 2r_{y/z}\,\mathrm{SP}y_i z_i)/n(n-1) \quad \text{in srswr} \qquad (3.11)$$

The following points should be noted:

1. Estimators of the type y_R^*, equation (3.6), are known as *ratio estimators,* obtained by the ratio method of estimation. Such ratio estimators should be distinguished from the *estimators of ratios* (of the type of equations (2.33), (2.48), or (3.5)), where both the numerator and the denominator are estimated from the sample. The two issues are related, but different.

2. The ratio estimator y_R^* is subject to the same observations as those for the estimation of a ratio (note 3, section 2.7). The ratio estimator y_R^* is biased generally: it becomes unbiased if the linear regression of y_0^* and z_0^* passes through the origin (0, 0). An estimator of the bias (this estimator is also generally biased) is

$$(r_{y/z}\, s^2_{z_0^*} - s_{y_0^* z_0^*})/Z \qquad (3.12)$$

3. The variance estimator of y_R^* is valid for large samples generally, say when $n \geqslant 30$.

• 4. *Correction for bias.* If the sample is drawn in the form of k independent sub-samples, each of the same size m (so that $n = mk$) and each giving unbiased estimators y_j^* and z_j^* ($j = 1, 2, \ldots, m$) of the universe totals Y and Z respectively, two estimators of the universe ratio Y/Z are:

$$r = y_0^*/z_0^* \qquad (3.13)$$

and

$$r' = \sum^m r_j/m \qquad (3.14)$$

where $y_0^* = \sum^m y_j^*/m;\quad z_0^* = \sum^m z_j^*/m;\quad r_j = y_j^*/z_j^*$

Noting that in large samples the bias of the estimator r' is m times that of the estimator r, an approximately unbiased estimator of the bias of r is

$$(r' - r)/(m-1)$$

and an almost unbiased estimator of the universe ratio (unbiased up to the second order of approximation) is

$$r'' = (mr - r')/(m-1) \qquad (3.15)$$

Thus an almost unbiased ratio estimator (Quenouille-Durbin-Murthy-Nanjamma estimator) of Y is

$$Z\,.\,r'' \qquad (3.16)$$

For any given sample design, an unbiased ratio-type estimator of the universe total Y (Nieto de Pascual-Rao estimator) is

$$y_R^{**} = Y_R^* + (y_0^* - r' z_0^*)/(m-1) \tag{3.17}$$

where

$$y_R^* = Z . r_{y/z} \tag{3.18}$$

For a simple random sample of n units out of the universe of N units, the ratio estimator of the universe total Y, corrected for bias, is

$$\begin{aligned} y_R^{**} &= y_R^* + (y_0^* - \bar{r} z_0^*)/(n-1) \\ &= y_R^* + N(\bar{y} - \bar{r}\bar{z})/(n-1) \end{aligned} \tag{3.19}$$

where

$$\bar{r} = \sum^n r_i/n = \sum^n (y_i/z_i)/n \tag{3.20}$$

There are other methods of debiasing the ratio estimators.●

5. In obtaining the ratio of two universe totals Y and X, the use of ratio estimators of the two totals is equivalent of using the unbiased estimators of the totals, for

$$\begin{aligned} r_{y/x} &= y_R^*/x_R^* \\ &= Zr_{y/z}/Zr_{x/z} = y_0^*/x_0^* \end{aligned} \tag{3.21}$$

We shall, however, see later in Parts II–IV that in stratified and multistage sampling, ratio estimators may be applied at different levels of aggregation.

Example 3.1

For the same data as for Example 2.3, where a simple random sample of 4 villages was taken from 30 villages, use the information given in Table 3.1 on the population data of a census, conducted five years previously, to obtain ratio estimates of the total numbers of households and persons in the 30 villages. The total population of the 30 villages in the previous census was 2815.

Here $N = 30$; $n = 4$; $Z = 2815$. The reader should verify the following: $\sum^n z_i = 419$; $\bar{z} = 104.75$; $\sum^n z_i^2 = 44\,301$; $\sum^n y_i z_i = 47\,597$; $\sum^n h_i z_i = 9102$; and the corrected sums of squares and products: $SSz_i = 410.75$; $SPy_iz_i = 354.75$; $SPh_iz_i = 93.50$. The other required computations have already been made for Example 2.3.

From equation (3.5), $r_{y/z} = \sum^n y_i/\sum^n z_i = 451/419 = 1.076372$, so that the ratio estimate of the current total population, using the previous census population, is, from equation (3.6),

$$\begin{aligned} y_R^* = Z . r_{y/z} &= 2815 \times 1.076372 \\ &= 3029.99 \text{ or } 3030 \end{aligned}$$

The estimated variance of y_R^* is, from equation (3.11), $30^2 \times 3.079132$, so that the estimated standard error of y_R^* is $30 \times 1.755 = 52.650$, and the estimated CV of y_R^* is 52.65/3030 or 1.74 per cent.

Table 3.1 Previous census population of the srs of 4 villages in Example 2.3

Village serial no.	19	24	18	15
Previous census population z_i	122	97	102	98

Similarly, the ratio estimate of the total number of households is $h_R^* = Z . r_{h/z} = 2815 \times 0.205251 = 577.78$ or 578, with estimated standard error of 24.345, and estimated CV of 4.21 per cent.

Note the following points:

1. The estimated average household size, using the ratio estimates, is the ratio estimate of the current total population divided by the ratio estimate of the total number of households, i.e. 3030 ÷ 578 = 5.24, the same figure as obtained by using simple unbiased estimates of totals in Example 2.3 (note 5 of section 3.2).
2. The ratio estimates of the total numbers of persons and households have smaller standard

Table 3.2 Computation of the ratio estimated, corrected for bias: data of Example 3.1

	Village serial no.				
	19	24	18	15	Average
Ratio h_i/z_i	0.204918	0.185567	0.225467	0.204082	0.2050085
Ratio y_i/z_i	1.040984	1.082474	1.039110	1.071429	1.0584992

errors than the unbiased estimates (in Example 2.3) and the ratio estimates are closer to the universe values.

•3. *Correction for bias.* To obtain ratio estimates of totals corrected for bias, we need the additional computations shown in Table 3.2.

The ratio estimate of the total current population, corrected for bias, is, from equation (3.19),

$$y_{\mathrm{R}}^{**} = y_{\mathrm{R}}^{*} + N(\bar{y} - \bar{r}_{y/z}\,\bar{z})/(n-1)$$
$$= 3048.71 \text{ or } 3049$$

Similarly, the ratio estimate of the total number of households, corrected for bias, is

$$h_{\mathrm{R}}^{**} = h_{\mathrm{R}}^{*} + N(\bar{h} - \bar{r}_{h/z}\,\bar{z})/(n-1)$$
$$= 578.034 \text{ or } 578$$

Here the correction is marginal. •

3.3 Regression method of estimation

Another method of using the ancillary information at the estimating stage which is more general than the ratio method is the regression method of estimation.

The estimated linear regression coefficient β^* for the regression of y_0^* on z_0^* is

$$\beta^* = \mathrm{SP}y_i^* z_i^* / \mathrm{SS}z_i^* \tag{3.22}$$

$$= \mathrm{SP}y_i z_i / \mathrm{SS}z_i \quad \text{in srswr} \tag{3.23}$$

The regression estimator of the universe total Y is defined as

$$y_{\mathrm{Reg}}^* = y_0^* + \beta^*(Z - z_0^*) \tag{3.24}$$

where Z is the known universe total of the ancillary variable.

The variance of y_{Reg}^* in large samples is

$$\sigma^2_{y^*_{\mathrm{Reg}}} = \frac{N^2}{n}\sigma^2(1-\rho^2) \tag{3.25}$$

where ρ is the correlation coefficient between the variables y and z.

An estimator of the variance of y_{Reg}^*, valid in large samples, is

$$s^2_{y^*_{\mathrm{Reg}}} = \frac{\sum^{n} [(y_i^* - y_0^*) - \beta^*(z_i^* - z_0^*)]^2}{n(n-2)} \tag{3.26}$$

$$= \frac{N^2[\mathrm{SS}y_i - (\mathrm{SP}y_i z_i)^2/\mathrm{SS}z_i]}{n(n-2)} \quad \text{in srswr} \tag{3.27}$$

Notes

1. The regression method of estimation is more general than the ratio method: the two become equivalent only when the (linear) regression of y on z passes through the origin (0, 0). The ratio method is, however, simpler to compute and is preferred when the regression line is expected to pass through the origin, i.e. when y is expected to be proportional to z.

2. Although the regression method is more general than the ratio method, it is not much used in practice in large-scale surveys for two reasons: first, the computation of the estimates is

more complex, and secondly, the gain in efficiency is not very marked in many cases as the regression line passes either through the origin or very close to it.

3. When the regression of y on z is perfectly linear, i.e. when the correlation coefficient $\rho = 1$, the variance of the regression estimator becomes zero; and when y and z are uncorrelated, the variance is the same as that of the unbiased estimator.

•4. In some textbooks, in the formula for the estimated variance of the regression estimator, the divisor $n(n-1)$ is used instead of $n(n-2)$. Although in large samples the difference in these two estimators is negligible, the later divisor has been suggested here, as in the regression method two estimators – the y-intercept on z and the (linear) regression coefficient of y on z – are to be computed from the sample, leading to the loss of two degrees of freedom; it is also standard in regression theory and is known to give an unbiased estimator of the error variance in the universe regression equation if the universe is infinite and the regression is linear (Cochran, section 7.3).

•5. In practice, the estimated variance of the regression estimator, namely, $s^2_{y^*_{\text{Reg}}}$ can be greater than that of the ratio estimator, namely, $s^2_{y^*_{\text{R}}}$; in fact, the greatest value the ratio $s^2_{y^*_{\text{Reg}}}/s^2_{y^*_{\text{R}}}$ can take is $(n-1)/(n-2)$, when the y-intercept on z happens to equal zero (that is, the regression line of y on z passes through the origin), so that $\beta^* = \sum^n y_i/\sum^n z_i$, and the numerators of the variance formulae (3.11) and (3.27) are identical.

•6. If the regression of y on z is non-linear, the regression estimator is subject to a bias of the order $1/n$, so that the ratio of the bias to the standard error is small for large samples. The bias is equal to –covariance (β^*, z_0^*).•

Example 3.2

For the same data as for Example 3.1, obtain the regression estimates of the total numbers of households and persons, using the data on the previous census population, along with their standard errors. Here $N = 30$; $n = 4$; $y_0^* = 3382.5$; $h_0^* = 645$; $z_0^* = 30 \times 104.75 = 3142.5$; $Z = 2815$; $\text{SS}y_i = 324.75$; $\text{SS}h_i = 29$; $\text{SS}z_i = 410.75$; $\text{SP}y_iz_i = 354.75$; $\text{SP}h_iz_i = 93.5$.

The estimated (linear) regression coefficient of y on z is, from equation (3.23),

$$\beta^* = \text{SP}y_iz_i/\text{SS}z_i = 0.863664$$

The regression estimate of the current total population is, from equation (3.24),

$$\begin{aligned} y^*_{\text{Reg}} &= y_0^* + \beta^*(Z - z_0^*) \\ &= 3382.5 + 0.863664\,(2815 - 3142.5) \\ &= 3099.65 \end{aligned}$$

or 3100, and the estimated variance of y^*_{Reg} is, from equation (3.27), $30^2 \times 2.295774$, so that the estimated standard error of y^*_{Reg} is $30 \times 1.515 = 45.45$, with the estimated CV of 1.47 per cent, somewhat less than that for the ratio estimate (1.74 per cent), and much smaller than that for the unbiased estimate (4.61 per cent).

Similarly, the regression estimate of the total number of households, using the previous census population, is 570 with an estimated standard error 29.46 (and an estimated CV of 5.16 per cent). Note that in this case, the estimated CV though less than that for the unbiased estimate (7.00 per cent), is larger than that for the ratio estimate (4.21 per cent).

Further reading

Cochran, section 6.1–6.9; Hansen *et al.*, Vol. I, sections 4C and 11.2, Vol. II, sections 4.11–4.19; Kish, Chapters 6 and 12; Murthy, sections 10.1–10.5, 10.9, 10.10, 10.11a–b, 11.2, and 11.3; Raj, sections 5.1–5.17; Sukhatme and Sukhatme, sections 4.1–4.10, 4.12–4.16, and 5.1–5.9; Yates, sections 6.8, 6.9, 6.12, 7.8, and 7.12.

Exercises

1. A simple random sample of 2055 (= n) farms, drawn from the universe of 75 308 (= N) farms to obtain the total number of cattle, gave the following data:

Sample total number of cattle,

$$\sum^n y_i = 25\,751$$

Sample total area of the 2055 farms,

$$\sum^n z_i = 62\,989 \text{ acres}$$

Actual total area of the 75 308 farms,

$$Z = 2\,353\,365 \text{ acres}$$

The uncorrected sums of squares and products were

$$\sum^n y_i^2 = 596\,737 \qquad \sum^n z_i^2 = 2\,937\,851$$

$$\sum^n y_iz_i = 1\,146\,391$$

Obtain the ratio estimate of the total number of cattle, along with the standard error, and compare the results of the unbiased estimate (United Nations *Manual,* Example 7.i).

2. For the same sample as in exercise 1, additional information obtained from a census was available on the number of cattle (w) five years ago in the 75 308 farms.

Sample total number of cattle in the 2055 farms five years ago, $\overset{n}{\Sigma}\, w_i = 23\,642$; $\overset{n}{\Sigma}\, w_i^2 = 504\,150$; $\overset{n}{\Sigma}\, y_i w_i = 499\,172$.

Actual total number of cattle in the 75 308 farms five years ago, $W = 882\,610$.

Obtain the ratio estimate of the total number of cattle and its standard error (United Nations *Manual,* Example 4.ii).

3. For the same data as for exercise 1, obtain the regression estimate of the total number of cattle and its standard error (United Nations *Manual,* Example 9.i).

4. For the same data as for exercise 3.2, obtain the regression estimate of the total number of cattle and its standard error (United Nations *Manual,* Example 9.ii).

5. For the same sample as for exercise 4, Chapter 2, use the previous census population figures, given in Appendix IV, to obtain ratio estimates of the universe characteristics, with the standard errors.

6. Using the same data as for exercise 5, obtain regression estimates of the universe characteristics, with the standard errors.

4 Systematic Sampling

4.1 Introduction

One operationally convenient method of selecting a one in k sample from a list of units is to select first a random number between 1 and k and then select the unit with this serial number and every kth unit afterwards: thus, to take a 5 per cent sample of households during a population census, one would first choose a random number between 1 and 20 (here the sampling fraction is 0.05, so $k = 1/0.05 = 20$); if the random number (also called the *random start*) is 12, then households numbered 12, 32, 52, 72, 92, 112, and so on will constitute a 5 per cent systematic sample. This procedure is known as *systematic sampling* and ensures that each universe unit has the same chance of being included in the sample: the constant k is known as the *sampling interval* and is generally taken as the integer nearest to N/n, the inverse of the sampling fraction. This method has several advantages. First, it is operationally convenient, especially when, as in multi-stage sample designs, information on the lower stage units, especially the ultimate and the penultimate stages (such as households or families or farms) is not available at the central office and the enumerators have to list these units and to draw samples from them. Second, N, the total number of universe units, need not be known beforehand and a systematic sample may be selected along with the listing of the universe units or with a census, if the sampling fraction is fixed beforehand. Third, a systematic sample is spread out more evenly over the universe, so that it is likely to produce a sample that is more representative and more efficient than a simple random sample.

There are two main disadvantages of systematic sampling. First, variance estimators cannot be obtained from a single systematic sample; and second, a bad arrangement of the units may produce a very inefficient sample.

Systematic sampling often suggests itself when there is a sequence of units occurring naturally in space (trees in a forest) or time (landing of fishing crafts on the coast).

In this chapter, we deal with the selection and estimating procedure of systematic samples.

4.2 Linear and circular systematic sampling

To illustrate, let us consider the universe of six households, a, b, c, d, e and f, with respective sizes of 8, 6, 3, 5, 4, and 4. The *linear systematic sample* of 2 households from this universe consists of one of the following: ad, be, and cf, with the respective average household sizes of 6.5, 5.0, and 3.5. The average of the sample averages is 5 (the universe average), showing that the sample mean is an unbiased estimator of the universe mean, and so also for the totals.

Linear systematic sampling has one limitation when the sampling interval is not an integer. In the above example, if the universe had consisted of 5 households a, b, c, d and e, the sampling interval $\frac{5}{2}$ is not an integer, and we can take it either as 2 or 3. Taking it as 2, the possible samples are ace, and bd; and taking it as 3, the possible samples are ad, be, and c. Thus the sample size does not remain fixed in linear systematic sampling when the sampling interval is not an integer. Although unbiased estimators of totals and averages could still be obtained in such a case by using modified formulae, a more satisfactory and simpler procedure is that of *circular systematic sampling,* by selecting first a random number between 1 and N and then selecting this and every kth unit (where k is the integer nearest to N/n) in a cyclical manner until n sample units are selected. In the above example of sampling 2 households systematically from 5 households, the possible circular systematic samples with 2 as the sampling interval are ac, bd, ce, da and eb; with 3 as the sampling interval, the possible circular systematic samples are ad, be, ca, db and ec (Figure 4.1). The sample size thus remains

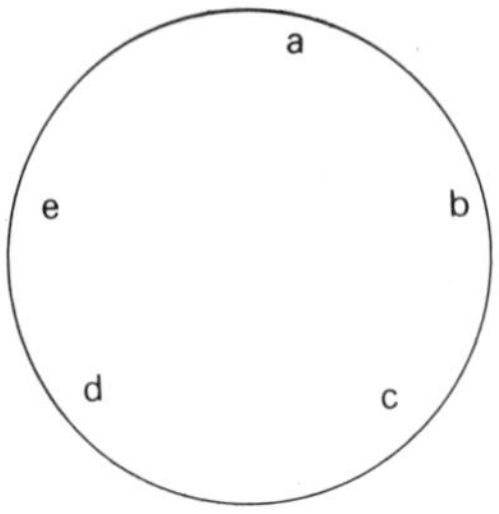

Figure 4.1 Diagrammatic representation of circular systematic sampling.

constant. The usual estimators of totals and averages would also be unbiased. The circular systematic sample reduces to linear systematic sampling when N/n is an integer: it is thus more general, and is to be preferred.

Note that in the first example, unlike the selection of 2 units by simple random sampling without replacement, the combinations ab, ac, bc, bd, bf, cd, ce, de, df and ef do not occur in systematic sampling.

4.3 Sample mean and variance

4.3.1 Mean and variance estimators

From a universe comprising $N = nk$ units, it is proposed to select a systematic sample of 1 in k. The universe units are shown schematically in Table 4.1, with k columns and n rows, the columns 1 to k showing the random starts (or clusters), each column containing n units. A random number is chosen between 1 and k, and the set of n units in the selected column constitutes one systematic sample. Or, in other words, every one of the k columns representing the k possible systematic samples has an equal (= $1/k$) chance of being selected.

Let y_{rj} denote the value of the unit with the serial number $r + (j - 1)k$ in the universe ($r = 1, 2, \ldots, k$; $j = 1, 2, \ldots, n$). If r is the chosen random number, then the sample mean is

$$\bar{y}_r = \frac{1}{n}\sum_{j=1}^{n} y_{rj} \tag{4.1}$$

and the universe mean is

$$\bar{Y} = \frac{1}{nk}\sum_{r=1}^{k}\sum_{j=1}^{n} y_{rj} = \frac{1}{k}\sum^{k} \bar{y}_r \tag{4.2}$$

i.e. the mean of the sample means.

Table 4.1 Serial numbering of the universe units showing the systematic samples of n units from N (= nk) units

Row number	*Random start (cluster number)* 1	2	...	r	...	k
1	1	2	...	r	...	k
2	$1+k$	$2+k$	...	$r+k$	...	$2k$
.	.	.	.	.	.	.
.	.	.	.	.	.	.
.	.	.	.	.	.	.
j	$1+(j-1)k$	$2+(j-1)k$	...	$r+(j-1)k$	...	jk
.	.	.	.	.	.	.
.	.	.	.	.	.	.
.	.	.	.	.	.	.
n	$1+(n-1)k$	$2+(n-1)k$	...	$r+(n-1)k$	...	nk

It can be shown that the sample mean $\bar{y}_r$ is an unbiased estimator of the universe mean $\bar{Y}$.

Noting that

$$\sigma^2 = \frac{1}{nk}\sum_{r=1}^{k}\sum_{j=1}^{n}(y_{rj}-\bar{Y})^2$$

$$= \frac{1}{k}\sum_{r=1}^{k}{}_{j}\,(\bar{y}_r-\bar{Y})^2 + \frac{1}{nk}\sum_{r=1}^{k}\sum_{j=1}^{n}(y_{rj}-\bar{y}_r)^2 \tag{4.3}$$

$$= \sigma_b^2 + \sigma_w^2 \tag{4.4}$$

where σ_b^2 is the *between-sample variance* and σ_w^2 is the *within-sample variance,* defined respectively by the two terms on the right-hand side of equation (4.3), then the sampling variance of the sample mean is

$$\sigma_b^2 = \sigma^2 - \sigma_w^2 \tag{4.5}$$

The sampling variance of the sample mean can also be expressed in terms of the intraclass correlation coefficient between pairs of sample units in a column

$$\rho_c = \frac{\sum_{r=1}^{k}\sum_{j'\neq j=1}^{n}(y_{rj}-\bar{Y})(y_{rj'}-\bar{Y})}{kn(n-1)\sigma^2} \tag{4.6}$$

so that

$$\sigma_b^2 = \frac{\sigma^2}{n}[1+(n-1)\rho_c] \tag{4.7}$$

Since $\sigma^2 \geqslant \sigma_b^2$, we note that ρ_c must lie between $-1/(n-1)$ and 1.

4.3.2 Comparison of systematic and simple random sampling

As the sampling variance of the sample mean in srs is σ^2/n in sampling with replacement and $[(N-n)/(N-1)]\ \sigma^2/n$ in sampling without replacement, the relative efficiency of systematic sampling compared to srs with replacement is

$$\frac{1}{1+(n-1)\rho_c} \tag{4.8}$$

and compared to srs without replacement is

$$\frac{N-n}{N-1} \cdot \frac{1}{1+(n-1)\rho_c} \tag{4.9}$$

4.3.3 Discussion of results

Expression (4.5) shows that the sampling variance of the sample mean in systematic sampling will be reduced if the within-sample variance is increased: this condition will be satisfied when the units within each systematic sample are as heterogenous as possible with respect to study variable. This is also seen from Expression (4.7), for σ_b^2 will be zero when ρ_c takes the minimum value of $-1/(n-1)$, i.e. when there obtains the highest degree of negative correlation between pairs of units within each systematic sample. When $\rho_c = -1/(N-1)$, srswor and systematic sampling will be equally efficient; systematic sampling will be more or less efficient than srswor according as ρ_c is less than or greater than $-1/(N-1)$. But when $\rho_c = 1$, systematic sampling will be most inefficient, for then one unit will provide as much information as n units.

In practice, it is difficult to know what value ρ_c will take in units occurring in a natural sequence. When the units can be re-arranged before sampling such that units homogeneous in respect of the study variable (or an ancillary variable that has a high degree of association with the study variable) are put together this will ensure heterogeneity within each systematic sample. Such a re-arrangement need not be physical in the sense that units that were neighbouring before the re-arrangement need not remain so afterwards.

Thus, in the socio-economic inquiry of rural households in the Indian National Sample Survey (1964–5), in a sample village the households were classified according to size, namely, (I) 1–4 persons, and (II) 5 and above, and according to the means of livelihood, namely, (1) agricultural labour, (2) self-employed in agriculture, and (3) others. They were then re-arranged according to the following combinations of the household size and means of livelihood: I–3; I–2; I–1; II–1; II–2; and II–3. This ensured that small and large households with different means of livelihood were proportionately represented in a systematic sample of households.

4.3.4 Periodicity of units

If a universe has periodic variations, then a systematic sample whose interval coincides with the length of the period will be extremely inefficient, but if the interval is made to be an odd multiple of half the period, a very efficient systematic sample will result. If, for example, sales in markets are high on Saturdays and low on Tuesdays, a systematic sample of one in six working days that includes only Saturdays or Tuesdays will be very inefficient. In such cases, instead of selecting a simple random sample, it would be better to take advantage of the knowledge by appropriately choosing the interval in systematic selection so as to break step with periodicity by including both the high and the low phases.

4.4 Estimation of variance

In general, it is not possible to estimate from one sample the variance of the estimators obtained from the systematic sample. (This is because the procedure does not ensure the inclusion of each of the ${}^{N}C_2$ pairs of units of the universe at least in one of the samples, which is a necessary condition for the existence of an unbiased estimator of a variance.) If, however, two or more systematic samples are taken, each with a separate random start, the combined estimator and its variance are obtained from the fundamental theorems of section 2.7.

In practice, moreover, the variance estimators for the simple random sampling (without replacement) may be used: these estimators will be unbiased only if the arrangement of the units is at random.

Example 4.1

(a): From village number 8 in zone I (Appendix IV), select a circular systematic sample of 5 households and for the 24 households in the village obtain estimates of the total number of persons, total monthly income and average monthly income per person; also obtain estimates of standard errors of these estimates, assuming the data came from a simple random sample.

(b): Draw with a fresh random start a second circular systematic sample of 5 households from the village, and obtain estimates with standard errors for the same characteristics.

(c): From the two circular systematic samples, obtain combined estimates of the characteristics along with their standard errors.

(a) The random start (i.e. the random number between 1 and 24) chosen was 14; the sampling interval is 24/5 or 5 to the nearest integer, so that the five sample households are those with serial numbers 14, (14 + 5) = 19, (19 + 5) = 24, 5 and (5 + 5) = 10. The data for these five households are given in Table 4.2.

$$s^2_{y_0^*} = (1 - f)N^2\ \mathrm{SS}y_i/n(n-1)$$
$$= 0.79167 \times 24^2 \times 9.2/20$$
$$= 0.79167 \times 24^2 \times 0.46$$
$$= 24^2 \times 0.364168$$

Therefore, an estimated standard error of y_0^* is

$$s_{y_0^*} = 24 \times 0.603 = 14.472;$$

and an estimated CV of y_0^* is

$$s_{y_0^*}/y_0^* = 16.75 \text{ per cent.}$$

The estimated total monthly income of all the 24 households in the village is

$$x_0^* = N\bar{x} = 24 \times \$38.6 = \$926.40 \text{ or } \$926$$

and estimated variance of x_0^* is

$$s^2_{x_0^*} = (1 - f)N^2\ \mathrm{SS}x_i/n(n-1)$$
$$= 0.79167 \times 24^2 \times 26.95$$
$$= 24^2 \times 21.335506$$

Therefore, an estimated standard error of x_0^* is

$$s_{x_0^*} = 24 \times \$4.619 = \$110.86$$

Table 4.2 Size and monthly income of the circular systematic sample of 5 households from 24 households in village no. 8 in zone I (Appendix IV)

Household serial no.	5	10	14	19	24
Number of persons (y_i)	4	5	4	1	4
Monthly income (\$) (x_i)	40	52	39	20	42

Here $N = 24$ and $n = 5$ so that $f = n/N = 0.2083$. Hence we have $\sum^n y_i = 18$; $\bar{y} = \sum^n y_i/n = 18/5 = 3.6$; $\sum^n x_i = \$193$; $\bar{x} = \sum^n x_i/n = \$193/5 = \$38.6$; $\sum^n y_i^2 = 74$; $\mathrm{SS}y_i = 9.2$; $\sum^n x_i^2 = 7989$; $\mathrm{SS}x_i = 539.2$; $\sum^n y_i x_i = 764$; $\mathrm{SP}y_i x_i = 69.2$.

The estimated total number of persons in the village is given by

$$y_0^* = N\bar{y} = 24 \times 3.6 = 86.4 \text{ or } 86$$

The estimated variance of y_0^* is, from equation (2.51),

and an estimated CV of x_0^* is 11.97 per cent.

The estimated average monthly income per person is given by the estimated total monthly income divided by the estimated total number of persons, i.e.

$$r = x_0^*/y_0^* = \$38.6/3.6 = \$10.72$$

An estimated variance of r is, from equation (2.53),

$$s_r^2 = \frac{(1-f)(\mathrm{SS}x_i + r^2\mathrm{SS}y_i - 2r\ \mathrm{SP}x_i y_i)}{n(n-1)\bar{y}^2} = 0.153267$$

Therefore, an estimated standard error of r is $s_r = s_r$ = \$0.392 and an estimated CV of r is 3.66 per cent.

(b) The second circular systematic sample is drawn by selecting another random start, 51/24, remainder 3; so the selected households are numbered 3, 8, 13, and 23. The data for these households are given in Table 4.3.

The reader should verify the following results: $y_0^* = 115.2$ persons; $s_{y_0^*} = 12.45$ persons; $x_0^* =$ \$1243.20; $s_{x_0^*} =$ \$125.86; $r =$ \$10.79; $s_r =$ \$0.200.

Table 4.3 Size and monthly income of the second circular systematic sample of 5 households selected from 24 households in village no. 8 in zone I (Appendix IV)

Household serial no.	3	8	13	18	23
Number of persons (y_i)	3	6	6	5	4
Monthly income (\$) (x_i)	32	62	65	53	47

(c) From the fundamental theorem of section 2.7, the combined estimate of total number of persons is $\frac{1}{2}(86.4 + 115.2) = 100.8$ or 101 persons, with an estimated standard error of, from equation (2.8), $\frac{1}{2}|86.4 - 115.2| = 14.4$ and an estimated CV of 14.29 per cent.

The combined estimate of the total monthly income is

$$\tfrac{1}{2}(\$926.4 + \$1243.2) = \$1084.8$$

with an estimated error

$$\tfrac{1}{2}|\$926.4 - \$1243.2| = \$158.4$$

and an estimated CV of 14.60 per cent.

The combined estimate of average monthly income per person is

$$\$1084.8/100.8 = \$10.76$$

an estimated variance of which is

$$\frac{[(926.4-1243.2)^2+(10.76)^2(86.4-115.2)^2 - 10.76(926.4-1243.2)(86.4-115.2)]}{[4 \times (100.8)^2]} = 0.1176$$

so that an estimated standard error is \$0.333 and the estimated CV is 3.09 per cent.

Further reading

Cochran, Chapter 8; Hansen *et al.*, Vol. I, section 11.8; Hendricks, Chapters VIII and IX; Kish, Chapter 4; Murthy, Chapter 5; Sukhatme and Sukhatme, Chapter IX; Yates, sections 3.6, 6.20, and 7.18.

Exercise

1. Out of 24 villages in an area, two linear systematic samples of 4 villages each were selected. The total area under wheat in each of these sample villages is given in Table 4.4. Estimate the total area under wheat in the area (Murthy, Problem 7.3, adapted).

Table 4.4 Area under wheat of the two linear systematic samples of 4 villages each from 24 villages in an area

Linear systematic sample	*Sample village*			
	1	*2*	*3*	*4*
1	427	326	481	445
2	335	412	503	348

5 Varying Probability Sampling: Sampling with Probability Proportional to Size

5.1 Introduction

In simple random and systematic sampling, the only information required on the universe units prior to sampling is the serial listing. For systematic sampling, although it was indicated that a re-arrangement of the universe units according to the values of an ancillary variable might provide more efficient estimators, only the relative magnitudes rather than the actual values of the ancillary variable are required.

It would seem reasonable to suppose that if the values of an ancillary variable related to the study variable were known for all the N units, the information could be utilized in selecting the sample so as to provide estimators with greater efficiency than those from simple random or systematic sampling. In contrast to the sampling procedures considered so far that assign the same probability of selection to all the units of the universe, another procedure will now be outlined that utilizes the values of the ancillary variable such that unequal probabilities of selection are assigned to the universe units; the estimating procedure adopted ensures unbiased estimators of totals and averages with much greater efficiency, under favourable conditions, than those from simple random or systematic sampling.

In this chapter we shall deal with the most common type of sampling with varying (or unequal) probabilities, namely, *sampling with probability proportional to 'size'* (pps), the 'size' being the value of the ancillary variable. We shall often refer to a sample so selected as a pps sample, and, unless otherwise specified shall deal with pps sampling with replacement.

5.2 Selection of one unit with probability proportional to size

Consider the universe of 6 households with respective household sizes given in Table 5.1. Let us draw one household from this universe with probability proportional to household size in order to estimate the total size (total number of persons): if, of course, the sizes of the households were known, we would not need to draw a sample to estimate the total size, but this will illustrate the logic of pps sampling. If households are selected with pps, the probability of selection of any household is its size divided by the total size of the universe (30); these probabilities are shown in the last column of Table 5.1. Note that the total of the probabilities is 1, as it should be.

In simple random sampling, the probability of selection of a unit at any draw is $1/N$, where N is the total number of universe units, and the unbiased estimator of the universe total Y for the study variable is obtained from any draw by multiplying the value of the unit drawn (y_i) by the total number of universe units (N), or in other words, on dividing y_i by the probability of selection ($= 1/N$, a constant in srs) of the unit i. Similarly, in varying probability sampling, the

Table 5.1 Selection of one household with probability proportional to size from a hypothetical universe of six households in Table 2.1

Household no. i	*Household identification*	*Household size* y_i	*Probability of selection* $\pi_i = y_i / \sum_i^N y_i$
1	a	8	8/30
2	b	6	6/30
3	c	3	3/30
4	d	5	5/30
5	e	4	4/30
6	f	4	4/30
Total	–	30	1

unbiased estimator of Y is obtained from the ith draw on dividing the value of the ith unit (y_i) by the probability of selection π_i (which varies from unit to unit in the universe); thus the unbiased estimator of the universe total Y in varying probability sampling is

$$y_i^* = y_i / \pi_i \tag{5.1}$$

If the values y_i of all the universe units were known before sampling and sampling is carried out with probability proportional to y_i, i.e.

$$\pi_i = y_i \Big/ \sum^N y_i \tag{5.2}$$

then the unbiased estimator y_i^* given in estimating equation (5.1) becomes

$$y_i^* = y_i / \pi_i = \sum^N y_i = Y$$

the universe total. In our example, if one draw of pps sampling gives household c, the probability of selection of which is $\frac{3}{30}$, then the unbiased estimator of the total universe size from this sampled household is $3 \div \frac{3}{30} = 30$, the actual universe total.

If, instead of drawing a unit with probability proportional to its actual value, we had drawn it with probability proportional to an ancillary variable, whose size (z_i) is related to the unit value (y_i) by the exact relation

$$z_i = \beta y_i$$

where β is a positive constant, the probability of selection

$$\text{pp}z_i = z_i \Big/ \sum^N z_i = \beta y_i \Big/ \beta \sum^N y_i = y_i \Big/ \sum^N y_i = \text{pp}y_i$$

remains the same, and would give the same results as probability proportional to the value of the study variable, i.e. there would be no sampling error.

The foregoing gives the clue to the determining factor for selection with pps. We cannot, of course, know the actual 'sizes' of the study variable, but if we can find an ancillary variable, the values of which are known to be roughly proportional to the values of the study variable, then we may select the units with probability proportional to the values of the ancillary variable in order to obtain estimators with greater efficiency than those obtained from srs. The ancillary variable chosen should be such that its values are known prior to sampling and the two are linearly related with the regression line passing through the origin (0, 0). If there is a perfect positive correlation between the study- and the ancillary-variables but the regression line does not pass through the origin (0, 0), sampling with pps of the ancillary variable will not necessarily be more efficient than srs.

Some examples of study- and ancillary-variables are as follows:

Study variable	*Ancillary variable*
Current population	Previous census population Geographical area (less suitable)
Current number of births	Previous census population
Current total income	Previous census population
Area under a crop	Total geographical area or cultivated area
Factory production	Number of workers

5.3 Estimating procedures

If we draw a sample of n units ($n \geqslant 2$) with replacement out of the universe of N units, with the initial probability of selection of the ith unit ($i = 1, 2, \ldots, N$)

$$\pi_i = z_i \Big/ \sum^{N} z_i = z_i/Z \tag{5.2}$$

where z_i is the value ('size') of an ancillary variable, the ith selected sample unit ($i = 1, 2, \ldots, n$) having the value of the study variable y_i will provide an unbiased estimator of the universe total Y by the estimating equation

$$y_i^* = \frac{y_i}{\pi_i} = Z\frac{y_i}{z_i} \tag{5.1}$$

By the fundamental theorem of section 2.7, a combined unbiased estimator of Y is

$$y_0^* = \frac{1}{n}\sum^{n} y_i^* = \frac{Z}{n}\sum^{n}\frac{y_i}{z_i} \tag{5.3}$$

The sampling variance of y_0^* is

$$\sigma^2_{y_0^*} = \frac{1}{n}\sum^{N}\left(\frac{Y_i}{\pi_i} - Y\right)^2 \pi_i = \frac{1}{n}\left(\sum^{N}\frac{Y_i^2}{\pi_i} - Y^2\right) \tag{5.4}$$

an unbiased estimator of which is

$$s^2_{y_0^*} = \sum^{n}(y_i^* - y_0^*)^2/n(n-1) \tag{5.5}$$

An unbiased estimator of the universe mean $\bar{Y} = Y/N$ is

$$\bar{y}_0^* = y_0^*/N \tag{5.6}$$

with an unbiased variance estimator

$$s^2_{\bar{y}_0^*} = s^2_{y_0^*}/N^2 \tag{5.7}$$

From the same sample, estimators of the total and the mean of another study variable may be computed from the estimating formulae of the above types. An unbiased estimator of the covariance of the estimators y_0^* and x_0^* of the two universe totals Y and X is

$$s_{y_0^* x_0^*} = \sum^n (y_i^* - y_0^*)(x_i^* - x_0^*)/n(n-1) \tag{5.8}$$

A consistent, but generally biased, estimator of the universe ratios $R = Y/X$ of two universe totals Y and X is the ratio of the sample estimators

$$r = y_0^*/x_0^* \tag{5.9}$$

and an estimator of the variance of r is

$$\begin{aligned} s_r^2 &= (s_{y_0^*}^2 + r^2 s_{x_0^*}^2 - 2r s_{y_0^* x_0^*})/x_0^{*2} \\ &= (\mathrm{SS}y_i^* + r^2\, \mathrm{SS}x_i^* - 2r\, \mathrm{SP}y_i^* x_i^*)/n(n-1)\, x_0^{*2} \end{aligned} \tag{5.10}$$

Notes

1. Theoretical proofs are given in Appendix II, section A2.3.7.
2. ● Pps sampling can be made without replacement, but the estimating formulae are rather complicated. Two relatively simple estimators are, however, mentioned for the case of two sample units ($n = 2$).

   (a) *Ordered (Raj) estimator.* Suppose the two units selected (in order) with pps and without replacement have the respective values y_1 and y_2 and the probabilities of selection π_1 and π_2. Then an unbiased ordered estimator of the universe total Y is

   $$\frac{1}{2}\left[(1+\pi_1)\frac{y_1}{\pi_1} + (1-\pi_1)\frac{y_2}{\pi_2}\right]$$

   with an unbiased variance estimator

   $$\frac{1}{4}(1-\pi_1)\left(\frac{y_1}{\pi_1} - \frac{y_2}{\pi_2}\right)^2$$

   (b) *Unordered (Murthy) estimator.* If y_1 and y_2 are the values of the two sample units selected with pps and without replacement with probabilities of selection π_1 and π_2 respectively, then an unbiased unordered estimator of Y is

   $$\frac{1}{2-\pi_1-\pi_2}\left[(1-\pi_2)\frac{y_1}{\pi_1} + (1-\pi_1)\frac{y_2}{\pi_2}\right]$$

   with an unbiased variance estimator

   $$\frac{(1-\pi_1)(1-\pi_2)(1-\pi_1-\pi_2)}{(2-\pi_1-\pi_2)^2}\left[\frac{y_1}{\pi_1} - \frac{y_2}{\pi_2}\right]^2$$

   With the advent of electronic computers and the use of sample designs with two sample first-stage units in each stratum (after the strata have been formed in a desirable manner, to be detailed later), the unordered estimator can be employed increasingly in the future.

   The unordered estimator is more efficient than the ordered, and both are more efficient than the unbiased pps 'with replacement' estimator. ●
3. Note that the selection of the sample units with probability proportional to 'size' is equivalent to the selection with probability proportional to the ratios of the sizes to their (a) total or (b) average. For in the latter cases, the selection probabilities are:

   (a) $\pi_i' = \dfrac{z_i/Z}{\sum^N (z_i/Z)} = \dfrac{z_i}{Z} = \pi_i$ (b) $\pi_i'' = \dfrac{z_i/\bar{Z}}{\sum^N (z_i/\bar{Z})} = \dfrac{z_i}{Z} = \pi_i$

Thus in pps sampling it is not necessary to know the 'sizes' if the ratios of these sizes to their total or average are known.

4. Pps sampling can be made with a suitable function of the value of ancillary variable. For example, in surveys on fruit count, the selection of branches with probability proportional to the fourth power of the girth may be more efficient than ppg^3, ppg^2, or ppg ('g' indicating girth) or srs (Murthy, Section 6.6); in a stratified two-stage design, selection with probability proportional to the square root of the number of ssu's will be reasonably close to the optimum and more efficient than probability proportional to the number of ssu's or srs if the costs vary substantially with both the number of fsu's and the number of ssu's per fsu (Hansen *et al.*, vol. I, section 8.14, vol. II, section 8.3).

5.4 Procedures for the selection of sample units with pps

5.4.1 Selection from a list

For selection of sample units with pps (and replacement) from a list, two methods are generally used.

(a) *Cumulative method.* This method entails cumulation of the sizes of the units in the universe.

$$z_1; z_1 + z_2;\ z_1 + z_2 + z_3; \ldots; \sum^{i} z_i; \ldots; \sum^{N} z_i = Z$$

Random numbers are then drawn between 1 and Z. If the random number is greater than $\sum^{i-1} z_i$, but less than or equal to $\sum^{i} z_i$, then the ith unit is selected. The procedure is continued till the required number n of sample units have been drawn.

(b) *Selection of a pair of random numbers* (*Lahiri's method*). The other method does not need cumulation of the sizes of the universe units. A pair of random numbers is chosen, the first between 1 and N, and the other between 1 and z_{max}, where z_{max} is the maximum value of z_i, obtained on inspection. If for any pair of random numbers chosen, the first number is i and the second number is $\leqslant z_i$, then the ith unit is selected; otherwise it is rejected, i.e. no selection is made, but a fresh pair of random numbers chosen. This is continued until the required total number of sample units n have been drawn. To minimize the number of rejections, a very large z_{max} may be split up into more than one part, and the original unit selected whenever one of the split units is drawn. Both these methods of procedure will be illustrated with examples.

5.4.2 Selection from a map

A third procedure is available for selection of geographical area units from a map with probability proportional to area (ppa). Figure 5.1 of the map of 16 fields in a village gives an example. A pair of random numbers is chosen, the first between 1 and the length of the village (in our example, 12, in a certain unit) and the second between 1 and the breadth of the village (say 9). The selected pair of random numbers fixes a point on the map, and the field on which it falls is selected; if the point falls outside the village area, it is, of course, rejected. For example, let the pair of random numbers be 06 and 86 (remainder after division of 86 by 9 is 5); this point (6, 5) is plotted on the map and is seen to fall in field

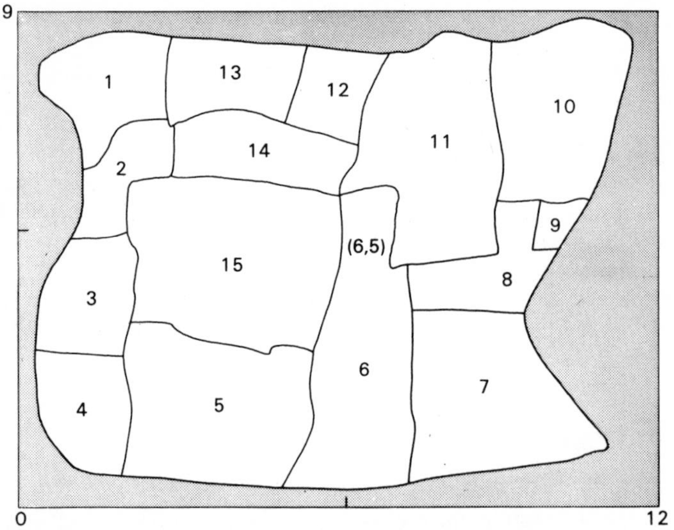

Figure 5.1 Selection of a farm with probability proportional to area.

number 6, which is selected. The procedure ensures selection with probability proportional to area, and does not require the values of the areas for selection, if a map is available; for estimation of the universe values, however, the areas of the selected fields and also the total area should be known.

• 5.4.3 Pps systematic sampling

In this method of selection, the cumulative sizes $\sum^{i} z_i$ are first obtained ($i = 1, 2, \ldots, N$ for the universe). If n is the sample size, the sampling interval I is the integer nearest to Z/n. If r is the number chosen at random from 1 to I, the units corresponding to the numbers $(r + jI)$, $j = 0, 1, 2, \ldots, (n-1)$ are selected. The ith universe unit will be selected if

$$\sum^{i-1} z_i < r + jI \leqslant \sum^{i} z_i$$

for some value of j between 0 and $(n-1)$. If Z/n is not an integer, a pps circular systematic sample can be obtained by selecting a random start from 1 to Z and then proceeding cyclically with the integer nearest to Z/n as the interval.

The estimator for the universe total Y has the same form as that for ppswr sampling, and is unbiased, but as in simple systematic sampling (Chapter 4), a single pps systematic sample cannot provide an unbiased variance estimator. The variance can be estimated unbiasedly by selecting two or more pps systematic samples, each with a separate random start. This method will provide more efficient estimators than simple pps sampling when the units are arranged in ascending or descending order of y_i/z_i; in practice, the y_i values will not be known and those for a previous period or of a related variable will have to be used.•

Example 5.1

From village number 8 in zone I (Appendix IV), given the sizes of the 24 households in the village, select 5 households with probability proportional to size (and with replacement) and from the data on total monthly income and food cost of these 5 households, obtain for the whole village estimates of the total monthly income, the total monthly food cost, and the proportion of income spent on food with their standard errors.

In this example, we shall illustrate the selection of the pps sample by cumulating the sizes. The cumulative sizes of the 24 households are shown in Table 5.2. As the total size of the 24 households is 109, we choose random numbers between 1 and 981 (the highest three digit number which is a multiple of 109); if the three-digit random number is greater than 109, it is divided by 109, and the remainder taken. The first three-digit random number is 213, which leaves a remainder of 104

Table 5.2 Cumulative household sizes and selection of households with probability proportional to size: village no. 8 in zone I (Appendix IV)

Serial no.	*Size*	*Cumulative size*	*Serial no.*	*Size*	*Cumulative size*
1	5	5	13	6	63
2	3	8	14	4	67
3	3	11	15	5	72
4	7	18	16	6	78
5	4	22	17	3	81
6	4	26	18	5	86
7	6	32	19	1	87
8	6	38	20	3	90
9	4	42	21	5	95
10	5	47	22	6	101
11	3	50	23	4	105
12	7	57	24	4	109

Table 5.3 Pps sample of 5 households: data of Table 5.2

Random number		
Three-digit	*Remainder after division by 109*	*Serial number of selected household*
213	104	23
290	72	15
953	81	17
908	36	8
464	28	7

Table 5.4 Monthly total income and food cost of 5 sample households selected with probability proportional to size in village no. 8, zone I (Appendix IV) and computation of estimates

Household serial no.	*Household sample no. i*	*Size* y_i	*Selection probability* $\pi_i = y_i/\sum_1^N y_i$	*Total monthly* *Income* x_i	*Food cost* w_i	$x_i^* = x_i/\pi_i$	$w_i^* = w_i/\pi_i$	x_i^{*2}	w_i^{*2}	$x_i^* w_i^*$
						109 x	109 x	109² x	109² x	109² x
7	1	6	6/109	$61	$27	10.1667	4.5	103.3611	20.25	37.7500
8	2	6	6/109	30	62	10.3333	5.0	106.7778	25.00	51.6667
15	3	5	5/109	58	25	11.6000	5.0	134.5600	25.00	58.0000
17	4	3	3/109	35	21	11.6667	7.0	136.1111	49.00	81.6667
23	5	4	4/109	47	22	11.7500	5.5	138.0625	30.25	64.6250
Total						109 x 55.5167	109 x 27.0	109² x 648.8725	109² x 149.50	109² x 301.7084
Mean						109 x 11.1033 (x_0^*)	109 x 5.4 (w_0^*)			

when divided by 109; from Table 5.2, this random number is seen to be greater than the cumulative size up to the 22nd household (101), but less than the cumulative size up to the 23rd household (105): therefore the 23rd household is selected.

The procedure of selection of five sample households with pps is shown in Table 5.3.

The data on the selected five households and the required computations are shown in Table 5.4.

Here $N = 24$; $n = 5$. The probability of selection of a household is

$$\pi_i = y_i / \sum^N y_i = y_i/109$$

where y_i is the known size of the ith household. The corrected sums of squares and products are:

$$SSx_i^* = 109^2 \times 2.451704$$
$$SSw_i^* = 109^2 \times 3.7$$
$$SPx_i^*w_i^* = 109^2 \times 1.91822$$

An unbiased estimate of the total monthly income in the 24 households is, from estimating equation (5.3),

$$x_0^* = \frac{1}{n}\sum x_i^*$$
$$= \frac{1}{5} \times 109 \times 55.5167$$
$$= 109 \times 11.1033 = \$1210.26$$

An unbiased estimate of the variance of x_0^* is, from equation (5.5),

$$s_{x_0^*}^2 = SSx_i^*/n(n-1)$$
$$= 109^2 \times 2.451704/20$$
$$= 109^2 \times 0.1225852$$

Therefore the estimated standard error of x_0^* is

$$s_{x_0^*} = 109 \times 0.34012 = \$38.16$$

and the estimated CV of x_0^* is 3.15 per cent.

Similarly, the estimated total monthly food cost is

$$w_0^* = \sum^n w_i^*/n = 109 \times 5.4 = \$588.6$$

with estimated variance

$$s_{w_0^*}^2 = 109^2 \times 0.185$$

i.e. with estimated standard error

$$s_{w_0^*} = 109 \times 0.43012 = \$46.88$$

and estimated CV of 7.96 per cent.

The estimated proportion of income spent on food is, from equation (5.9),

$$r = w_0^*/x_0^*$$
$$= \$588.6/\$1210.26$$
$$= 0.4863 \text{ or } 48.63 \text{ per cent.}$$

The estimated variance, s_r^2, of r is, from equation (5.10),

$$\frac{109^2(3.7 + 0.4863^2 \times 2.451704 - 2 \times 0.4863 \times 1.9182}{109^2 \times 11.1033^2 \times 20}$$

$$= 0.120704/123.283271 = 0.0009791$$

so that the estimated standard error of r is s_r = 0.03129 and the estimated CV of r is 6.43 per cent.

Example 5.2

For the thirty villages listed in Appendix IV, the population data obtained from a census conducted five years previously are available and given in

Table 5.5 Previous census population of the 30 villages: data of Appendix IV

Zone no.	*Village serial no.*	*Continuous serial no.*	*Previous census population*	*Zone no.*	*Village serial no.*	*Continuous serial no.*	*Previous census population*
I	1	1	69	II	6	16	84
	2	2	81		7	17	85
	3	3	110		8	18	102
	4	4	80		9	19	122
	5	5	92		10	20	102
	6	6	65		11	21	86
	7	7	72	III	1	22	78
	8	8	108		2	23	112
	9	9	106		3	24	97
	10	10	80		4	25	117
II	1	11	72		5	26	106
	2	12	102		6	27	115
	3	13	73		7	28	110
	4	14	84		8	29	104
	5	15	98		9	30	103
						Total	2815

Table 5.5. Select four sample villages with probability proportional to their previous census population, and on the basis of the current population and number of households in these four sample villages, obtain for the thirty villages estimates of the current total population, number of households, and average household size, along with their standard errors.

We shall follow the second procedure for selecting the pps sample which does not require cumulation of the sizes. As there are 30 villages and the maximum previous census population in any village is 122, we take five-digit random numbers, the first two digits referring to the village serial number, and the last three digits to the village census population. The first five-digit number is 06 723; the last three digits, divided by 122, leaves a remainder of 1, which is less than the census population of village serial number 6, namely, 65, so this village is selected. The second five-digit random number if 65 511; the first two-digits, on division by 30, leaves a remainder of 5, and the last three digits, on division by 122, leaves a remainder of 31; which is less than the census population of the village serial number 5, namely, 92, so this village too is selected. The third five-digit random number is 01 932, the last three digits, on division by 122, leaves a remainder of 78, which is greater than the census population of village serial number 1, namely, 69, so this random number is rejected. We continue in this manner until four villages have been selected, as shown in Table 5.6.

Table 5.6 Pps sample of 4 villages: data of Table 5.5

*Random number for**		*Actual census population*	*Accept/ reject*
Village	*Census population*		
06	723 (R 1)	65	Accept
65 (R 5)	511 (R 23)	92	Accept
01	932 (R 78)	69	Reject
71 (R 11)	508 (R 20)	80	Accept
48 (R 18)	222 (R 100)	102	Accept

* R indicates remainder after division by 30 and 122 respectively.

Note: We shall illustrate with this example the procedure of selection of a pps systematic sample. As the interval $I = Z/n = 2815/4 = 703.75$ is not an integer, we first select a random start between 1 and Z (i.e. 2815). Let this be 1938; then our random numbers are 1938; $1938 + 704 = 2642$; $(1938 + 2 \times 704 =) 3346 - 2815 = 531$; and $531 + 704 = 1235$. From Table 5.5 the reader may verify that the cumulative previous census population is 1873 up to village no. 11 in zone II and 1951 up to village no. 1 in zone III; the random number 1938 therefore corresponds to the latter village, which is selected. Similarly the other three sample villages would be village no. 8 in zone III, village no. 7 in zone I, and village no. 5 in zone II.

The data on the four sample villages and the required computations are shown in Table 5.7: the computational procedures are somewhat different from those Example 5.1, and are generally to be preferred.

Here $N = 30$; $n = 4$; $Z = \sum_1^N z_i = 2815$. The corrected sums of squares and products are $SSy_i^* = 88\,945.57$; $SSh_i^* = 16\,714.07$; and $SPy_i^*h_i^* = 37\,524.44$.

An unbiased estimate of the present total population in the thirty villages is, from equation (5.3),

$$y_0^* = \frac{1}{n}\sum y_i^* = 3239.33 \quad \text{or} \quad 3239$$

an unbiased estimate of variance of which is, from equation (5.5),

$$s_{y_0^*}^2 = SSy_i^*/n(n-1) = 7412.1308$$

so that the estimated standard error of y_0^* is

$$s_{y_0^*} = 86.09$$

and the estimated CV is 2.66 per cent.

Similarly, an unbiased estimate of the total number of households in the thirty villages is

$$h_0^* = \frac{1}{n}\sum h_i^* = 672.95 \quad \text{or} \quad 673$$

an unbiased estimate of the variance of which is $s_{h_0^*}^2 = 1392.8392$, so that the estimated standard error of h_0^* is $s_{h_0^*} = 37.32$ and the estimated CV is 5.55 per cent.

The estimated average household size is, from equation (5.9),

$$\begin{aligned} r &= y_0^*/h_0^* \\ &= 3239.33/672.95 \\ &= 4.8136 \quad \text{or} \quad 4.81 \end{aligned}$$

Table 5.7 Present population and number of households of 4 sample villages selected with probability proportional to previous census population and computation of estimates for 30 villages

Village serial no. (1)	Sample village no. i (2)	Size (previous census population) z_i (3)	Reciprocal of probability $1/\pi_i = Z/z_i$ (4)	Present population y_i (5)	Present number of households h_i (6)	$y_i^* = y_i/\pi_i$ (7)	$h_i^* = h_i/\pi_i$ (8)
5	1	92	30.5978	112	24	3426.95	734.35
6	2	65	43.3077	77	17	3334.69	736.23
11	3	72	39.0972	78	15	3049.58	586.46
18	4	102	27.5980	114	23	3146.17	634.75
					Total	12 957.39	2691.79
					Mean	3239.33 (y_0^*)	672.95 (h_0^*)

Sample village no. (2)	y_i^{*2} (9)	h_i^{*2} (10)	$y_i^* h_i^*$ (11)
1	11 743 986.30	539 269.92	2 516 580.73
2	11 120 157.40	542 034.61	2 455 098.82
3	9 299 938.18	343 935.33	1 788 456.69
4	9 898 385.67	402 907 56	1 997 031.41
Total	42 062 467.55	1 828 147.42	8 757 167.65

As an unbiased estimate of the covariance of y_0^* and h_0^* is

$$s_{y_0^* h_0^*} = \mathrm{SP} y_i^* h_i^* / n(n-1) = 3127.0367,$$

the estimated variance of r is, from equation (5.10),

$$s_r^2 = (s_{y_0^*}^2 + r^2 s_{h_0^*}^2 - 2r s_{y_0^* h_0^*})/h_0^{*2}$$
$$= 0.0211556$$

so that the estimated standard error of r is $s_r = 0.1454$ and the estimated CV 3.02 per cent.

Note that although this particular sample has not provided units with relatively large sizes, the CV's are much smaller than those for a simple random sample (Example 2.3).

5.5 Special cases of crop surveys

5.5.1 Introduction

In a survey designed to estimate the total area under any particular crop and its total yield, sampling of fields (or farms or plots) with *probability proportional to total (geographical) area* (ppa) introduces simplifications in the estimating procedures in addition to possible improvements in the efficiency of the estimators. If a map of fields (or farms or plots) is available, selection with ppa may be made by the procedure described in section 5.4.2.

5.5.2 Area surveys of crops

Suppose n fields are selected with ppa

$$\pi_i = a_i/A$$

where

$$A = \sum^{N} a_i$$

and where a_i is the area of the ith field ($i = 1, 2, \ldots, N$ for the universe, and $i = 1, 2, \ldots, n$ for the sample). Let y_i denote the area under a particular crop for the ith sample field, then an unbiased estimator of the total area under the crop in the universe is, from estimating equation (5.3),

$$y_0^* = \sum^{n} y_i^*/n = \sum^{n} y_i/n\pi_i$$

$$= A \sum^{n} a_i p_i/n\pi_i = A \sum^{n} p_i/n = A\bar{p} \qquad (5.11)$$

where $p_i = y_i/a_i$ is the proportion of the area of the ith field under the particular crop, which varies from 0 to 1.

An unbiased estimator of the variance of y_0^* is, from equation (5.5),

$$s_{y_0^*}^2 = A^2 s_{\bar{p}}^2 = A^2 \text{ SS}p_i/n(n-1) \qquad (5.12)$$

where $\text{SS}p_i$ is the corrected sum of squares of p_i.

An unbiased estimator of the overall proportion under the crop is given by the estimated total area under the crop divided by the total geographical area, i.e.

$$A\bar{p}/A = \bar{p} = \sum^{n} p_i/n \qquad (5.13)$$

Thus a simple (unweighted) average of the sample proportions under the crop gives an unbiased estimator of the universe proportion.

An unbiased estimator of the variance of $\bar{p}$ is, from equation (5.12),

$$s_{\bar{p}}^2 = \text{SS}p_i/n(n-1) \qquad (5.14)$$

Note: If the crop is such that it either occupies the whole of a field or no part of it, or if the fields are small enough for this assumption to hold, such that the proportion of the total area under the crop (p_i) is either 1 (whole) or 0 (none), the estimator for the variance of the total area under the crop in equation (5.12) reduces to

$$s_{y_0^*} = A^2\bar{p}(1-\bar{p})/(n-1) \qquad (5.15)$$

For let r of the n sample units have $p_i = 1$, and the rest $(n-r)$ have $p_i = 0$. Then $\sum^{n} p_i = \sum^{r} 1 = r$, also $\sum^{n} p_i^2 = \sum^{r} 1 = r$, so that $\bar{p} = \sum^{n} p_i/n = r/n$, and $\text{SS}p_i = \sum^{n} p_i^2 - n\bar{p}^2 = r - n\bar{p}^2 = n\bar{p} - n\bar{p}^2 = n\bar{p}(1-\bar{p})$. Substituting this value of $\text{SS}p_i$ in equation (5.12), we obtain equation (5.15).

Example 5.3

Ten farms selected with probability proportional to total area from the universe of 100 farms gave the following proportion of area under a crop:

(p_i): 0.20; 0.25; 0.10; 0.30: 0.15; 0.25; 0.20; 0.25; 0.10; 0.20

Estimate the total area under the crop for the 100 farms and its CV; the total geographical area is 16 124 acres.

Here $N = 100$; $n = 10$; $A = 16\,124$; $\sum^{n} p_i = 2.00$; $\sum^{n} p_i^2 = 0.4175$; and $\text{SS}p_i = 0.0175$.

The estimated proportion of area under the crop is, from equation (5.13),

$$\bar{p} = \frac{1}{n}\sum p_i = 2.00/10 = 0.20$$

with estimated variance (from equation (5.14))

$$s_{\bar{p}}^2 = 0.0175/90 = 0.00019444$$

so that $s_{\bar{p}} = 0.01394$.

An unbiased estimate of the total area under the crop is (equation (5.11)) $y_0^* = A.\bar{p} = 16\ 124 \times 0.20 = 3224.8$ or 3225, with an estimated standard error of y_0^* (equation (5.12)) of

$$s_{y_0^*} = As_{\bar{p}} = 16\ 124 \times 0.01394 = 224.77$$

and the estimated CV is 6.97 per cent.

As the t-value corresponding to probability 0.05 and degrees of freedom $(n - 1 =)$ 9 is 2.262, the 95 per cent probability limits to the total area under the crop are $y_0^* \pm 2.262 \times s_{y_0^*}$ or 3225 ± 508 or 2717 and 3733 acres.

5.5.3 Yield surveys of crops

Similar considerations apply for estimating the average yield of a crop per unit of area. If $r_i = x_i/a_i$ is the yield per unit are in the ith sample field (obtained on harvesting the crop and measuring the yield x_i), then an unbiased estimator of the average yield per unit area is (from an equation of the type (5.13))

$$\bar{r} = \frac{1}{n}\sum^{n} r_i \qquad (5.16)$$

i.e. the simple (unweighted) average of the yields per unit area in the different fields. Also,

$$s_{\bar{r}}^2 = \text{SS}r_i/n(n-1) \qquad (5.17)$$

An unbiased estimator of the total yield X is (from an equation of the type 5.11)

$$x_0^* = A\bar{r} \qquad (5.18)$$

and

$$s_{x_0^*}^2 = A^2 s_{\bar{r}}^2 \qquad (5.19)$$

A generally biased but consistent estimator of the average yield per unit of crop area is

$$r' = x_0^*/y_0^* = \bar{r}/\bar{p} \qquad (5.20)$$

an estimator of the variance of which is

$$s_{r'}^2 = (s_{x_0^*}^2 + r'^2 s_{y_0^*}^2 - 2r' s_{y_0^* x_0^*})/y_0^{*2} \qquad (5.21)$$

where $s_{y_0^* x_0^*}$ is an unbiased estimator of the covariance of y_0^* and x_0^*, given by

$$s_{y_0^* x_0^*} = A^2 \ \text{SP}p_i r_i/n(n-1) \qquad (5.22)$$

Note: In the far less common situation when the areas under a particular crop are known for all the fields in the universe, sample fields can be selected with probability proportional to crop area, and an unbiased estimator of the average yield per unit of crop area is given by an equation of the type (5.16), namely

$$\bar{r}'' = \sum r_i''/n \qquad (5.23)$$

and an unbiased estimator of the total yield of the crop by $Y\bar{r}''$, where $r_i'' = x_i/y_i$. Variance estimators of these two estimators are given by estimating equations of the types (5.17) and (5.19) respectively.

5.6 Ratio method of estimation

As with simple random sampling, so also with pps sampling, the ratio method of estimation may be used to improve the efficiency of estimators. The principle is the same and will be illustrated with examples.

Note, first, however, that if sampling is with pps, and the ratio method is used with the help of the sizes themselves, the ratio estimator of the universe total becomes the same as the unbiased estimator from the pps sampling. For if y is the study variable and z the size variable, then with the usual notations,

$$y_0^* = \frac{1}{n}\sum^n y_i^* = \frac{Z}{n}\sum^n \frac{y_i}{z_i} \tag{5.3}$$

$$z_0^* = \frac{1}{n}\sum^n z_i^* = \frac{Z}{n}\sum^n \frac{z_i}{z_i} = Z \tag{5.24}$$

The ratio estimator of the total Y, using the size variable z, is therefore, from an estimating equation of the type (3.6),

$$\begin{aligned} y_R^* &= Zy_0^*/z_0^* \\ &= Zy_0^*/Z = y_0^* \end{aligned}$$

i.e. the unbiased estimator from the pps sample.

A corollary of the above is that the simple (i.e. unweighted) mean of the ratios y_i/z_i becomes the unbiased estimator of the universe ratio Y/Z, for the estimator of Y/Z is

$$\frac{y_0^*}{Z} = \frac{Z\sum^n (y_i/z_i)/n}{Z} \quad \frac{\sum^n r}{n} \tag{5.25}$$

where $r_i = y_i/z_i$. This result has special applications in agricultural crop surveys (see section 5.5).

Although it is known that for sampling with pps to be more efficient than simple random sampling, the size variable should have a high, positive correlation with the study variable, and the linear regression line of the study variable on the size should pass through the origin, in a multi-subject inquiry the size variable chosen (as a necessary compromise owing to the conflicting desiderata of a number of variables) may be such that the the above conditions are not fulfilled in respect of a particular study variable; in other situations, the required information on the desired size may not be available at the time of the sample selection (see exercise 3 at the end of this chapter). In these cases, the ratio method of estimation may be used in order to improve the efficiency of the estimators obtained from the pps sampling.

Thus, if w is the ancillary variable used for the ratio estimation and z is the size variable used for pps sampling, then with the usual notations

$$w_0^* = \frac{1}{n}\sum^n w_i^* = \frac{Z}{n}\sum^n \frac{w_i}{z_i} \tag{5.26}$$

The ratio estimator of the universe total Y is then, using an estimating equation of the type (3.6),

$$y_R^* = W(y_0^*/w_0^*) = Wr \tag{5.27}$$

where

$$W = \sum^N w_i \qquad r = y_0^*/w_0^* \tag{5.28}$$

The variance estimator of y_R^* is, from equation (3.10),

$$s_{y_R^*}^2 = W^2 s_r^2 = (s_{y_0^*}^2 + r^2 s_{w_0^*}^2 - 2rs_{y_0^* w_0^*}) \tag{5.29}$$

Note: For the ratio method of estimation to be efficient, the selection probability should be appropriate for both y and w.

Example 5.4

A sample of 4 villages, drawn with probability proportional to area from the list of 30 villages given in Appendix IV, gave the data on the present population (Table 5.8). Obtain an unbiased estimate of the present total population in the 30 villages. Given the previous census population of the 30 villages in Table 5.6, obtain the ratio estimate of the present total population and compare the two estimates. The total area of the 30 villages is 270.0 km² and the total previous census population, 2815.

data on the previous census were available, we could use these to improve our estimates. We first obtain the unbiased estimate of the total previous census population, from equation (5.26),

$$w_0^* = \frac{1}{n}\sum w_i^* = 3137.68 \quad \text{or} \quad 3138$$

An unbiased estimate of the variances of w_0^* is

$$s_{w_0^*}^2 = SSw_i^*/n(n-1)$$
$$= 224\,942.3142$$

Table 5.8 Present and previous census population of 4 villages selected with probability proportional to area and computation of estimates for 30 villages

Village serial no.	*Sample village no.* i	*Area (km²)* z_i	*Reciprocal of probability* $1/\pi_i = Z/z_i$	*Present population* y_i	$y_i^* = y_i/\pi_i$	y_i^{*2}
(1)	(2)	(3)	(4)	(5)	(6)	(7)
7	1	4.5	60.0000	88	5 280.00	27 878 400.00
13	2	5.8	46.5517	80	3 724.14	13 869 218.74
22	3	10.0	27.0000	83	2 241.00	5 022 081.00
30	4	10.2	26.4706	105	2 779 41	7 725 119.95
				Total	14 024.55	54 494 818.69
				Mean	3 506.14 (y_0^*)	

Sample village no. i	*Previous census population* w_i	$w_i^* = w_i/\pi_i$	w_i^{*2}	$y_i^* w_i^*$
(2)	(8)	(9)	(10)	(11)
1	72	4320.00	18 662 400.00	22 809 600.00
2	73	3398.27	11 548 238.99	12 655 633.24
3	78	2106.00	4 435 236.00	4 719 546.00
4	103	2726.47	7 433 638.66	7 577 977.98
	Total	12 550.74	42 079 513.65	47 762 757.22
	Mean	3137.68 (w_0^*)		

Here $N = 30$; $n = 4$; $Z = \sum^N z_i = 270.0$ km²; $W = \sum^N w_i = 2815$.

An unbiased estimate of the present total population of the 30 villages is

$$y_0^* = \frac{1}{n}\sum y_i^* = 3506.14 \quad \text{or} \quad 3506$$

An unbiased estimate of the variance of y_0^* is

$$s_{y_0^*}^2 = SSy_i^*/n(n-1) = 443\,565.2459$$

so that the estimated standard error of y_0^* is $s_{y_0^*} =$ 666.01 and the estimated CV is 19.00 per cent.

These are the estimates we would obtain if no other information were available. If, however, the

Also,

$$s_{y_0^* w_0^*} = SPy_i^*w_i^*/n(n-1)$$
$$= 313\,175.4731.$$

As the ratio of the two unbiased estimates y_0^* and w_0^* is

$$r = y_0^*/w_0^*$$
$$= 3506.14/3137.68 = 1.11743$$

the ratio estimate of the present total population, using the previous census population is, from equation (5.27),

$$y_R^* = Wy_0^*/w_0^* = Wr$$
$$= 2815 \times 1.11743$$
$$= 3145.57 \quad \text{or} \quad 3146$$

From equation (5.29), the estimated variance of y_R^* is 24 514.8162, so that the estimated standard error is 156.57 and the estimated CV is 4.98 per cent.

Note the tremendous improvement in the estimates by the ratio method of estimation as compared to the unbiased estimate.

Further reading

Kish, Chapter 7; Murthy, Chapter 6 and section 15.5c; Raj, sections 3.12–3.25; Sukhatme and Sukhatme, Chapter II and section 4.17; Yates, sec'ons 3.9, 6.16, 7.15, 8.9, and 10.8.

Exercises

1. A sample of 10 villages was drawn with probability proportional to the 1951 Census population in a *tehsil* (a *tehsil* is an administrative sub-division) in India. The data on the 1951 Census population, and the sample data on cultivated area, are given in Table 5.9. Estimate the total cultivated area in the *tehsil* and its CV, given the total population in the *tehsil* in 1951 as 415 149 (Murthy, Problem 6.2).

Table 5.9 Cultivated area of 10 sample villages, selected with probability proportional to 1951 Census population in a *tehsil* in India

Village serial no.	*1*	*2*	*3*	*4*	*5*	*6*	*7*	*8*	*9*	*10*
1951 census population	5511	865	2535	3523	8368	7357	5131	4654	1146	1165
Cultivated area (acres)	4824	924	1948	3013	7678	5506	4051	4060	809	1013

Table 5.10 Area under a crop and the proportion to total area of 5 sample farms, selected with probability proportional to the total area

Farm No.	3	18	28	34	35
Total area	52	110	300	410	430
Area under a crop	10	24	59	72	103
Proportion of area under the crop	0.1923	0.2182	0.1967	0.1756	0.2395

Table 5.11 Total cultivated area in 1931 and area under wheat in 1936 and 1937 for a sample of 34 villages selected with probability proportional to cultivated area in Lucknow sub-division, India

Village serial no.	*Total cultivated area, 1931 (acres)*	*Area under wheat in 1936 (acres)*	*Area under wheat in 1937 (acres)*	*Village serial no.*	*Total cultivated area, 1931 (acres)*	*Area under wheat in 1936 (acres)*	*Area under wheat in 1937 (acres)*
1	401	75	52	18	186	45	27
2	634	163	149	19	1767	564	515
3	1194	326	289	20	604	238	249
4	1770	442	381	21	701	92	85
5	1060	254	278	22	524	247	221
6	827	125	111	23	571	134	133
7	1737	559	634	24	962	131	144
8	1060	254	278	25	407	129	103
9	360	101	112	26	715	192	179
10	946	359	355	27	845	663	330
11	470	109	99	28	1016	236	219
12	1625	481	498	29	184	73	62
13	827	125	111	30	282	62	79
14	96	5	6	31	194	71	69
15	1304	427	399	32	439	137	100
16	377	78	79	33	854	196	141
17	259	78	105	34	824	255	265

2. From 35 farms with a total geographical area of 5759, a sample of 5 farms drawn with probability proportional to the total area of the farms gave the data shown in Table 5.10. Estimate the average and the total area under the crop with respective standard errors (Sampford, pp. 124–125).

3. Table 5.11 shows for 1937 the area under wheat in 34 villages in Lucknow sub-division (India) selected out of 170 villages with probability proportional to the cultivated area as recorded in 1931.

(a) Given that the total cultivated area in 1931 in the 170 villages was 78 019 acres, obtain an unbiased estimate of the total area under wheat and its standard error.

(b) After the sample selection and enumeration, data on the area under wheat in 1936 became available. Using this information, and given that the total area under wheat in 1936 was 21 288 acres, obtain the ratio estimate of the area under wheat in 1937 and its standard error (Sukhatme and Sukhatme, Example 4.4).

4. Comment on the following observation relating to the unbiased method of estimation in pps sampling. 'Here is a method of estimation that sort of contradicts itself by allotting weights to the selected units that are inversely proportional to their selection probabilities. The smaller is the selection probability of a unit (the greater is the desire to avoid selecting the unit) the larger is the weight that it carries (when selected)'.

(*Hints:* No contradiction is involved in the unbiased method of estimation in pps sampling, the logic of which has been explained in section 5.2. As a pps sample assigns higher probabilities of selection to large-sized units, it is sometimes said to give rise to a biased sample; the bias is later removed by introducing another bias, in the opposite direction, in the estimating procedure by assigning weighting factors that are inversely proportional to the selection probabilities.)

6 Choice of Sampling Units: Cluster Sampling

6.1 Introduction

In the cases of single-stage sampling so far considered, we have illustrated sampling procedures such as simple random sampling, systematic sampling, and varying probability sampling, with different sampling units such as villages, farms, households, and persons. When the recording units occur in clusters or groups, there may be some advantage in selecting a sample of clusters and completely surveying all the recording units within the selected clusters. This is known as *cluster sampling*. Thus, in a household inquiry, a sample of villages may be selected from the total list of villages and all the households in these sample villages surveyed; or in a demographic inquiry, where the elementary, recording unit is the individual for sex, age etc., all the members of the households may be surveyed in a sample of households; and so on. There may thus be a hierarchy of recording (and sampling) units.

The cluster may refer to naturally occurring groups such as individuals in a household, or households or farms in a village, or all the sheep in a flock; clusters may also be formed artificially by grouping together units that are neighbouring or can be surveyed together conveniently. When the total geographical area of the universe is subdivided into smaller areas and a sample of areas is taken, the sampling plan is known as *area sampling*.

No new principles are involved in cluster sampling for the estimation of universe totals, averages, ratios etc., or for their variance estimators: the estimation procedure will of course depend on whether the sample of clusters is selected with equal or varying probabilities. However, the expressions for the universe variances of the estimators and the sample estimators of variances may be formulated differently in order to facilitate the choice of the sampling unit – the cluster or its elements – and its size.

In this chapter, we shall consider the criteria for the choice of the sampling units, and for simplicity, only clusters of equal size (i.e. each with an equal number of elementary units) selected by simple random sampling.

We shall see that when, as it generally happens, the elementary units within a cluster tend to be similar in respect of the study variable, then cluster sampling will be less efficient than a direct (unrestricted) simple random sample of the elementary units, given the same total number of the units in the sample. However, cluster sampling reduces the costs and labour of travel, identification, contact, data collection etc., and may sometimes also reduce non-sampling errors and biases in data (section 25.7). Apart from the latter consideration, the question therefore arises of balancing the general increase in sampling error against the decrease in costs. This leads to the problem of determining the optimal size of cluster, i.e. the number of elementary units it should contain. This problem is considered in the next chapter.

Notes

1. Cluster sampling, as defined, may be used with single- as well as multi-stage designs, and unstratified as well as stratified designs. A single-stage cluster sampling may be considered as the case where all the second-stage units (i.e. the elementary units) in the selected first-stage units (i.e. clusters) are surveyed. Cluster sampling and two-stage sampling, to be considered in Part III, have many common considerations. The term 'cluster sampling' is sometimes used to denote any multi-stage design; the difference in definitions is therefore well worth bearing in mind. The similarity in formulation of the theory in section 6.2 with that of systematic sampling in section 4.3 may also be noted.
2. Cluster sampling in general refers to geographical units; and a sample of households is not generally described as a cluster sample of household members.

6.2 Simple random sampling

6.2.1 Universe parameters

Let the universe consist of N mutually exclusive and exhaustive clusters, and in each cluster let there be the same number of M_0 elementary units, so that the total number of units in the universe is NM_0 (note that until now we had denoted the total number of universe units by N).

Let Y_{ij} denote the value of the study variable for the jth unit ($j = 1, 2, \ldots, M_0$) in the ith cluster ($i = 1, 2, \ldots, N$). The total value of the study variable in the ith cluster is

$$Y_i = \sum_{j=1}^{M_0} Y_{ij} \tag{6.1}$$

and the cluster mean

$$\overline{Y}_i = Y_i/M_0 \tag{6.2}$$

The universe total of the study variable is

$$Y = \sum_{i=1}^{N} Y_i = \sum_{i=1}^{N} \sum_{j=1}^{M_0} Y_{ij} \tag{6.3}$$

and the overall universe mean per unit

$$\overline{Y} = Y/NM_0 = \sum^{N} \overline{Y}_i/N \tag{6.4}$$

is also the mean of the cluster means.

The universe variance

$$\sigma^2 = \frac{\sum_{i=1}^{N} \sum_{j=1}^{M_0} (Y_{ij} - \overline{Y})^2}{NM_0} = \sigma_w^2 + \sigma_b^2 \tag{6.5}$$

where

$$\sigma_w^2 = \frac{\sum_{i=1}^{N} \sum_{j=1}^{M_0} (Y_{ij} - \overline{Y}_i)^2}{NM_0} \tag{6.6}$$

is the *within-cluster variance*, and

$$\sigma_b^2 = \frac{\sum_{i=1}^{N} (\overline{Y}_i - \overline{Y})^2}{N} \tag{6.7}$$

is the *between-cluster variance.*

The between-cluster variance can also be expressed in terms of the intraclass correlation coefficient ρ_c between pairs of units within the clusters (or 'intra-cluster' correlation coefficient), which is defined by

$$\rho_c = \frac{\sum_{i=1}^{N} \sum_{j' \neq j=1}^{M_0} (Y_{ij} - \overline{Y})(Y_{ij'} - \overline{Y})}{NM_0(M_0 - 1)\sigma^2} \tag{6.8}$$

and it can be shown that

$$\sigma_b^2 = \sigma^2 [1 + (M_0 - 1)\rho_c]/M_0 \tag{6.9}$$

Proofs are given in Appendix II, section A2.3.8.

6.2.2 Sample estimators

If n clusters are selected by srs with replacement and all the elementary units in the sample clusters surveyed, the total number of units in the sample is nM_0. If y_{ij} denotes the value of the study variable for the jth unit ($j = 1, 2, \ldots, M_0$) in the ith sample cluster ($i = 1, 2, \ldots, n$), the total value of the study variable in the ith sample cluster is

$$y_i = \sum_{j=1}^{M_0} Y_{ij} \tag{6.10}$$

and the mean of the ith cluster is

$$\overline{y}_i = y_i/M_0 = \overline{Y}_i \tag{6.11}$$

The overall mean per unit in the sample is also the mean of the sample cluster means

$$\overline{y}_c = \frac{\sum_{i=1}^{n} \sum_{j=1}^{M_0} Y_{ij}}{nM_0} = \frac{\sum^{n} \overline{y}_i}{n} \tag{6.12}$$

and it is an unbiased estimator of the universe mean $\overline{Y}$.

The sampling variance of $\overline{y}_c$ is

$$\sigma_{\overline{y}_c}^2 = \sigma_b^2/n \tag{6.13}$$

An unbiased estimator of σ_b^2 is the sample estimator

$$s_b^2 = \sum_{i=1}^{n} (\overline{y}_i - \overline{y}_c)^2/(n-1) \tag{6.14}$$

An unbiased estimator of $\sigma_{\overline{y}_c}^2$ is therefore

$$s_{\overline{y}_c}^2 = s_b^2/n \tag{6.15}$$

From equation (6.9), the sampling variance of the sample estimator $\overline{y}_c$ in terms of the intraclass correlation coefficient ρ_c is

$$\sigma_{\overline{y}_c}^2 = \sigma^2 [1 + (M_0 - 1)\rho_c]/nM_0 \tag{6.16}$$

$$= \sigma_{\overline{y}_{srs}}^2 [1 + (M_0 - 1)\rho_c] \tag{6.17}$$

where

$$\sigma_{\overline{y}_{srs}}^2 = \sigma^2/nM_0 \tag{6.18}$$

is the sampling variance of the mean of nM_0 units selected directly by srs (with replacement). Equation (6.17) is the fundamental formula in cluster sampling.

An unbiased estimator of the within-cluster variance σ_w^2 is the sample estimator

$$s_w^2 = \sum_{i=1}^{n} \sum_{j=1}^{M_0} (y_{ij} - \bar{y}_i)^2 / n(M_0 - 1) \tag{6.19}$$

From equations (6.5) and (6.9),

$$\rho_c = 1 - M_0 \sigma_w^2 / (M_0 - 1)\sigma^2 \tag{6.20}$$

A sample estimator of ρ_c is

$$\hat{\rho}_c = 1 - M_0 s_w^2 / (M_0 - 1)(s_b^2 + s_w^2) \tag{6.21}$$

6.3 Discussion of the results

The following points emerge from the preceding formulae:

1. As $\sigma^2 \geqslant \sigma_w^2$ (from equation (6.5)), from equation (6.20) one sees that ρ_c lies between $-1/(M_0 - 1)$ (when $\sigma_w^2 = 0$) and 1 (when $\sigma_b^2 = 0$).

2. From equation (6.17), it can be seen that
   (i) If $M_0 = 1$, i.e. if each cluster consists of one elementary unit, then there is no clustering, and the variance formula for cluster sampling becomes the same as that for srs.
   (ii) If $\rho_c = 0$ (i.e. when the characteristic is distributed randomly on the ground, and there is no intra-cluster correlation), then the variance formula for cluster sampling becomes the same as that for srs; a sample of one cluster of size M_0 will then provide as much information on the study variable as an srs of M_0 selected directly.
   (iii) If $\rho_c = 1$ (the maximum value), i.e. if all the units in the cluster have the same value for the study variable such that the greatest degree of homogeneity obtains in a cluster, then $\sigma_{\bar{y}_c}^2 = M_0 \sigma_{\bar{y}_{srs}}^2$; i.e. the variance for a cluster sampling will be M_0 times that for an srs. Cluster sampling will then be extremely inefficient.
   (iv) The factor $(M_0 - 1)\rho_c$ gives a measure of the relative change in the sampling variance due to sampling clusters instead of sampling the elementary units directly; for example, if clusters of $(M_0 =)$ 100 persons each are formed and $\rho_c = 0.01$, then $1 + (M_0 - 1)\rho_c = 1.99$ or 2 approximately, so that the variance of cluster sampling will be twice that of an srs of individuals. Thus, a relatively small value of the intracluster correlation coefficient, multiplied by the size of the cluster, could lead to a substantial increase in variance.
   (v) ρ_c will be *negative* for the sex and age composition of members of a household (*see* Note 4, and exercises 3 and 4 at the end of this chapter); cluster sampling of households will then be more efficient than an srs of persons, and the cost-efficiency of cluster sampling greater still.
   (vi) In general, however, ρ_c is positive, and decreases with the cluster size M_0, but the factor $(M_0 - 1)\rho_c$ increases with increasing cluster size, so that cluster sampling becomes less efficient than srs, and increasingly inefficient as the cluster size increases.

3. The comparison of the efficiencies of estimators from a direct (unrestricted) srs of the elementary units and an srs cluster sample of the units will be misleading for two reasons. First, unrestricted simple random samples are not generally realized in practice; and second, even when a cluster sample is

taken (either single- or multi-stage), special procedures of sampling (such as stratification, pps selection of clusters) and of estimation (such as the ratio method of estimation) are often used to increase the efficiency of the estimators.

6.4 Notes

1. Similar considerations apply to unequal size clusters, selected with equal or varying probabilities, with $\overline{M}$, the average size of a cluster, replacing M_0 in equation (6.17) and others.

2. The above formulation also holds for the estimation of proportion of the units in the universe possessing a certain attribute. Defining, as in section 2.13, our study variable so that it takes the value 1, if the unit has the attribute, the value zero otherwise, and if in the ith sample cluster, M_i' units possess the attribute, then the cluster proportion of units possessing the attribute is $M_i'/M_0 = P_i = p_i$; P_i and p_i denote respectively the universe proportion and the sample estimator in the ith sample cluster.

   An unbiased estimator of the universe proportion P is

$$\overline{p}_c = \sum^{n} p_i/n \qquad (6.22)$$

   the mean of the sample cluster proportions.
   The unbiased estimator of the variance of $\overline{p}_c$ is

$$s^2_{\overline{p}_c} = s_b^2/n = \sum^{n} (p_i - \overline{p}_c)^2/n(n-1) \qquad (6.23)$$

3. For proportions, $P(1 - P)/n$ would be the sampling variance if the units were selected directly from an srs. But design-based estimates of variance should be computed in every case. If, for example, an srs of clusters is taken and the ratio method of estimation used for the birth rate, the three variance estimators, namely from srs of clusters with and without ratio method of estimation, and the binomial variance, could be compared, which would show the loss of efficiency due to clustering on the one hand, and the gain due to using the ratio method of estimation.

4. When the study variable is a zero-one or yes-no characteristic, and a design-based estimate of its variance has been obtained, the value of ρ_c can be computed from equation (6.17). Thus, in Example 2.5, we have seen that the design-based estimate of the variance of proportion of males is $s_r^2 = 0.0005391082$, whereas the estimated variance of the proportion had the sample persons been selected directly is $s_p^2 = 0.00206545$. From equation (6.17) and using the sample estimates of variances, we have

$$1 + (M_0 - 1)\rho_c = s_r^2/s_p^2 = 0.5391$$

Taking M_0, the average number of persons per household from the sample, namely, 5.1, we have

$$\rho_c = (0.5391 - 1)/4.1 = -0.11.$$

In this case, the efficiency of sampling households to that of sampling persons directly is $s_p^2/s_r^2 = 2.61$ or 261 per cent (*cf.* exercise 3a at the end of this chapter).

As another example, suppose that a simple random sample (with replacement) of 50 areal units (clusters) each of 300 persons estimated the birth rate at 0.045 per person with variance 0.00000395081, computed by the method of section 2.9.1. Had the 50 x 300 = 15 000 sample persons been selected directly from the universe of persons, then under some simplifying assumptions, from equation (2.74), the same birth rate would have an estimated variance 0.00000283689. Using equation (6.17), the intraclass correlation coefficient is then estimated at 0.00013.

5. In the different geographical strata with clusters of about 300 persons in Cameroon (1960–5), the value of the intraclass correlation coefficient for the birth rate ranged from −0.0013 to 0.0054, with a central value of 0.0013, and for the death rate from −0.0005 to 0.0101, with a central value of 0.0025 (Scott, 1967 and 1968). For the proportion of adult males employed, the coefficient was 0.1 for census enumeration areas (with an average of 1000 persons each) in the Ghana Census of 1960 (Scott, 1967): this indicates that in this case, cluster sampling of enumeration areas would be very inefficient and a sub-sample of households in selected enumeration areas should be taken. If for practical reasons (see section 26.13), a multi-subject survey covering both a demographic and a labour force inquiry is conducted, a possible solution would be to enumerate completely the selected geographical areas for population, births, and deaths, and to select a sub-sample of households for an 'in-depth' inquiry into the size and characteristics of the labour force.

•6. *Variance function.* The intra-cluster correlation coefficient, and therefore the variance of the cluster means, are not explicit functions of the size of the cluster; that is to say, the functions cannot be derived mathematically but depend on how the characteristic is in fact distributed on the ground. This makes it difficult to estimate the sampling variance for a sample of clusters of any one size, given the variance of an equivalent sample of cluster of any other particular size. However, in area sampling for crops in India, the U.K., and the U.S.A., the within-cluster variance has been seen generally to follow the form aM_0^g where a and g are constants to be evaluated from the data (and g is a positive number less than unity). From this and using the sample estimator of the total variance, an expression for the between-cluster variance can be obtained in terms of M_0, if there are at least two values of the within-cluster variance for estimating a and g. Regarding, however, the total universe as a single cluster of NM_0 units, the total variance becomes the same as the within-cluster variance in the finite universe and is equal to $a(NM_0)^g$, i.e. a and g can be estimated from a survey in which only one value of M_0 is used: this formulation may not hold for clusters with large M_0s.•

Further reading

Cochran, Chapter 9; Deming (1966), Chapter 3C; Hansen *et al.*, Vol. I, Chapter 6A–D, Vol. II, Chapter 6; Hendricks, Chapters VI and VII; Kish, sections 5.1, 5.2, 5.4, 6.1, and 6.2; Murthy, Chapter 8; Raj, sections 6.1 and 6.2; Sukhatme and Sukhatme, Chapter VI; Yates, section 2.9.

Exercises

1. From the data and results of exercise 7, Chapter 2, estimate the value of intra-class correlation coefficient for the proportion of persons absent.

2. A bed of white pine seedlings contained six rows, each 434 ft long. Data for four types of sampling

units into which the bed could be divided are shown in Table 6.1 along with estimates of cost (in terms of length of a row that could be covered in 15 minutes). Obtain the optimum sampling unit after comparing the relative cost-efficiencies of the different sampling units (Cochran, pp. 236–237).

Table 6.1 Data for four types of sampling unit

Types of unit	*Relative size of unit*	*Total number of units*	*Universe variance per unit*	*Length of a row (ft) that can be covered in 15 minutes*
u	M_u	N_u	S_u^2	
1-ft row	1	2604	2.537	44
2-ft row	2	1302	6.746	62
1-ft bed	6	434	23.094	78
2-ft bed	12	217	68.558	108

(*Hint:* First compute the relative cost of measuring one unit, in this example, in terms of time required to count one unit: C_u = relative size of unit/length of a row (ft) that can be covered in 15 minutes; then compute the relative net precision of each unit which is defined as being inversely proportional to the variance obtained for fixed cost, namely $M_u^2/(C_u S_u^2)$ is the universe variance per unit. For details, *see* Cochran).

3. (a) Consider households all of size 4, consisting of a couple and two children, and assume that the sex of a child is binomially distributed with the proportion of male children being one-half. Show by computing the between- and within-cluster variances or otherwise that the intraclass correlation coefficient between sexes of different members of the households is $-\frac{1}{6}$, and that the efficiency of sampling households (clusters of persons) to that of sampling persons directly for estimating the sex-ratio is 200% (Sukhatme and Sukhatme, section 6.3).

•(b) Compute separately the values of the intraclass correlation coefficient between sexes of different members of households with 1 child, 3 children, and 4 children respectively in addition to a couple. Generalize the results for a household with M_0 members (a couple and the rest of their children) to show that the intraclass correlation coefficient between the sexes of different members of the households is $-1/{}^{M_0}C_2$.•

(*Hint:* Generalize the following statement for a household of size 4; 'The correlation between the sexes of husband and wife is −1, but for every other of the remaining five pairs is zero, since the sex of the husband or wife will not determine the sex of their children, nor will the sex of one child determine the sex of another. The average value of the correlation between sexes of different members in a household of 4 consisting of a husband, a wife and two children is therefore $-\frac{1}{6}$, Sukhatme and Sukhatme, Section 6.3. Also Deming (1966) pp. 209–11.) (*Note:* As sampling of individuals is rarely done in the field for demographic inquiries, this and the following exercise have relevance mainly to sampling (of punched cards) after the survey.)

• 4. Consider households all of size 5, consisting of a couple and three children (none adult). Show that the intraclass correlation between the age compositions of household members (taking only two broad age-groups of adults and children) is $-\frac{1}{5}$, so that the efficiency of sampling households to that of sampling persons directly to obtain the age composition of the population is 500%. Generalize this result to households with M_0 members (a couple and the rest children, none of whom are adult) to show that the intraclass correlation coefficient is

$$[1 + {}^{(M_0-2)}C_2 - 2(M_0 - 2)]/{}^{M_0}C_2$$ •

7 Size of Sample: Cost and Error

7.1 Introduction

Two important questions on the designing of any sample inquiry are the total cost of the survey and the precision of the main estimates. Both these are related to the size of the sample, given the variability of the data, the type of sampling and the method of estimation. Obviously, the larger the sample, the smaller will be the sampling error, i.e. the greater the precision of the estimates, but the higher will also be the cost. The survey should be so designed as to provide estimates with minimum sampling errors (i.e. with maximum precision) when the total cost is fixed, and to result in the minimum total cost when the precision is preassigned: a sample size fulfilling these conditions is called the *optimal sample size*. We shall see later in Chapter 25 that other considerations such as the existence of non-sampling errors and biases in data have also to be taken into account: these generally increase with sample size beyond a certain point.

The size of a sample will be determined by the objective of the inquiry, and the permissible margin of error in the estimates. For example, during the depression of the Thirties in the U.S.A., when it was not known whether the unemployed numbered five million or fifteen million, the first sample did not necessarily have to be large to be useful; at present, however, larger samples with more sensitive measures are required to estimate if the unemployment rate (the number unemployed as a percentage of the total number in labour force) increases from say 3 to $3\frac{1}{4}$ per cent (Kish, 1971).

For simplicity, we shall in this chapter consider simple random sampling. In general, under conditions in which pps sampling and the ratio method of estimation are applied, these require smaller samples than an srs with the same efficiency.

7.2 Simple random sampling

7.2.1 Sampling for proportions

We have seen in section 2.13 that the sampling variance of a sample proportion p in an srs of n units is

$$\sigma_p^2 = P(1 - P)/n \tag{7.1}$$

which depends only on the universe proportion P and the sample size n. The coefficient of variation of p is

$$e = \frac{\sigma_p}{P} = \sqrt{\frac{1 - P}{nP}} \tag{7.2}$$

from which

$$n = \frac{1 - P}{Pe^2} \tag{7.3}$$

This determines the sample size n required for any given coefficient of variation e.

As the CV per unit is

$$\mathrm{CV} = \sqrt{\left[\frac{P(1-P)}{P}\right]} = \sqrt{\frac{1-P}{P}} \tag{7.4}$$

the required sample size n can also be expressed as

$$n = \left[\frac{\text{CV of one unit in universe}}{\text{desired CV of sample estimator}}\right]^2$$

$$= \left[\frac{\text{CV per unit}}{e}\right]^2 \tag{7.5}$$

The sample size can also be determined such that the universe proportion P would lie within a given margin of error d on both sides of the sample estimator with a certain probability $(1 - \alpha)$, which is equivalent to saying that the acceptable risk that the universe proportion P will lie outside the limits $p \pm d$ is α. For this, we use the assumption (which holds when n is large and the proportion P is not too small) that the sample proportion is normally distributed with mean P and standard deviation σ_p. Then the $(1 - \alpha)$ per cent probability limits of the universe proportion P are

$$p \pm t_\alpha \sigma_p \tag{7.6}$$

where t_α is the value of the standard normal deviate that cuts off a total area α from both the tails taken together in the normal curve. Some illustrative values of t_α are given in Appendix III, Table 2; for example, for $\alpha = 0.05$ (1 in 20), $1 - \alpha = 0.95$ and $t = 1.96$ or approximately 2.

Setting the permissible margin of error

$$d = t_\alpha \sigma_p = t_\alpha \sqrt{[P(1-P)/n]} \tag{7.7}$$

and solving for n, we get

$$n = t_\alpha^2 P(1-P)/d^2 \tag{7.8}$$

In Appendix III, Table 3 have been tabulated the values of CV per unit in the universe for some values of the universe proportion P, and in Appendix III, Table 4, the required sample sizes (n) corresponding to given values of the universe CV per unit and the desired CV of the sample estimator. Given a value of P, the universe CV per unit can be obtained from equation (7.4) or read off Appendix III, Table 3; and the required sample size n can then be obtained from equation (7.5) or from Appendix III, Table 4, given the desired CV of the sample estimator e.

The margin of error d can be expressed as

$$d = t_\alpha \sigma_p = t_\alpha P e \tag{7.9}$$

Notes

1. As the CV of the estimated number of units possessing an attribute is the same as that of the estimated proportion of such units in the universe (exercise 6, Chapter 2), the same formulae for n will hold for the estimated number with the attribute as for the estimated proportion.
2. As the universe proportion P will not generally be known, an advance estimate p may be taken and used in the preceding equations.

3. As the value of $p(1-p)$ increases as p approaches $\frac{1}{2}$, a safer estimate of n is obtained on taking as an advance estimate that value of p which is nearer to $\frac{1}{2}$. For very small p, the advanced estimate should not be too rough; in this case, the Poisson approximation can be used, so that equation (7.8) is simplified to

$$n = t_\alpha^2 P/d^2 \tag{7.10}$$

and equation (7.3) to

$$n = 1/Pe^2 \tag{7.11}$$

(The Poisson distribution can be regarded as the limiting distribution of a binomial when P becomes indefinitely small and n increased sufficiently to keep nP finite (say m), but not necessarily large; both the mean and the universe variance per unit are then equal to m).

4. When sampling is without replacement, and the sampling fraction n/N is not negligible, a more satisfactory estimate of the sample size is

$$n' = \frac{n}{1 + n/N} \tag{7.12}$$

where n is obtained from the previous equations on the basis of assumption of sampling with replacement.

5. In practice, estimates of proportions and ratios such as birth and death rates are seldom obtained from a simple random sample of individuals.

Example 7.1

A survey is to be made of the prevalence of the common diseases in a large population. For any disease that affects at least 1 per cent of the individuals in the population, it is desired to estimate the total needed, assuming that the presence of the disease cent. What size of a simple random sample is needed, assuming that the presence of the disease can be recognized without mistake? (Cochran, exercise 4.3).

As the CV of a sample proportion is the same as the CV of the number of persons possessing the attribute, for $P = 0.01$, universe CV per unit = 9.95 (from equation (7.4)). As the desired CV is $e = 0.20$, the required sample size is, from equation (7.5), $n = 2475$. Or, from Appendix III, Table 3, CV = 995% for $P = 0.01$; and for CV 1000% (nearest to 995%) and $e = 20\%$, $n = 2500$.

Example 7.2

The (crude) birth rate is to be estimated in a country from a sample survey with 2.5 per cent CV. What is the sample size required if a rough estimate of the birth rate places it at 40 per 1000 persons?

Here $P = 0.04$, and the universe CV per unit is 4.9 (from equation (7.4)). As the desired CV is $e = 0.025$, $n = 38\,400$ persons (from equation (7.5)).

7.2.2 Sampling of continuous data

If σ^2 is the universe variance per unit for the study variable, and $\overline{Y}$ the universe mean, the universe coefficient of variation per unit in the universe is

$$\text{CV} = \sigma/\overline{Y} \tag{7.13}$$

The sampling variance of the sample estimator $\bar{y}$, obtained from an srs of n units (with replacement), is

$$\sigma_{\bar{y}}^2 = \sigma^2/n \tag{7.14}$$

The CV of the sample estimator $\bar{y}$ is therefore

$$\begin{aligned} e &= \sigma_{\bar{y}}/\overline{Y} \\ &= \sigma/\overline{Y}\sqrt{n} \\ &= (\text{universe CV per unit})/\sqrt{n} \end{aligned} \tag{7.15}$$

From equation (7.15), the sample size n required to obtain the sample mean with a given CV e is

$$n = (\text{universe CV per unit}/e)^2$$
$$= (\text{universe CV per unit/desired CV of sample mean})^2 \qquad (7.16)$$

In Appendix III, Table 4, have been given the required sample sizes (n) corresponding to some values of the universe CV per unit and the desired CV of the sample mean (e).

The sample size may also be determined such that the acceptable risk that the universe mean $\overline{Y}$ will lie outside the limits $\overline{Y} \pm d$ is α. For this, we assume that the sample mean $\bar{y}$ is normally distributed with mean $\overline{Y}$ and standard deviation $\sigma_{\bar{y}} = \sigma/\sqrt{n}$. The $(1 - \alpha)$ per cent probability limits of the universe mean $\overline{Y}$ are

$$\bar{y} \pm t_\alpha \sigma_{\bar{y}} \qquad (7.17)$$

where t_α is the value of the standard normal deviate that cuts off a total area α from both the tails taken together in the normal curve. Some illustrative values of t_α are given in Appendix III, Table 2.

Setting the permissible margin of error

$$d = t_\alpha \sigma_{\bar{y}} = t_\alpha \sigma/\sqrt{n} \qquad (7.18)$$

and solving for n, we get

$$n = (t_\alpha \sigma/d)^2 = \sigma^2/V \qquad (7.19)$$

where V is the desired variance of the sample mean. This determines the sample size n, given the values of σ, d and α.

Notes

1. If the universe total Y is to be estimated with margin of error D, the sample size required is

$$n = (Nt\sigma/D)^2 = (N\sigma)^2/V' \qquad (7.20)$$

where V' is the desired variance of the sample total. Note that the unbiased estimators of the universe mean and total have the same CV (exercise 2, Chapter 6), and equations (7.19) and (7.20) give the same value of n.

2. The above formulation assumes a knowledge of universe CV or σ. The universe CV remains remarkably stable over time and space and for characteristics of the same nature as the study variable; the CV for a previous study in the same or a different area for a related characteristic may therefore be taken. An advance estimate of σ may be taken from a pilot or some other study.

3. When sampling is without replacement, and the sampling fraction n/N is not negligible, a more satisfactory estimator of the sample size is

$$n' = \frac{n}{1 + n/N} \qquad (7.21)$$

This applies both to equations (7.19) and (7.20).

4. If a sample of size n gives a CV e for a sample estimator (mean or total), then to obtain the sample estimator with CV e'', the required sample size is

$$n'' = n(e/e'')^2 \qquad (7.22)$$

This follows from the relation

$$(\text{CV per unit})^2 = ne^2 = n''e''^2 \qquad (7.23)$$

which in turn follows from equation (7.15).

Example 7.3

The coefficients of variation per unit (an area of 1 mile square) obtained in a farm survey in Iowa, U.S.A. are given in Table 7.1. A survey is planned to estimate acreage items with a CV of 2.5 per cent, and the numbers of workers (excluding the unemployed) with a CV of 5 per cent. With simple random sampling, how many units are needed? How well would this sample be expected to estimate the number unemployed? (Cochran, exercise 4.5)

Example 7.4

In a village of 625 households, a simple random sample of 50 households were surveyed in order to estimate the average monthly household expenditure on toilet items. The estimate came out at \$0.88 with a standard error of \$0.10. Using this information, determine the sample size required to estimate the same characteristic in a neighbouring village such that the permissible margin of error at 95 per cent probability level is 10 per cent of the true value (adapted from Murthy, problem 4.5).

Table 7.1 Coefficients of variation per unit (area of 1 mile square) in a farm survey in Iowa, U.S.A.

	Acres in farms	*Acres in corn*	*Acres in oats*	*No. of family workers*	*No. of hired workers*	*No. of un-employed*
Estimated CV per unit (%)	38	39	44	100	110	317

The maximum CV of the items (other than the number unemployed) is that of the number of hired workers, for which the CV is 1.10, the desired CV of which is 0.05. From equation (7.16),

n = (CV per unit/desired CV of sample estimator)^2

$= (1.10/0.05)^2$

$= 22^2 = 484$

The acreage items are required with CV of 0.025. Taking the acreage item with the maximum CV, namely, acres in oats, the required sample size is, from equation (7.16),

$$n = (0.44/0.025)^2 = 17.6^2 = 310$$

so that with a sample size of 484, required for the number of workers, the desired CV of the acreage items will also be attained.

For the number unemployed, we get, from equation (7.15), the CV of the sample estimator from a sample of size 484,

$$e = \text{CV per unit}/\sqrt{n}$$

$$= 3.17/22$$

$$= 0.44 \text{ or } 14.4\%$$

We assume that the CV per unit is the same in both the villages. From the first village, the CV of the sample estimator based on 50 households is $e = s_{\bar{y}}/\bar{y} = \$0.10/\$0.88 = 0.1136$. The permissible margin of error of the estimator in the second village is

d' = 10 per cent of the true value at 95 per cent probability level, or

$$0.1\,\bar{Y}' = d' = t_{95\%}\sigma'_{\bar{y}} = 2\sigma'_{\bar{y}}$$

from equation (7.18); i.e. the desired CV for the sample estimator in the second village is

$$e' = \sigma'_{\bar{y}}/\bar{Y} = 0.1/2 = 0.05$$

From equation (7.22), the required sample size in the second village is

$$n' = n(e/e')^2$$

$$= 50 \times (0.1136/0.05)^2$$

$$= 50 \times 2.27^2 = 258$$

7.2.3 Sample sizes for sub-divisions of the universe

If estimates are required not only for the universe as a whole, but for sub-divisions such as geographical area, or sex and age groups of the population, obviously the sample size, obtained to estimate the overall universe value with a given precision, must be enlarged if estimators for the sub-divisions are required with the same precision as that of the overall universe estimators.

As a general rule, if estimators with variance V are required for each of the k universe sub-divisions, the sample size should be

$$n'' = kn \tag{7.24}$$

where n is the required sample size for the overall universe estimate with the same variance V.

If the proportion of the universe units in any sub-division p_i is known or could be estimated, the required sample size to estimate the average from the sub-division with variance V is

$$n_i = \sigma_i^2/p_i V \tag{7.25}$$

where σ_i^2 is the variance per unit in the ith sub-division. We have to take the maximum value of the right-hand side of equation (7.25) in order that it holds for all sub-divisions,

$$\begin{aligned} n &= \text{maximum } (\sigma_i^2/p_i V) \\ &= (\sigma^2/V) \text{ maximum } (1/p_i) \end{aligned} \tag{7.26}$$

as the σ_i^2s will, on an average, be slightly smaller than the universe variance σ^2. If the sub-divisions are approximately equal in size, then p_i may be taken to be equal to $1/k$, and this relation is used in equation (7.26).

Example 7.5

In Example 7.1, what size of sample is required if total cases are wanted separately for males and females, with the same precision? (Cochran, exercise 4.3).

Here $k = 2$, and assuming that males and females are about equal in number in the population, we have, from equation (7.24), $n'' = 2 \times 2475 = 4590$.

7.3 Estimation of variance

An estimate of the universe variance or CV may be made from pilot studies or from data of related characteristics. The relations between the range (h) and the variance of mathematical distributions can also be utilized for estimating the variance. Deming (1960) has given the following rules (Table 7.2) for estimating the variance from the range if the shape of the distribution is known or could be guessed.

Table 7.2 Rules for estimating the variance from the range (h) depending on the shape of the distribution

Type	*Mean*	*Variance*	*Standard deviation*	*Coefficient of variation*
Binomial	ph	$p(1-p)/h^2$	$h\sqrt{[p(1-p)]}$	$\sqrt{[(1-p)/p]}$
Rectangular	$\frac{1}{2}h$	$h^2/12$	$0.29h$	0.58
Right-triangle (I)	$\frac{1}{3}h$	$h^2/18$	$0.24h$	0.71
Right-triangle (II)	$\frac{2}{3}h$	$h^2/18$	$0.24h$	0.35
Symmetrical triangle	$\frac{1}{2}h$	$h^2/24$	$0.20h$	0.40
Normal	$\frac{1}{2}h$	$h^2/36$	$h/6$	$\frac{1}{3}$
		(Set $h = 6\sigma$)		

Right-triangle (I)

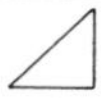
Right-triangle (II)

Example 7.6

The four-year colleges in the U.S.A. were divided into classes of four different sizes according to their 1952–3 enrolments. The standard deviations within each class are given in Table 7.3. If you know the class boundaries but not the value of σ, how well can you guess the σ values by using simple mathematical figures? No college has less than 200

Table 7.3 Actual and estimated standard deviations within each size class of four-year colleges in the U.S.A., 1952–3

	Size class of colleges 1	2	3	4
Number of students	up to 1000	1000–3000	3000–10 000	10 000+
Range (h)	800	2000	7000	40 000
Standard deviation				
(a) actual	236	625	2000	10 023
(b) estimated (= $0.29h$)	232	580	2030	11 600
(c) estimated	192	580	2030	9600
	($0.24h$)	($0.29h$)	($0.29h$)	($0.24h$)

students and the largest has about 50 000 students (Cochran, exercise 4.8).

Assuming a rectangular distribution within each class, the range (h) and the estimates of standard deviation (= $0.29h$) are given in Table 7.3. If for Size class 1, we assume the right triangle distribution (II), the estimated s.d. = $0.24h$ = 192; and if for Size class 4, we assume the right triangle distribution (I), estimated s.d. = $0.24h$ = 9600.

Note: The mathematical relations are not of much use if h is large or cannot be estimated closely. If h is large, the universe can be stratified (Part II) when, within a stratum, the shape of the distribution becomes simpler, closer to a rectangle, if stratification is effective and the mathematical relation between h and σ can be used for each stratum.

7.4 Cost considerations

We have seen how to obtain the sample size required to provide estimators with a given precision in simple random sampling. To determine the implied total cost of the survey, we take the simplest type of cost function

$$c = c_0 + nc_1 \quad (7.27)$$

where c_0 is the overhead cost, and c_1 the cost of surveying one sample unit.

On the other hand, if the total cost C is fixed, the sample size is determined from the above relation, namely,

$$n = (C - c_0)/c_1 \quad (7.28)$$

and the only thing left is to estimate the CV of the sample estimators from a sample of this size if prior information is available on the variability of data.

7.5 Balance of cost and error

With increase in sample size, the cost of the survey increases but the sampling error, and so the loss involved in basing any decision on the sample estimators, decreases; it is necessary to express this loss in monetary terms in order to find a balance between cost and error.

Taking an srs (with replacement) of n units, where the sample mean $\bar{y}$ is used to estimate the universe mean $\bar{Y}$, the error in the estimator $\bar{y}$ is

$$b = \bar{y} - \bar{Y}$$

If the loss due to this error b in the estimator $\bar{y}$ is taken as proportional to b^2, the expected loss for a given sample size is

$$L = lE(b^2) = l\sigma^2/n = l\sigma_{\bar{y}}^2 \quad (7.29)$$

where l is the constant of proportionality. With a total survey cost C, given by equation (7.27), the reasonable procedure would be to obtain the value of n that minimizes the total survey cost *plus* the loss involved with a sample of that size, namely, to minimize

$$C + L = c_0 + nc_1 + l\sigma^2/n \tag{7.30}$$

The minimizing value of n is obtained on partially differentiating equation (7.30) with respect to n, and equating the result to zero, when

$$n = \sigma\sqrt{\frac{l}{c_1}} \tag{7.31}$$ •

7.6 Cluster sampling

7.6.1 Sample size in cluster sampling

As cluster sampling is generally less efficient than an srs of the same size (Chapter 6), a larger total number of elementary units have to be included in a cluster sample in order to attain the same degree of precision as that of an unrestricted srs of the elementary units.

From equation (6.17) and section 6.2, it is seen that the relative change in variance due to sampling clusters instead of the elementary units directly is $(M_0 - 1)\,\rho_c \approx M_0\rho_c$, when M_0 (the cluster size) is large.

The required sample size due to sampling clusters is approximately

$$n_c = n(1 + M_0\rho_c) \tag{7.32}$$

where n is the sample size required for the given precision had the ultimate units been sampled directly.

Example 7.7

In Example 7.2, what would be the sample size required for estimating the birth rate with 2.5 per cent CV if clusters of 300 persons each are taken and the intraclass correlation coefficient is estimated at 0.001?

From equation (7.32), $n_c = 38\,400\,(1 + 300 \times 0.001) = 49\,920$, an increase of 30 per cent.

• 7.6.2 Optimal cluster size

The formulation in section 7.6.1 does not take into account the question of cost, for a cluster sample is less costly to enumerate than a direct sample of the elementary units. The optimal cluster size is so determined that the sampling variance is minimized for a given total cost or the total cost is minimized for a given precision.

A simple cost function with cluster sampling is

$$C = c_0 + nc_1 + nM_0c_2 \tag{7.33}$$

where c_0 is the overhead cost, c_1 the cost of travel, identification, contact etc., per cluster, and c_2 the cost of enumerating one ultimate unit. Generally, c_2 will be considerably smaller than c_1.

If the total cost is fixed at C, the value of n (the number of sample clusters) is (from equation 7.33)

$$n = (C - c_0)/(c_1 + M_0c_2) \tag{7.34}$$

The universe variance of the sample cluster mean $\bar{y}_c$ is, from equation (6.17),

$$\sigma^2_{\bar{y}_c} = \sigma^2_b/n = \sigma^2 [1 + (M_0 - 1)\rho_c]/nM_0$$
$$= \sigma^2_b(c_1 + M_0c_2)/(C - c_0) \text{ from equation (7.34)}$$
$$= c_1\sigma^2(1 + M_0c_2/c_1)[1 + (M_0 - 1)\rho_0]/M_0(C - c_0) \qquad (7.35)$$

If previous information is available from empirical or pilot studies, the values of the equation (7.35) can be computed and plotted for different values of M_0, to show the particular value of M_0 that minimizes the expression.

Putting

$$C - c_0 = C' = nc_1 + nM_0c_2 \qquad (7.36)$$

the optimal cluster size is

$$M_0 = \sqrt{\left[\frac{c_1}{c_2} \cdot \frac{1 - \rho_c}{\rho_c}\right]} \qquad (7.37)$$

In practice, instead of dealing with the costs c_1 and c_2, we might consider the man-days required to be spent on the different operations. If the enumerators work singly, c_1 may be 2 to 3 man-days, and if 30 to 50 persons can be enumerated per day in a demographic inquiry, c_2 will be from 1/30 to 1/50 man-day; the ratio c_1/c_2 will then range from 60 to 150. The optimal size of cluster for some typical values of the ratio c_1/c_2 and of ρ_c are given in Table 7.4.

Table 7.4 Optimum size of cluster (number of persons) in typical surveys on birth and death rates

c_1/c_2	*Intraclass correlation coefficient* ρ_c				
	0.001	0.002	0.003	0.004	0.005
50	224	158	129	112	100
75	274	194	158	137	122
100	316	224	183	158	141
125	373	264	215	187	158
150	387	274	223	194	173

Taking a typical value of the ratio c_1/c_2 as 100, and the value of the intraclass correlation coefficient at 0.001 for the birth rate, the optimal cluster size is 316 persons; for the death rate, taking the intraclass correlation coefficient at 0.003, the optimal cluster size is 183 persons. The value of n (number of sample clusters) is determined from equation (7.34), given the values of C, c_0, c_1, c_2, and the optimum value of M_0.

Note that optimal cluster size is generally very broad (except for very small values of the intraclass correlation coefficient), so that substantial deviations from the optimal cluster size would not affect the cost very much.●

Example 7.8

For a survey on birth rate, given that the total cost, neglecting overheads, is fixed at \$20 000, and the enumerator cost per month is \$300, what is the optimal size of sample if it is decided to select a cluster of persons, assuming that an enumerator has to spend, on average, two days in contacting the clusters and in other preliminary work; that he can enumerate an average of 40 persons a day; and that the intraclass correlation is estimated at 0.001?

Here C = \$20 000; 1 man-day = \$300/30 = \$10; c_1 = 2.5 man-days = \$25; c_2 = 1/40 man-day = \$0.25, so that c_1/c_2 = 100; ρ_c = 0.001. From Table 7.4, the optimal size of cluster is 316 persons. Taking M_0 at 300 persons, the total cost (neglecting overheads) is, from equation (7.33)

$$\$20\,000 = \$25n + \$0.25 \times 300n$$
$$= \$25n + \$75n = \$100n,$$

or n = 200 sample clusters. The total sample size is nM_0 = 200 x 300 = 60 000 persons.

7.7 General remarks

Estimates of sample size required to obtain measures with a given precision will often be found to be quite large, when derived on the basis of unrestricted simple random sampling. But the 'paean for large samples must be interrupted with caveats' (Kish, 1971). First, an unrestricted simple random sample is rarely used in practice; and, as noted earlier, special procedures (such as stratification, pps sampling etc.) and methods of estimation (such as ratio and regression methods), that require smaller sample sizes than srs, are used to provide estimators with the same efficiency. Second, small samples have proved useful, not only as pilot studies to full-scale surveys, but also in providing interim estimates as in the earlier 'rounds' of the Indian National Sample Survey where fairly accurate estimates of birth and death rates were obtained from only 3000 to 8000 sample households after special analytical techniques were applied (section 25.5.5): later, in 1958–9, the inquiry on population, births, and deaths was made to cover 2600 sample villages with 234 000 households and 1.2 million persons. Third, a country with inadequate resources can start from a small sample and with increasing resources build up a fully adequate sample; the Current Population Survey of the U.S.A., for example, started in 1943 with 68 primary areas which were enlarged to the present 449. Fourth, it is possible to combine smaller monthly or quarterly estimates into yearly estimates, and the yearly estimates into estimates covering longer periods, to provide estimates with acceptable precision. And finally, in the interest of true accuracy, it may sometimes be better to conduct a smaller sample with adequate control than try to canvass a much larger sample but with poor quality data (see also Chapter 25).

Further reading

Cochran, Chapter 4 and section 9.6; Deming (1960), Chapter 14; Hansen *et al.*, Vol. I, section 4.11 and Chapter 6D, Vol. II, section 4.9; Kish, section 2.6; Murthy, sections 4.8 and 8.3; Raj, sections 3.2 and 3.7; Sukhatme and Sukhatme, sections 1.11 and 1.12; Yates, section 4.31.

Exercises

1. Material for the construction of 5000 wells was issued in a district in India in 1944. The list of cultivators to whom the material was issued was available along with the proposed location of each well. A large part of the material was, however, reported to have been misused, having been diverted to other purposes. It is proposed to estimate the proportion of wells not actually constructed, by taking a simple random sample of wells with the permissible margin of error of 10 per cent, and the degree of assurance desired 95 per cent. Determine the size of the sample required to estimate the proportion of wells not constructed for different values of the universe proportion ranging from 0.5 to 0.9. (Sukhatme and Sukhatme, Example 1.3, modified to with replacement plan in our answer.)

2. An anthropologist wishes to know the percentage of people with blood-group 'O' in an island of 3200 persons with a margin of 5 per cent and 95 per cent degree of assurance. What is the required sample size? (Cochran, p. 74.)

3. From the estimates obtained from the srs of 20 households in Example 2.2, compute the sample size required to provide an estimate of the total number of persons in the universe within 10 per cent, apart from a chance of 1 in 20.

4. For the 30 villages listed in Appendix IV, if no village is assumed to have less than 60 or more than 150 persons, estimate the s.d.

8 Self-weighting Designs

8.1 Introduction

To obtain estimates of totals, the sample observations have to be multiplied by factors, variously called weights, multipliers, or weighting-, raising- or inflation-factors, which depend on the particular sample design and the method of estimation adopted. A sample design becomes self-weighting with respect to the particular linear estimator of a total (or an average) when the multipliers of the sample units are all equal: with only one multiplier, the tabulation becomes simpler, speedier and more economical, as the unbiased estimator of the universe total is obtained on multiplying the sample total by the constant, overall multiplier, and the estimator of the ratio of the universe totals of two variables is obtained on taking the ratio of the corresponding sample totals; the estimation of the variances, covariances, etc., is also simplified.

The problem of how to make a design self-weighting becomes particularly important in stratified multi-stage designs, but the principles are introduced in this chapter relating to unstratified single-stage samples.

8.2 Simple random sampling

For a simple random sample, the design is self-weighting with respect to the unbiased estimators of universe totals, averages, ratios. For with the notations used in Chapter 2, the unbiased estimator of the universe total Y, obtained from a simple random sample of n units out of the universe total of N units, is

$$y_0^* = N \sum^n y_i/n \tag{8.1}$$

where y_i ($i = 1, 2, \ldots, n$) is the value of the study variable for the ith sample unit. The weighting-factor for the ith sample unit is, therefore,

$$w_i = N/n \tag{8.2}$$

which, when multiplied by y_i and added for the n sample units, provide the unbiased estimator y_0^* of the universe total Y. The multiplier is constant for all the sample units.

Similarly, the unbiased estimator of the universe mean $\bar{Y}$ is given by the sample mean

$$\bar{y} = \sum^n y_i/n \tag{8.3}$$

where the multiplier is $1/n$, constant for all the sample units.

In the estimation of the ratio of two universe totals, the sample estimator is

$$r = \sum^n y_i \Big/ \sum^n x_i \tag{8.4}$$

Here, because the design is self-weighting with respect to the two totals, the multiplier does not enter in the ratio; it does not also enter into the estimating equation for the variance of the ratio.

8.3 Varying probability sampling

In sampling with varying probability, the single-stage sample cannot in general be made self-weighting.

The unbiased estimator of the universe total Y obtained from a pps sample of n units is

$$y_0^* = \sum^n y_i/n\pi_i \tag{8.5}$$

where π_i is the probability of selection of the ith unit.

The multiplier of the ith sample unit for estimating the universe total Y is

$$w_i = 1/n\pi_i \tag{8.6}$$

which will not be the same for all the sample units, and the design will not, therefore, be self-weighting.

However, if $\pi_i = z_i/Z$, where z_i is the 'size' of the ith universe unit and $Z = \sum^N z_i$ is the total 'size' of the universe, then

$$y_0^* = \sum^n y_i/n\pi_i = Z\sum^n \frac{(y_i/z_i)}{n} = Z\sum^n r_i/n \tag{8.7}$$

where $r_i = y_i/z_i$.

If the ratios $r_i = y_i/z_i$ could be observed and recorded easily in the field, so that the r_is could be considered as the values of a newly defined study variable, then the design will be self-weighting with respect to the unbiased estimators, for the multipliers for the r_i values are $w_i' = Z/n$, which are the same for all the sample units. This has practical uses in crop surveys, as we have seen in section 5.5.

Notes

1. A pps sample design can be made self-weighting at the tabulation stage by selecting a sub-sample of the sample units with probability proportional to the multipliers, but the sampling variance will be increased (see section 13.5).
2. We shall see later in Chapters 18 and 23 how in a multi-stage design the sample can be made self-weighting even with pps sampling at some stages.

8.4 Rounding off of multipliers

When the original design is not self-weighting, one of the following procedures may be adopted to reduce the number of multipliers and thus to achieve at least partial self-weighting at the tabulation stage.

1. The multipliers w_is may be replaced by their simple average.
2. The multipliers may be rounded off to some convenient numbers as the nearest multiples of ten, hundred etc.
3. The multipliers may be rounded off to a small number of weights by a *random process* which would retain the unbiased character of the estimators (Note 3(c) to section 12.3.8).
4. A sub-sample of n' ultimate units may be selected from the original sample with probability proportional to their multipliers (Note 1 to section 8.3).

The first two procedures will lead to biased estimates with possible decrease in variance. However, the bias in the first procedure will be negligible if in the sample the covariance between the sample values and the multipliers is small.

The third and the fourth procedures give unbiased estimators but with some increase in variance. In the third procedure, determination of optimum weights is very complicated if the number of such weights exceed three; on the other hand, the increase in variance will be large for a small number of weights.

Part II:

Stratified Single-stage Sampling

9 Stratified Sampling: Introduction

9.1 Introduction

If the universe is sub-divided into a number of sub-universes, called *strata,* and sampling is carried out independently in each stratum, the sampling plan is known as *stratified sampling.* Stratified sampling can be used in single- as well as multi-stage designs.

In this part, we shall illustrate the use of stratification in single-stage designs. The general theorems relating to stratified sampling will be dealt with in the present chapter; subsequent chapters in Part II will deal with stratified simple-random and varying-probability sampling; the size of the sample and allocation to different strata; formation of strata; and self-weighting designs.

9.2 Reasons for stratification

Stratified sampling is adopted in a number of situations:

1. When estimates are required for each sub-division of the universe separately, such as for geographical sub-divisions or for households in different social and economic groups in a household survey;
2. when estimates of universe characteristics are required with increased efficiency per unit of cost;
3. when a greater weightage is required to be given to some units that occur infrequently in the universe, such as households with very high incomes; and when the universe has a large variance, i.e. the units vary greatly in the values;
4. when different sampling procedures are to be adopted for different sub-universes, in which case the field work is easier to organize in the different strata, formed according to the nature of the available ancillary information required for sample selection. For example, the population in many African countries can be classified according to their modes of living into urban, rural-sedentary, and rural-nomad; and these can be considered to constitute different strata; the sampling units – at least for the first-stage in a multi-stage design – could be town-blocks for the urban stratum, villages for the rural-sedentary stratum, and tribal hierarchies for the rural-nomad stratum. In sampling human populations, people in institutions such as boarding houses, hospitals, and jails may be considered separately from those living in households. Sampling might be simple random in one stratum, with probability proportional to size in another, and so on. The geographical strata may be further subdivided, each sub-stratum, for example, being allotted to a separate supervisor.

Notes

1. That stratification may lead to a gain in efficiency per unit of cost may be seen if we consider a universe composed of strata that are, with respect to the study variable, internally homogeneous, but heterogeneous with respect to each other: in this case, a very small sample from each strata would provide estimates with relatively small sampling variances.

Taking an extreme case, consider the universe of six households with respective sizes 4, 4, 4, 5, 5 and 5. The reader will verify that the variance per unit in this universe is $\sigma^2 = 0.25$. If we were to draw a simple random sample of 4 households with replacement from this universe, the unbiased estimator of the total size is $N\bar{y}$ (where $N = 6$ and $\bar{y}$ is the sample mean, based on $n = 4$ sample units), with the sampling variance $N^2\sigma^2/n = 2.25$. Suppose, however, that the universe is subdivided into two strata, the first with the three households each of size 4, and the second with the other three households each of size 5; and an srs with replacement of 2 units is to be drawn from each stratum. Then, as we shall see later, the unbiased estimator of the total size is $(N_1\bar{y}_1 + N_2\bar{y}_2)$, where $N_1 = 3$, $N_2 = 3$, being respectively the total number of units in the two strata, and $\bar{y}_1$ and $\bar{y}_2$ the respective stratum means based on $n_1 = 2$ and $n_2 = 2$ sample units; the sampling variance of this unbiased estimator of the total is $N_1^2\sigma_1^2/n_1 + N_2^2\sigma_2^2/n_2 = 0$, as the variance per unit in the two strata $\sigma_1^2 = \sigma_2^2 = 0$.

2. Stratification may be carried out at different stages of sampling. The most common type of stratification is by administrative and geographical sub-divisions, such as by provinces, prefectures, counties, districts, and rural/urban categories. In household surveys, stratification is often carried out before the sample households are selected, by listing the households from which the sample is to be drawn, and recording such of their characteristics that may be readily obtained, e.g. size, and social and economic classes; the households are then stratified on the basis of the characteristics recorded and then sampled (with a different sampling fraction) in each stratum.
3. Strata may be formed of units that are not geographically contiguous. Thus in a rural socio-economic survey, the villages may be stratified according to their population as given by the most recent census; or in a crop-yield survey, the fields may be classified according as they are irrigated or not; in the demographic inquiry in Mysore (a former State in India), conducted by the United Nations and the Government of India in 1951–2, three strata were formed in the rural areas: rural hills area with large-scale anti-malarial operations, rural hills area without large-scale anti-malarial operations, and rural plains (tank-irrigated area).
4. The data from a sample design that is not stratified, or is stratified according to some variable other than that desired, may, under some circumstances, be treated as if coming from a sample stratified according to the desirable stratification variable. This technique known as *stratification after sampling* or the *technique of post-stratification,* is explained in Chapter 10 in connection with stratified simple random sampling.

9.3 Fundamental theorems in stratified sampling

In the following the fundamental theorems given in section 2.7 are extended to the case of stratified sample designs.

1. The universe is sub-divided into L mutually exclusive and exhaustive strata. In the hth stratum ($h = 1, 2, \ldots, L$) there are n_h ($\geqslant 2$) independent and unbiased estimators t_{hi} ($i = 1, 2, \ldots, n_h$) of the universe parameter T_h, a combined unbiased estimator of T_h is, from estimating equation (2.28), the arithmetic mean

$$\bar{t}_h = \sum_{i=1}^{n_h} t_{hi}/n_h \tag{9.1}$$

An unbiased estimator of the variance of $\bar{t}_h$ is, from equation (2.29),

$$\begin{aligned} s_{\bar{t}_h}^2 &= \sum_{i=1}^{n_h} (t_{hi} - \bar{t}_h)^2/n_h(n_h - 1) \\ &= \mathrm{SS}t_{hi}/n_h(n_h - 1) \end{aligned} \tag{9.2}$$

where

$$\begin{aligned} SSt_{hi} &= \sum_{i=1}^{n_h} (t_{hi} - \bar{t}_h)^2 \\ &= \sum_{i=1}^{n_h} t_{h_i}^2 - \left(\sum_{i=1}^{n_h} t_{hi}\right)^2 / n_h \end{aligned} \tag{9.3}$$

is the corrected sum of squares of t_{hi}.

An unbiased estimator of the universe parameter for all the strata combined

$$T = \sum_{h=1}^{L} T_h \tag{9.4}$$

is obtained on summing up the stratum estimators $\bar{t}_h$, namely,

$$t = \sum_{h=1}^{L} \bar{t}_h \tag{9.5}$$

and an unbiased estimator of the variance of t is obtained on summing the L unbiased estimators of the stratum variances $s_{\bar{t}_h}^2$ (sampling in one stratum is independent of sampling in another, so that the stratum estimators $\bar{t}_h$ are mutually independent)

$$s_t^2 = \sum_{h=1}^{L} s_{\bar{t}_h}^2 \tag{9.6}$$

For another study variable similarly defined, an unbiased estimator of the universe parameter

$$U = \sum_{h=1}^{L} U_h \tag{9.7}$$

is the sum of the stratum estimators $\bar{u}_h$ of U_h,

$$u = \sum_{h=1}^{L} \bar{u}_h \tag{9.8}$$

and an unbiased variance estimator of u is the sum of the L unbiased estimators of the stratum variances $s_{\bar{u}_h}^2$, i.e.

$$s_u^2 = \sum_{h=1}^{L} s_{\bar{u}_h}^2 \tag{9.9}$$

2. An unbiased estimator of the covariance of $\bar{t}_h$ and $\bar{u}_h$ for the hth stratum is, from equation (2.32),

$$\begin{aligned} s_{\bar{t}_h \bar{u}_h} &= \sum_{i=1}^{n_h} (t_{hi} - \bar{t}_h)(u_{hi} - \bar{u}_h)/n_h(n_h - 1) \\ &= SPt_{hi}u_{hi}/n_h(n_h - 1) \end{aligned} \tag{9.10}$$

where

$$\mathrm{SP}t_{hi}u_{hi} = \sum_{i=1}^{n_h} (t_{hi} - \bar{t}_h)(u_{hi} - \bar{u}_h)$$

$$= \sum_{i=1}^{n_h} t_{hi}u_{hi} - \left(\sum_{i=1}^{n_h} t_{hi}\right)\left(\sum_{i=1}^{n_h} u_{hi}\right) / n_h \tag{9.11}$$

is the corrected sum of products of t_{hi} and u_{hi}.

For all the strata combined, an unbiased estimator of the covariance between t and u is the sum of the stratum estimators $s_{\bar{t}_h \bar{u}_h}$, i.e.

$$s_{tu} = \sum_{h=1}^{L} s_{\bar{t}_h \bar{u}_h} \tag{9.12}$$

A consistent but generally biased estimator of the ratio of two universe parameters in the hth stratum $R_h = T_h/U_h$ is, from equation (2.33), the ratio of the sample estimators $\bar{t}_h$ and $\bar{u}_h$ in the stratum

$$r_h = \bar{t}_h/\bar{u}_h \tag{9.13}$$

with estimated variance (from equation (2.34))

$$s_{r_h}^2 = (s_{\bar{t}_h}^2 + r_h^2 s_{\bar{u}_h}^2 - 2r_h s_{\bar{t}_h \bar{u}_h})/\bar{u}_h^2 \tag{9.14}$$

For all the strata combined, a consistent but generally biased estimator of the ratio of two universe parameters $R = T/U$ is the ratio of the respective sample estimators t and u,

$$r = t/u \tag{9.15}$$

with estimated variance

$$s_r^2 = (s_t^2 + r^2 s_u^2 - 2r s_{tu})/u^2 \tag{9.16}$$

A generally biased but consistent estimator of the correlation coefficient ρ_h between the two study variables at the stratum level is, from equation (2.35),

$$\hat{\rho}_h = s_{\bar{t}_h \bar{u}_h} / s_{\bar{t}} s_{\bar{h}} \tag{9.17}$$

and an estimator of the correlation coefficient ρ of the two study variables at the overall level is

$$\hat{\rho} = s_{tu} / s_t s_u \tag{9.18}$$

Notes

1. The question of formation of strata, and the allocation of the total number of sample units into the different strata will be taken up later in Chapter 12.
2. Estimates are often required for higher levels of aggregation than strata, in addition to the overall universe estimators. For example, for the Indian National Sample Survey on Population, Births and Deaths (1958–9), the strata consisted of *tehsils* or groups of *tehsils* (a *tehsil* being an administrative unit comprising a number of villages and some small towns); a sample of villages was selected from each stratum and the population and the births and deaths during the preceding twelve months recorded. The country comprises a number of states, each state comprising a number of *tehsils*, and estimates were required not only for the country as a whole, but also for the states separately. For estimators for the whole country, estimating equations of tye types (9.5), (9.6), (9.12), (9.15) and (9.16) were used; for the estimators for the states, the same types of equations were used excepting that for any state, the values of only those strata that fell within the state were considered (see Example 10.3).

3. As noted in section 9.2, the selection probabilities might be different in different strata.
4. For other notes, e.g. relating to estimators of ratios, see the notes to section 2.7.

9.4 Setting of probability limits

We have seen in section 2.8 that for an unstratified single-stage sample, the $(100 - \alpha)$ per cent probability limits of a universe parameter T are set with the sample estimator $\bar{t}$ and its estimated standard error $s_{\bar{t}}$, thus:

$$\bar{t} \pm t'_{\alpha, n-1} s_{\bar{t}} \tag{2.40}$$

where $t'_{\alpha, n-1}$ is the 100α percentage point of the t-distribution for $(n - 1)$ degrees of freedom. It was also noted that for large n, the percentage points of the normal distribution could be used for those of the t-distribution.

The above holds for each stratum separately. Thus the 100α per cent probability limits of the universe parameter T_h for the hth stratum are

$$\bar{t}_h \pm t'_{\alpha, n_h-1} s_{\bar{t}_h} \tag{9.19}$$

the sample stratum estimator $\bar{t}_h$ being defined by equation (9.1) and its estimated standard error $s_{\bar{t}_h}$ by equation (9.2).

For the probability limits of the overall universe parameter T, the normal distribution may be used if the estimators of variances of the stratum estimators $\bar{t}_h$, namely $s^2_{\bar{t}_h}$, are based on not too few degrees of freedom and it is assumed that the overall universe estimator t, defined by equation (9.5), is normally distributed. Then

$$t \pm t'_{\alpha} s_t \tag{9.20}$$

are the $(100 - \alpha)$ per cent probability limits of the universe parameter T where s_t is defined by equation (9.5) and t'_{α} is the normal percentage point for probability α.

If, however, the variance estimators in the different strata $s^2_{\bar{t}_h}$ are based on small numbers of degrees of freedom, the normal distribution cannot be used. The t-distribution can be used by computing the effective number of degrees of freedom,

$$n'' = \frac{\left(\sum^{L} s^2_{\bar{t}_h}\right)^2}{\sum^{L} [s^4_{\bar{t}_h}/(n_h - 1)]} \tag{9.21}$$

which will lie between (the smallest of the) $n_h - 1$ and $\sum^{L} (n_h - 1)$. The $(100 - \alpha)$ per cent probability limits of T are

$$t \pm t'_{\alpha, n''} s_t \tag{9.22}$$

where $t'_{\alpha, n''}$ is the 100α percentage point of the t-distribution of n'' degrees of freedom.

9.5 Selection of two units from each stratum

The computation of estimates is considerably simplified when, in each stratum, two units are selected with replacement out of the total number of units. Noting that $n_h = 2$ in this case,

$$\bar{t}_h = \tfrac{1}{2}(t_{h1} + t_{h2}) \tag{9.23}$$

where t_{h1} and t_{h2} are the two unbiased estimators of T_h from the two selected sample units.

An unbiased estimator of the variance of t_h is

$$s^2_{\bar{t}_h} = \tfrac{1}{4}(t_{h1} - t_{h2})^2 \tag{9.24}$$

so that

$$s_{\bar{t}_h} = \tfrac{1}{2}|t_{h1} - t_{h2}| \tag{9.25}$$

An unbiased estimator of the covariance of $\bar{t}_h$ and $\bar{u}_h$ is

$$s_{\bar{t}_h\bar{u}_h} = \tfrac{1}{4}(t_{h1} - t_{h2})(u_{h1} - u_{h2}) \tag{9.26}$$

Also,

$$r_h = t_h/u_h = (t_{h1} + t_{h2})/(u_{h1} + u_{h2}) \tag{9.27}$$

with a variance estimator

$$s^2_{r_h} = \frac{(t_{h1} - t_{h2})^2 + r_h^2(u_{h1} - u_{h2})^2 - 2r_h(t_{h1} - t_{h2})(u_{h1} - u_{h2})}{(u_{h1} + u_{h2})^2} \tag{9.28}$$

The overall estimators (for all the strata taken together) are:

$$t = \sum_{h=1}^{L} \bar{t}_h = \tfrac{1}{2}\sum_{1}^{L}(t_{h1} + t_{h2}) \tag{9.29}$$

$$s_t^2 = \sum_{h=1}^{L} s^2_{\bar{t}_h} = \tfrac{1}{4}\sum_{h=1}^{L}(t_{h1} - t_{h2})^2 \tag{9.30}$$

$$s_{tu} = \sum_{h=1}^{L} s_{\bar{t}_h\bar{u}_h} = \tfrac{1}{4}\sum_{h=1}^{L}(t_{h1} - t_{h2})(u_{h1} - u_{h2}) \tag{9.31}$$

Also,

$$r = t/u = \sum_{h=1}^{L}(t_{h1} + t_{h2}) \Big/ \sum_{h=1}^{L}(u_{h1} + u_{h2}) \tag{9.32}$$

with a variance estimator

$$s_r^2 = \frac{\sum_{h=1}^{L}[(t_{h1} - t_{h2})^2 + r^2(u_{h1} - u_{h2})^2 - 2r(t_{h1} - t_{h2})(u_{h1} - u_{h2})]}{\sum_{h=1}^{L}(u_{h1} + u_{h2})^2} \tag{9.33}$$

Thus the sums and differences of the two estimates obtained from the two (first-stage) sample units in the different strata supply the required data for computing the overall estimators and their variances, not only for the totals but also for the ratios. This procedure has been followed in the Indian National Sample Survey since its inception in 1950, and some standard errors of ratios and ratio estimates were published in 1956; the procedure is sometimes attributed to Keyfitz, who also suggested this in 1957.

Note: Another unbiased estimator of the variance of t is

$$\frac{1}{4}\left(\sum_{h=1}^{L} t_{h1} - \sum_{h=1}^{L} t_{h2}\right)^2 \tag{9.34}$$

and the corresponding estimated standard error

$$\frac{1}{2}\left|\sum_{h=1}^{L} t_{h1} - \sum_{h=1}^{L} t_{h2}\right| \tag{9.35}$$

These are simpler to compute, but being based on only one degree of freedom will be less efficient than the estimators defined earlier which are based on an effective number of degrees of freedom, given by equation (9.21), that will lie between 1 and $(L - 1)$. Another unbiased estimator of the covariance of t and u can be defined similarly, and so also the estimated variance of r.

Further reading

Cochran, section 5.1; Hansen *et al.*, Vol. I, Chapter 5A; Murthy, sections 7.1–7.3; Raj, sections 4.1–4.2; Sukhatme and Sukhatme, section 3.1; Yates, section 3.3.

10 Stratified Simple Random Sampling

10.1 Introduction

In this chapter we will consider the estimating methods for totals, means, ratios of study variables and their variances when a simple random sample has been drawn in each stratum. The methods of estimation of proportion of units in a category, the use of ratio estimators and stratification after sampling will also be considered.

10.2 Estimation of totals, means, ratios, and their variances

10.2.1 Structure of a stratified srs

Consider the total universe to be divided on certain criteria into L strata, and let the hth stratum ($h = 1, 2, \ldots, L$) contain N_h total number of units. The values of two study variables for the ith unit ($i = 1, 2, \ldots, N_h$) in the hth stratum will be denoted respectively by Y_{hi} and X_{hi}. The total number of units in the universe is

$$N = \sum_{h=1}^{L} N_h \tag{10.1}$$

In each stratum, a simple random sample is taken with replacement (or without replacement, but with a small sampling fraction, 10 per cent or less). Let the number of sample units in the hth stratum be n_h and the values of two study variables for the ith selected sample unit ($i = 1, 2, \ldots, n_h$) in the hth stratum be denoted respectively by y_{hi} and x_{hi}. The total number of sample units is

$$n = \sum_{h=1}^{L} n_h \tag{10.2}$$

10.2.2 Estimators of totals, means, and ratio of two totals

An unbiased estimator of the stratum total Y_h, obtained from the ith sample unit in the hth stratum is, from equation (2.42),

$$y_{hi}^* = N_h y_{hi} \tag{10.3}$$

and the combined unbiased estimator for the stratum total Y_h, obtained from the n_h sample units in the hth stratum, is, from equation (2.43) or (9.1),

$$y_{ho}^* = \sum_{i=1}^{n_h} y_{hi}^*/n_h = N_h \bar{y}_h \tag{10.4}$$

where $\bar{y}_h = \sum^{n_h} y_{hi}/n_h$ is the mean of the y_{hi} values in the hth stratum.

Universe totals, means, and ratios of two totals and their sample estimators are defined in Table 10.1. These follow from the results of section 2.9 for any stratum and of section 9.3 for all the strata combined. The sample estimators of

Table 10.1 Some universe parameters and their sample estimators for a stratified simple random sample

	For the hth stratum		For all strata combined		
	Universe parameter (a)	Sample estimator (b)	Universe parameter (c)	Sample estimator (d)	
For a study variable:					
Total	$Y_h = \sum_{i=1}^{N_h} Y_{hi}$	$y^*_{h0} = \sum_{i=1}^{n_h} y^*_{hi}/n_h = N_h\bar{y}_h$	$Y = \sum_{h=1}^{L} Y_h$	$y = \sum_{h=1}^{L} y^*_{h0}$	(10.5)
Mean	$\bar{Y}_h = Y_h/N_h$	$\bar{y}_h$	$\bar{Y} = Y/N$	y/N	(10.6)
For two study variables:					
Ratio of totals	$R_h = Y_h/X_h$	$r_h = y^*_{h0}/x^*_{h0} = \bar{y}_h/\bar{x}_h$	$R = Y/X$	$r = y/x$	(10.7)

totals and means are unbiased (theoretical proofs are given in Appendix II, section A2.3.7); those for ratios are consistent, but generally biased.

10.2.3 Sampling variances and their estimators for a stratum

In the hth stratum, the variance per unit is

$$\sigma_h^2 = \sum_{i=1}^{N_h} (Y_{hi} - \bar{Y}_h)^2/N_h \qquad (10.8)$$

and the variance of the sample mean $\bar{y}_h$ in a simple random sample (with replacement) of n_h units is, from equation (2.17),

$$\sigma_{\bar{y}_h}^2 = \sigma_h^2/n_h \qquad (10.9)$$

so that the variance of the sample estimator y^*_{h0} of the stratum total Y_h is

$$\sigma_{y^*_{h0}}^2 = N_h^2\sigma_{\bar{y}_h}^2 = N_h^2\sigma_h^2/n_h \qquad (10.10)$$

An unbiased estimator of the variance of the estimator y^*_{h0} is, from equations (2.44) or (9.2),

$$\begin{aligned} s_{y^*_{h0}}^2 &= \sum_{i=1}^{n_h} (y^*_{hi} - y^*_{h0})^2/n_h(n_h - 1) \\ &= SSy^*_{hi}/n_h(n_h - 1) \\ &= N_h^2 \sum_{i=1}^{n_h} (y_{hi} - \bar{y}_h)^2/n_h(n_h - 1) \\ &= N_h^2 SSy_{hi}/n_h(n_h - 1) \end{aligned} \qquad (10.11)$$

An unbiased estimator of the variance of the sample mean $\bar{y}_h$ is, from equation (2.46),

$$s_{\bar{y}_h}^2 = s_{y^*_{h0}}^2/N_h^2 \qquad (10.12)$$

An unbiased estimator of the covariance of y^*_{ho} and x^*_{ho} (x^*_{ho} being defined similarly for another study variable) is from equation (2.47) or (9.10).

$$s_{y^*_{ho}x^*_{ho}} = \sum_{i=1}^{n_h} (y^*_{hi} - y^*_{ho})(x^*_{hi} - x^*_{ho})/n_h(n_h - 1)$$
$$= \mathrm{SP}y^*_{hi}x^*_{hi}/n_h(n_h - 1)$$
$$= N_h^2 \sum_{i=1}^{n_h} (y_{hi} - \bar{y}_h)(x_{hi} - \bar{x}_h)n_h(n_h - 1)$$
$$= N_h^2\,\mathrm{SP}y_{hi}x_{hi}/n_h(n_h - 1) \tag{10.13}$$

An estimator of the variance of r_h is, from equation (9.14) or (2.49),

$$s^2_{r_h} = (s^2_{y^*_{ho}} + r_h^2 s^2_{x^*_{ho}} - 2r_h s_{y^*_{ho}x^*_{ho}})/x^{*2}_{ho}$$
$$= (\mathrm{SS}y_{hi} + r_h^2\,\mathrm{SS}x_{hi} - 2r_h\,\mathrm{SP}y_{hi}x_{hi})/n_h(n_h - 1)\bar{x}_h^2 \tag{10.14}$$

10.2.4 Sampling variances and their estimators for all strata combined

The variance of the estimator y of the universe total Y is

$$\sigma_y^2 = \sum^{L} \sigma^2_{y^*_{ho}} = \sum^{L} N_h^2\sigma_h^2/n_h \tag{10.15}$$

which is estimated unbiasedly by the sum of the unbiased variance estimators of y^*_{ho} from equation (9.6), namely, by

$$s_y^2 = \sum^{L} s^2_{y^*_{ho}} \tag{10.16}$$

The sampling variance of the estimator of the universe mean is

$$\sigma_y^2/N^2 \tag{10.17}$$

which is estimated unbiasedly by

$$s_y^2/N^2 \tag{10.18}$$

Theoretical proofs of the above are given in Appendix II, section A2.3.9.

An unbiased estimator of the covariance of the estimators y and x is, from equation (9.12),

$$s_{yx} = \sum^{L} s_{y^*_{ho}x^*_{ho}} \tag{10.19}$$

A variance estimator of r is, from equation (9.16)

$$s_r^2 = (s_y^2 + r^2 s_x^2 - zrs_{yx})/x^2 \tag{10.20}$$

Notes

1. *Sampling without replacement.* In srs without replacement, the sample estimators of a universe total, mean, and ratio of two universe totals remain the same as for srs with replacement. The sampling variance of $y = \sum^{L} y^*_{ho}$ is, however, of the form

$$\sigma_y^2 = \sum^{L} \frac{N_h^2\sigma_h^2}{n_h}\,\frac{(N_h - n_h)}{(N_h - 1)} = \sum^{L} \frac{N_h^2 S_h^2}{n_h}(1 - f_h) \tag{10.21}$$

where

$$S_h^2 = \frac{N_h}{N_h - 1}\sigma_h^2 \tag{10.22}$$

and

$$f_h = n_h/N_h \qquad (10.23)$$

σ_y^2 is estimated unbiasedly by

$$s_y^2 = \sum^{L} \frac{N_h^2 s_h^2}{n_h}(1 - f_h) \qquad (10.24)$$

where

$$s_h^2 = \sum_{i=1}^{n_h} (y_{hi} - \bar{y}_h)^2/(n_h - 1) \qquad (10.25)$$

The sampling variance of the universe mean and its unbiased estimator are obtained on dividing the respective expressions for the universe total by N^2.

Theoretical proofs are given in Appendix II, section A2.3.9.

The unbiased covariance estimator s_{yx} is similarly defined and so also the variance estimator of the ratio $r = y/x$.

2. The estimator of the overall universe mean $\bar{Y}$ is, from equation (10.6*d*)

$$\bar{y} = y/N = \sum_{h=1}^{L} y_{ho}^*/N = \sum_{h=1}^{L} N_h \bar{y}_h/N \qquad (10.26)$$

i.e. the weighted average of the sample stratum means ($\bar{y}_h$), the weights being N_h/N, the proportion of the total number of universe units contained in each stratum. This will be the same as the simple unweighted mean of the values of the study variables (which is the same as the simple unweighted mean of the sample stratum means),

$$\sum_{h=1}^{L} \sum_{i=1}^{n_h} y_{hi}/n = \sum_{h=1}^{L} n_h \bar{y}_h/n \qquad (10.27)$$

only when $\qquad N_h/N = n_h/n \quad$ or $\quad n_h/N_h = n/N \qquad$ (10.28)

i.e. when the sampling fraction is the same in all the strata, in which case

$$n_h = (n/N)N_h \qquad (10.29)$$

Thus the allocation of the number of sample units to the different strata is in proportion to the total number of units in each stratum: this is known as *proportional allocation*. It leads to a self-weighting design for a simple random sample.

3. For other notes relating for example to estimators of ratios, see the notes to section 2.7.

Example 10.1

From village 8 in Zone I (Appendix IV), assume that for a family budget inquiry, a preliminary listing of all the 24 households has been made along with information on size. Divide these 24 households into two strata – one with households of sizes 1–5 persons and the other with sizes 6 persons and above. Select a simple random sample (with replacement) of 3 households from the first stratum and of 2 households from the second stratum, and on the basis of the data on income and food cost of these sample households, estimate for all the 24 households the total income, total food cost, income per household and per person, food cost per household and per person and the proportion of income spent on food.

The details of the selected households in the two strata with the required information are given in Table 10.2. The required computations are shown in a summary form in Table 10.3 and the final results in Table 10.4.

Table 10.2 Size, total monthly income, and food cost in the srs of households in village no. 8 in zone I (Appendix IV)

Household serial no.	*Household size*	*Total monthly income* ($)	*Monthly food cost* ($)
Stratum I (households of sizes 1–5); total number of households = 17			
2	3	33	21
9	4	39	24
11	3	35	22
Stratum II (households of sizes 6 and above); total number of households = 7			
16	6	62	31
22	6	61	29

Table 10.3 Computation of the required estimates for a stratified srs: data of Table 10.2

	Stratum I ($h = 1$)	*Stratum II* ($h = 2$)	*Combined*
(1) Number of units: total (N_h)	17	7	24 (N)
(2) Number of units: sample (n_h)	3	2	5 (n)
For total monthly income			
(3) Total income in sample households $\left(\sum_1^{n_h} x_{hi}\right)$	$107	$123	
(4) Mean income per household $\left(\bar{x}_h = \sum_1^{n_h} x_{hi}/n_h\right)$	$35.67	$61.50	$43.20*'† ($x/n$)
(5) Estimated total income $(x^*_{h0} = N_h\bar{x}_h)$	$606.34	$430.5	$1036.84 ($x$)
(6) $\sum_1^{n_h} x^2_{hi}$	3835	7565	
(7) $\left(\sum_1^{n_h} x_{hi}\right)^2/n_h$	3816.3333	7564.5	
(8) SSx_{hi} = row (6) − row (7)	18.6667	0.5	
(9) $SSx_{hi}/n_h(n_h - 1)$	3.1111	0.25	
(10) $s^2_{x^*_{ho}} = N^2_h \times$ row (9)	899.1111	12.25	911.36 (s^2_x)
(11) $s_{x^*_{ho}}$ = sq. root of row (10)	$29.99‡	$3.5‡	$30.19* ($s_x$)
(12) CV of x^*_{h0} = row (11)/row (5)	4.9%‡	0.81%‡	2.91%*
For Food Cost			
(13) $\left(\sum_1^{n_h} w_{hi}\right)$	$67	$60	
(14) Mean food cost per household $\bar{w}_h = \sum_1^{n_h} w_{hi}/n_h$	$22.33	$30	$24.57[a,d] ($w/N$)

	Stratum I (h = 1)	Stratum II (h = 2)	Combined
(15) Estimated total food cost ($w^*_{h0} = N_h \overline{w}_h$)	\$379.67	\$210	\$589.67 (w)
(16) $\sum_1^{n_h} w_{hi}^2$	1501	1802	
(17) $\left(\sum_1^{n_h} w_{hi}\right)^2/n_h$	1496.3333	1800	
(18) SSw_{hi} = row (16) – row (17)	4.3333	2	
(19) $SSw_{hi}/n_h(n_h-1)$	0.777778	1	
(20) $s^2_{w^*_{h0}} = N_h^2$ x row (19)	224.7777788	49	273.777778 (s_w^2)
(21) $s_{w^*_{h0}}$ = sq. root of row (20)	15.00‡	7‡	16.55* (s_w)
(22) CV of w^*_{h0} = row (21)/row (15)	3.9%‡	3.33%‡	2.81%*
(23) Proportion of income spent on food $r_h = w^*_{h0}/x^*_{h0}$ = row (15)/row (5)	0.626166‡	0.487800‡	0.568718 (r)*
(24) r_h^2	0.3092661‡	0.2379488‡	0.3234402 (r^2)*
(25) $2r_h$	1.243332‡	0.975600‡	1.137436($2r$)*
(26) $\sum_1^{n_h} x_{hi}w_{hi}$	2399	3691	
(27) $\left(\sum_1^{n_h} x_{hi}\right)\left(\sum w_{hi}\right)/n_h$	2389.666667	3690	
(28) $SPx_{hi}w_{hi}$ = row (26) – row (27)	9.333333	1	
(29) $SPx_{hi}w_{hi}/n_h(n-1)$	1.555556	0.5	
(30) $s_{x^*_{h0}w^*_{h0}} = N_h^2$ x row (29)	449.555556	24.5	474.055556 (s_{xw})
(31) $r_h^2 \times s^2_{x^*_{h0}}$	349.993432‡	0.291480‡	294.770278 ($r^2 x s_x^2$)*
(32) $2r_h s_{x^*_{h0}} w^*_{h0}$	558.946009‡	23.90220‡	539.209802 ($2rs_{xw}$)*
(33) row (20) + row (31) – row (32)	15.825201‡	25.389286‡	29.338254*
(34) x^{*2}_{h0}	367 648.1956‡	185 330.25‡	1 075 037.1856 (x^2)*
(35) $s^2_{r_h}$ = row (33)/x^{*2}_{h0}	0.0000430444‡	0.000136995‡	0.0000272904* (s_r^2)
(36) s_{r_h} = sq. root of row (34)	0.00656‡	0.0117‡	0.00522 (s_r)*
(37) CV of r_h = row (36)/row (23)	1.05%‡	2.40%‡	0.91%*
(4) Average income per household $\overline{x}_h = \sum_1^{n_h} x_{hi}/n_h$	\$35.67‡	\$61.50‡	\$43.20 (x/N)*
(9) $s^2_{\overline{x}_h} = s^2_{x^*_{h0}}/N_h^2$	3.1111‡	0.25‡	1.582222 (s_x^2/N^2)*
(38) $s_{\overline{x}_h}$	\$1.76‡	\$0.50‡	\$1.26 (s_x/N)*
(14) Average food cost per household ($\overline{w}_h$)	\$22.33‡	\$30‡	\$24.57 (w/N)*
(19) $s^2_{\overline{w}_h} = s^2_{w^*_{h0}}/N_h^2$	0.777778‡	1‡	0.475309 (s_w^2/N^2)*
(39) $s_{\overline{w}_h}$	\$0.88	\$1	\$0.69 (s_w/N)*
(40) Total number of persons	65	44	109
(41) Average income per head = row (3)/row (41)	\$9.33‡	\$9.78‡	\$9.51*
(42) Estimated S.E. of average income per head = row (11)/row (40)	\$0.46‡	\$0.08‡	\$0.28*

Table 10.3 *continued*

	Stratum I (h = 1)	Stratum II (h = 2)	Combined
(43) Average food cost per head = row (15)/ raw (40)	\$5.84‡	\$4.47‡	\$5.41★
(44) Estimated S.E. of average food cost per head = row (21)/row (40)	\$0.23‡	\$0.16‡	\$0.15★

★ Not additive, i.e. the combined estimate is not obtained by adding the stratum estimates.
† Obtained by dividing the value of x by N (= 24).
‡ Computations not necessary if separate estimates are not required for each stratum.
§ Obtained by dividing the value of w by N (= 24)

Table 10.4 Estimates and standard errors of income and food costs, computed from the data of Tables 10.2 and 10.3: Stratified srs of households

Item	*Estimate*	*Standard error*	*CV (%)*
1. *Monthly income*			
(i) Total	\$1037	\$30.19	2.91
(ii) Per household	\$43.20	\$1.26	2.91
(iii) Per person	\$9.51	\$0.28	2.91
2. *Monthly Food cost*			
(i) Total	\$590	\$16.55	2.81
(ii) Per household	\$24.57	\$0.69	2.81
(iii) Per person	\$5.41	\$0.15	2.81
3. *Proportion of income spent on food*	0.5687	0.00522	0.91

Table 10.5 Population and number of households in each of the two sample villages in the three zones (Appendix IV) and computation of stratum estimates: stratified srs of villages

Zone	*Number of villages* Total	*Number of villages* Sample	*Village serial no.*	*Sample village no.*	*Number of households* In sample village	*Number of households* Stratum estimate	*Population* In sample village	*Population* Stratum estimate
h	N_h	n_h		i	x_{hi}	$x^*_{hi} = N_h x_{hi}$	y_{hi}	$y^*_{hi} = N_h y_h$
(1)	(2)	(3)	(4)	(5)	(6)	(7)	(8)	(9)
I	10	2	2	1	18	180	82	820
			10	2	22	220	88	880
					Total	400		1700
					Mean (= stratum estimate of total)	200 (x^*_{ho})		850 (y^*_{ho})
					Difference	−40 (d_{xh})		−60 (d_{yh})
					½ |Difference| (= estimated standard error)	20 ($s_{x^*_{ho}}$)		30 ($s_{y^*_{ho}}$)
II	11	2	5	1	20	220	105	1155
			11	2	18	198	94	1034
					Total	418		2189
					Mean (= stratum estimate of total)	209 (x^*_{ho})		1094.5 (y^*_{ho})
					Difference	22 (d_{xh})		121 (d_{yh})
					½ |Difference| (= estimated standard error)	11 ($s_{x^*_{ho}}$)		60.5 ($s_{y^*_{ho}}$)
III	9	2	2	1	21	189	121	1089
			3	2	18	162	105	945
					Total	351		2034
					Mean (= stratum estimate of total)	175.5 (x^*_{ho})		1017 (y^*_{ho})
					Difference	27 (d_{xh})		144 (d_{yh})
					½ |Difference| (= estimated standard error)	13.5 ($s_{x^*_{ho}}$)		72 ($s_{y^*_{ho}}$)

Example 10.2

From each of the three zones (Appendix IV), select a simple random sample (with replacement) of two villages each, and on the basis of the data on the number of households and of persons in these sample villages, estimate for three zones separately, and also for the three zones combined, the total numbers of households and of persons, and the average household size with standard errors.

Here the zones are the strata. The information for the selected sample villages is given in Table 10.5. As two sample villages are selected from each zone, the simplified procedures of computation of section 9.5 will be used, where the sums and the differences of the values of the study variables for the two sample units will provide the required estimates. The required computations are shown in Tables 10.5–10.7 and the final results in Table 10.8

Table 10.6 Computation of estimated total population and number of households and their standard errors: data of Table 10.5

Zone	*Number of households*			*Population*		
	Estimate	*Standard error*	*Variance*	*Estimate*	*Standard error*	*Variance*
h	x^*_{ho}	$s_{x^*_{ho}}$	$s^2_{x^*_{ho}}$	y^*_{ho}	$s_{y^*_{ho}}$	$s^2_{y^*_{ho}}$
I	200.0	20.0	400.00	850.0	30.0	900.00
II	209.0	11.0	121.00	1094.5	60.5	3660.25
III	175.5	13.5	182.25	1017.0	72.0	5184.00
All zones combined	584.5 (x)	26.52* (s_x)	703.25 (s_x^2)	2961.5 (y)	98.71* (s_y)	9744.25 (s_y^2)

* Not additive.

Table 10.7 Computation of the estimated average household size and sampling variance: data of Tables 10.5 and 10.6

Zone h (1)	*Average household size* $r_h = y^*_{ho}/x^*_{ho}$ (2)	$s_{y^*_{ho}x^*_{ho}} = \frac{1}{4}d_{yh}d_{xh}$ (3)	$r_h^2 s^2_{x^*_{ho}}$ (4)	$2r_h s_{y^*_{ho}x^*_{ho}}$ (5)	$s^2_{y^*_{ho}} + r^2 s^2_{x^*_{ho}} - 2r_h s_{y^*_{ho}x^*_{ho}}$ (6)	$s^2_{r_h} =$ col. (6)$/x^{*2}_{ho}$ (7)
I	4.2500	600.0	7225.0000	5100.0000	3025.0000	0.07562500
II	5.2368	665.5	3318.3161	6970.1808	8.3853	0.00015621
III	5.7949	972.0	6120.1190	11 265.2865	38.8334	0.00126081
All zones combined	5.0667 (r)	2237.5 (s_{yx})	18 053.4120 ($r^2s_x^2$)	22 673.4825 ($2rs_{yx}$)	5124.1795 ($s_y^2 + r^2s_x^2 - 2rs_{yx}$)	0.01499876 (s_r^2)

Table 10.8 Estimates and standard errors, computed from the data of Table 10.5: Stratified srs of villages

	Zone I	*Zone II*	*Zone III*	*All zones combined*
1. *Total population*				
(a) Estimate	850.0	1094.5	1017.0	2961.5
(b) Standard error	30.0	60.5	72.0	98.7
(c) Coefficient of variation (%)	3.53	5.53	7.08	3.33
2. *Total number of households*				
(a) Estimate	200.0	209.0	175.5	584.5
(b) Standard error	20.0	11.0	13.5	26.5
(c) Coefficient of variation (%)	10.00	5.26	7.69	4.54
3. *Average household size*				
(a) Estimate	4.25	5.24	5.79	5.07
(b) Standard error	0.2750	0.0125	0.0355	0.1225
(c) Coefficient of variation (%)	6.47	0.24	0.61	2.42

Example 10.3

In the Indian National Sample Survey (1958–9), the inquiry into population, births and deaths in the rural areas was based on a single-stage stratified design. Rural India was divided into 218 strata (composed of *tehsils* or groups of *tehsils*, a *tehsil* being an administrative unit consisting of villages and a few towns), each stratum containing approximately equal populations as in the Census of 1951. The total period of survey of one year was divided into six sub-rounds each of two months' duration. A total of 2616 sample villages were covered, with 436 sample villages in each sub-round. In each stratum, two enumerators collected the information independently, each surveying a sub-sample of six villages, i.e. one village each in a sub-round. The allocation of the total sample of 2616 villages to the different States of India was made on the basis of various factors, including the total 1951 Census population, and the allocations were rounded to multiples of 12. In each stratum, the sample villages were selected systematically with a random start after arranging the tehsils in a particular order so as to increase the efficiency of the estimators. In the selected villages, all the households were surveyed in regard to demographic characteristics, including births and deaths during the 365 days preceding the enumeration. For any sub-round, the estimating equations will take the forms in Example 10.2 as there were two sample villages in each stratum in a sub-round. (For further details, see Murthy, Chapter 15, and Som, De, Das, Pillai, Mukkerjee, and Sarma, 1961.)

Some estimates and their standard errors are given in Table 10.9 for the first sub-round (July–August 1958). The estimated rate of growth of population was 2 per cent per annum in 1958–9. The registration system for births and deaths is still defective in India, and this was the first national survey which showed that population in India was growing at a rate much faster than the annual 1.3 per cent rate recorded during the 1941–1951 census decade, which was used in the first two five-year plans covering the period 1951–61. This was later confirmed by the Census of 1961 and the third five-year plan took account of this accelerated rate of growth of population.

Note that such a large-scale sample survey (with 2616 sample villages, about 234 000 sample households and 1.2 million sample persons) used the simple design described in this chapter, the data of which the reader should by now be able to analyse by himself.

Table 10.9 Estimates and standard errors of some demographic parameters: Indian National Sample Survey, July–August 1958 (404 sample villages with 36 365 households)

Parameter	*Estimate*	*Standard error*	*CV (%)*
Birth rate (per 1000 persons)	38.61	0.74	1.92
Death rate (per 1000 persons)	18.97	0.73	3.85
Growth rate (per cent)	1.96	0.08	4.08
Percentage of population employed	45.55	0.71	1.56
Household size	5.06	0.04	0.79

• Example 10.4

For crop surveys starting from 1937 in the Bengal province of pre-independence India and in the West Bengal State of India since 1948, the design was stratified single-stage. The whole area is cadastrally surveyed (i.e. surveyed for tax purposes) so that village maps are available showing each plot. A number of sample grids were selected, the size varying in different years. (A 'grid' is a square mesh of a plane formed by two sets of lines perpendicular to each other, each line of each set being at a constant interval from the adjacent lines.)

With the help of the village maps, the field enumerators identified the individual plots wholly or partially included in each grid, indicating against each plot the fraction of it that was sown with a particular crop. From these data and the total area of each plot (which was known), the proportion of the total area of the grid occupied by the crop was calculated. Let this proportion be denoted by p_{hi} for the ith sample grid in the hth stratum ($h = 1, 2, \ldots, L$; $i = 1, 2, \ldots, n_h$).

If A_h is the total area of the hth stratum, an unbiased estimator of the area under the crop in the stratum, obtained from the hith sample grid, is

$$y^*_{hi} = A_h p_{hi}$$

and the combined unbiased estimator, obtained from all the n_h sample grids in the stratum is, from equation (10.5.b),

$$y^*_{h0} = \sum_{i=1}^{n_h} y^*_{hi}/n_h = A_h \sum_{i=1}^{n_h} p_{hi}/n_h = A_h \bar{p}_h$$

where $\bar{p}_h$ is the simple (unweighted) average of the p_{hi} values in the hth stratum.

An unbiased variance estimator of y^*_{h0} is given by

Table 10.10 Estimated area under Aus paddy (1962–3); Aman paddy, and Jute (1963–4): West Bengal.

Crop	*Number of effective grids of 2.25 acres*	*Estimated area*	
		In thousand acres	*CV (%)*
Aus paddy	52 439	1352	0.94
Aman paddy	54 303	9668	0.32
Jute	52 439	1074	0.94

equation (10.11).

For the whole universe, an unbiased estimator of the total area under the crop is

$$y = \sum^{L} y^*_{ho}$$

an unbiased variance estimator of which is given by equation (10.16).

In surveys for estimating crop areas, the sampling fraction was generally 1 in 140 to 250, but in estimating crop yields it was necessarily much smaller, about 1 in 6 million – for details, see Mahalanobis, 1944, 1946, and 1968. (For the crop surveys recently being conducted in India by the National Sample Survey and the Indian Council of Agricultural Research, the design is stratified multi-stage.)

Some results for the crop survey in the State of West Bengal, India (1962–4) are given in Table 10.10. The total geographical area of West Bengal is about 20 million acres (1 acre = 0.404 hectare), and square grids of size 2.25 acres each were the sampling units in each stratum.●

10.3 Estimation of proportion of units

For the estimation of the proportion of the universe units that fall into a certain class or possess a certain attribute, the extension from the case of unstratified simple random sampling (dealt with in section 2.13) to stratified simple random sampling is straightforward.

Let N'_h be the number of units possessing the attribute, and N_h the total number of units in the hth stratum; a simple random sample of n_h units from the N_h units show that n'_h of the sample of n_h units possess the attribute. We have already seen in section 2.13 that the sample proportion

$$p_h = r_h/n_h \qquad (10.30)$$

is an unbiased estimator of the proportion

$$P_h = N'_h/N_h \qquad (10.31)$$

The sampling variance of the estimator p_h is

$$P_h(1 - P_h)/n_h \qquad (10.32)$$

and an unbiased estimator of this is

$$p_h(1 - p_h)/(n_h - 1) \qquad (10.33)$$

The overall (i.e. for all the strata combined) universe proportion is

$$P = N'/N = \sum N'_h / \sum N_h = \sum N_h P_h / \sum N_h \qquad (10.34)$$

and an unbiased sample estimator of P is

$$p = \sum N_h p_h / N \qquad (10.35)$$

The sampling variance of p is

$$\frac{1}{N^2}\sum \frac{N_h^2 P_h(1 - P_h)}{n_h} \tag{10.36}$$

an unbiased estimator of which is

$$s_p^2 = \Sigma N_h^2 s_{p_h}^2 / N^2$$
$$= \frac{1}{N^2}\sum \frac{N_h^2 p_h(1 - p_h)}{(n_h - 1)} \tag{10.37}$$

Note: p is a weighted average of the p_hs, the weights being N_h/N. If a constant sampling fraction is taken in each stratum, i.e. $n_h/N_h = n/N$, then

$$p = \Sigma r_h / n = r/n \tag{10.38}$$

where $r = \Sigma r_h$ is total number of sample units possessing the attribute.

10.4 Ratio method of estimation: Combined and separate ratio estimators

As for unstratified simple random sampling (Chapter 3), so also for stratified simple random sampling, the use of ancillary information may increase the efficiency of estimators: we shall illustrate the most common use of ancillary information, namely, the ratio method of estimation.

Following the methods of section 3.2, and using the same types of notations as for the study variables, the ratio estimator of Y_h, the stratum total of the study variable in the hth stratum, using the ancillary information, is, from equation (3.6),

$$y_{hR}^* = Z_h r_{1h}$$
$$= Z_h (y_{ho}^* / z_{ho}^*)$$
$$= Z_h (\bar{y}_h / \bar{z}_h) \tag{10.39}$$

where

$$r_{1h} = y_{ho}^* / z_{ho}^* \tag{10.40}$$

A consistent but generally biased estimator of the variance of y_{hR}^* is, from equation (3.11),

$$s_{y_{hR}^*}^2 = Z_h^2 s_{r_{1h}}^2 = (s_{y_{ho}^*}^2 + r_{1h}^2 s_{z_{ho}^*}^2 - 2r_{1h} s_{y_{ho}^* z_{ho}^*} \tag{10.41}$$

$$= N_h^2 (\mathrm{SS}y_{hi} + r_{1h}^2 \mathrm{SS}z_{hi} - 2r_{1h} \mathrm{SP}y_{hi}z_{hi}) / n_h(n_h - 1) \quad \text{in srswr} \tag{10.42}$$

For the ratio estimator of the total of the study variable $Y = \sum^{L} Y_h$, two types of ratio estimators could be used:

1. *Separate ratio estimator.* Here, the ratio method of estimation is first applied at the stratum levels to obtain ratio estimators of the stratum totals y_{hR}^* (equation (10.39)); these stratum ratio estimators are then summed up to provide the ratio estimator of the overall total Y; thus

$$y_{RS} = \sum^{L} y_{hR}^* = \sum^{L} Z_h \bar{y}_h / \bar{z}_h \tag{10.43}$$

(the additional subscript 'S' in y_{RS} standing for 'separate' ratio estimator).

The variance estimator of $y_{\mathbf{RS}}$ is the sum of the estimated variances of the stratum ratio estimators; thus

$$s^2_{y_{\mathbf{RS}}} = \sum^L s^2_{y^*_{h\mathbf{R}}} \quad (10.44)$$

$s^2_{y^*_{h\mathbf{R}}}$ being given by equation (10.42).

2. *Combined ratio estimator.* Here the ratio method of estimation is applied to the estimators of the overall totals; thus

$$y_{\mathbf{RC}} = Zr = Zy/z \quad (10.45)$$

(the additional subscript 'C' in $y_{\mathbf{RC}}$ standing for 'combined' ratio estimator); y and z are the unbiased estimators of the universe totals Y and Z respectively, defined by equations of the type (10.5(d)), i.e.

$$z = \sum^L z^*_{h0} = \sum^L N_h \bar{z}_h \quad (10.46)$$

and

$$r = y/z \quad (10.47)$$

The variance estimator of $y_{\mathbf{RC}}$ is (from equation (3.10))

$$s^2_{y_{\mathbf{RC}}} = Z^2 s^2_r = (s^2_y + r^2 s^2_z - 2rs_{yz}) \quad (10.48)$$

where s_{yz} is the unbiased estimator of the covariance of y and z, given by an equation of the type (10.19), i.e.

$$s_{yz} = \sum^L s_{y^*_{h0} z^*_{h0}}$$

$$= \sum^L N_h^2 \mathrm{SP} y_{hi} z_{hi} / n_h(n_h - 1) \quad (10.49)$$

Notes:

1. The combined ratio estimator $y_{\mathbf{RC}}$ does not require a knowledge of the stratum totals Z_h of the ancillary variable, but only of the overall total Z.
2. The separate ratio estimator will be more efficient than the combined ratio estimator if the sample in each stratum is not too small and the universe stratum ratios $R_{1h} = Y_h/Z_h$ vary considerably from stratum to stratum. If the sample in each stratum is small, the separate ratio estimator will be subject to a large bias.
3. If stratification is made with respect to an ancillary variable, then the ratio method of estimation using the same ancillary variable is not likely to further improve the efficiency of estimators.
4. The ratio method of estimation can be applied at intermediate levels of aggregation between the individual strata and the overall universe; in the example given in note 2 of section 9.3, the ratio method could be used not only at the stratum and country levels, but also at the level of the states. However, the ratio estimate obtained for the country or for a similar higher level of aggregation by applying the ratio method of estimation at the lower levels is likely to contain a bias that may be relatively large in relation to the standard error.

Example 10.5

For the same data as for Example 10.2, given the additional information on the previous census population in Table 10.11, obtain ratio estimates of the present total population and the number of households for the three zones separately and also for all the zones combined.

The required computations are given in Tables 10.11–10.13 and the final estimates in Table 10.14. First, unbiased stratum estimates of the previous census population are computed in Table 10.11. For the ratio estimates of the present total population for the three zones separately, we use equation (10.39); for all the strata combined, the separate ratio estimate is obtained using equation (10.43), and the combined ratio estimate by using equation (10.45), and the respective variance estimates by equations (10.44) and (10.48). The required computations are given in Table 10.12. Similar computations for the ratio estimates of the total number of households are given in Table 10.13.

Compared with the results of Example 10.2, the ratio estimates are seen to be more efficient, especially for total population. The combined ratio estimate is more efficient than the separate ratio estimate for estimating the population.

Table 10.11 Previous census population in each of the two sample villages, selected in Example 10.2 in each of the three zones and computation of stratum estimates: Stratified srs of villages

Zone	*Number of villages*		*Village serial no.*	*Sample village no.*	*Previous census population*	
	Total	*Sample*			*In sample village*	*Stratum estimate*
h	N_h	n_h		i	z_{hi}	$z^*_{hi} = N_h z_{hi}$
(1)	(2)	(3)	(4)	(5)	(6)	(7)
I	10	2	2	1	81	810
			10	2	80	800
					Total	1610
					Mean (= stratum estimate of total)	805 (z^*_{ho})
					Difference	10 (d_{zh})
					$\frac{1}{2}$ |Difference| (= estimated standard error)	5 ($s_{z^*_{ho}}$)
II	11	2	5	1	98	1078
			11	2	86	946
					Total	2024
					Mean (= stratum estimate of total)	1012 (z^*_{ho})
					Difference	132 (d_{zh})
					$\frac{1}{2}$ |Difference| (= estimated standard error)	66 ($s_{z^*_{ho}}$)
III	9	2	2	1	112	1008
			3	2	97	873
					Total	1881
					Mean (= stratum estimate of total)	940.5 (z^*_{ho})
					Difference	135 (d_{zh})
					$\frac{1}{2}$ |Difference| (= estimated standard error)	67.5 ($s_{z^*_{ho}}$)

Table 10.12 Computation of ratio estimates of population and their variances, using the previous census population: data of Tables 10.5–10.7 and 10.11

Zone (h)	Z_h	$r_{1h} = y^*_{ho}/z^*_{ho}$	$y^*_{hR} = Z_h r_{1h}$	$s_{y^*_{ho}z^*_{ho}} = \frac{1}{4}d_{yh}d_{zh}$	$s^2_{y^*_{hR}} = s^2_{y^*_{ho}} + r^2_{1h}s^2_{z^*_{ho}} - 2r_{1h}s_{y^*_{ho}z^*_{ho}}$	$s_{y^*_{hR}}$
(1)	(2)	(3)	(4)	(5)	(6)	(7)
I	863	1.0559	911.24	−150	1244.6431	35.28
II	1010	1.0815	1092.34	3993	116.8691	10.81
III	942	1.0813	1018.62	4860	0.5988	0.77
Total	2815		3022.20 (y_{RS})		1362.1110 ($s^2_{y_{RS}}$)	36.91 ($s_{y_{RS}}$)
Combined	2815 (Z)	1.0734 (r_1)	3023.25 ($y_{RC} = Zr_1$)	8703 (s_{yz})	725.5130 ($s^2_{y_{RC}} = s^2_y + r^2_1 s^2_z - 2r_1 s_{yz}$)	26.98 ($s_{y_{RC}}$)

Table 10.13 Computation of ratio estimates of number of households and their variances, using the previous census population: data of Tables 10.5–10.7 and 10.11–10.12

Zone (h)	Z_h	$r_{2h} = x^*_{ho}/z^*_{ho}$	$x^*_{hR} = Z_h r_{2h}$	$s_{x^*_{ho} z^*_{ho}} = \frac{1}{4} d_{xh} d_{zh}$	$s^2_{x^*_{hR}} = s^2_{x^*_{ho}} + r^2_{2h} s^2_{z^*_{ho}} - 2 r_{2h} s_{x^*_{ho} z^*_{ho}}$	$s_{x^*_{hR}}$
(1)	(2)	(3)	(4)	(5)	(6)	(7)
I	863	0.24845	214.41	−100.00	426.3882	20.65
II	1010	0.20652	208.59	726.00	6.9157	2.63
III	942	0.18660	175.78	911.25	0.8128	0.90
Total	2815		598.78 (x_{RS})		437.1167 $(s^2_{x_{RS}})$	20.84 $(s_{x_{RS}})$
Combined	2815 (Z)	0.21179 (r_2)	596.18 $(x_{RC} = Z r_2)$	1537.25 (s_{xz})	452.9819 $(s^2_{x_{RC}} = s^2_x + r^2_2 s^2_z - 2 r_2 s_{xz})$	21.28 $(s_{x_{RC}})$

Table 10.14 Ratio estimates of population and number of households, using the previous census population, computed from the data of Tables 10.5 and 10.11: Stratified srs of villages

Zone	*Ratio estimates of population*			*Ratio estimates of number of households*		
	Estimated number	*Standard error*	*CV* (%)	*Estimated number*	*Standard error*	*CV* (%)
I	911.2	35.28	3.89	214.4	20.65	9.63
II	1092.3	10.81	0.99	208.6	2.64	1.26
III	1018.6	0.77	0.076	175.8	0.90	0.51
All zones:						
Separate ratio estimate	3022.2	36.91	1.22	598.8	20.84	3.48
Combined ratio estimate	3023.2	26.98	0.89	596.2	21.28	3.57

Example 10.6

For the Demographic Sample Survey in Chad (1964), the universe of population (excluding nomads) was divided into nine strata – one for the urban areas and eight for rural areas, where the stratification criterion was chosen on the basis of the most dominant ethnic group. In each rural stratum, the sub-universe was classified according to the administrative division (*Préfectures*, *Sous-préfectures*, and *Cantons*) and the size of the village. The rural strata were composed of a number of *Sous-préfectures.* Each rural stratum was divided into three sub-strata according to the size of the village: (a) with up to 199 persons, (b) with 200 to 499 persons, and (c) with 500 or more persons, according to the previous administrative census. Primary units were constructed in sub-stratum (a) by grouping small villages, in (b) by considering the villages as the primary units, and in (c) by grouping the localities of a village so that they contain an average of 300 persons: a systematic sample of 1 in 20 primary units was chosen for the survey, and the population, and births and deaths during the preceding 365 days recorded. The relatively large sampling fraction was chosen in order to provide reliable estimates at the stratum level.

Note: Sampling in rural sub-stratum (c) was stratified two-stage, to be considered in Part IV.

The sample comprized 101 thousand persons in the rural areas and 11 thousand persons in the urban: the overall sampling fraction was 5 per cent. Using the previous administrative census data, the ratio estimates of population were 2356 ± 87.1 thousand in the rural sector and 2530 ± 91.1 thousand in the country. On the basis of these results, the 95 per cent limits were placed at 2268 and 2444 thousand for the rural population and 2439 and 2629 thousand for the total population of the country (excluding nomads). (For further details, *see* the report by Behmoiras).

• 10.5 Ratio of ratio estimators of totals of two study variables

Ratio estimators of totals of two study variables may be used in estimating their ratio. However, it becomes the same as the ratio of the corresponding unbiased estimators of totals if computed at the stratum level, or for all the strata combined if the combined ratio estimator is used (see section 3.2, note 5).

If separate ratio estimators are used for the totals of two study variables namely

$$y_{\mathrm{RS}} = \sum^{L} y^{*}_{h\mathrm{R}} = \sum^{L} Z_h r_{1h} = \sum^{L} Z_h \bar{y}_h / \bar{z}_h \tag{10.43}$$

$$x_{\mathrm{RS}} = \sum^{L} x^{*}_{h\mathrm{R}} = \sum^{L} Z_h r_{2h} = \sum^{L} Z_h \bar{x}_h / \bar{z}_h \tag{10.50}$$

as respective estimators of the universe totals Y and X, then an estimator of the universe ratio

$$R = Y/X \tag{10.7.c}$$

is

$$r' = y_{\mathrm{RS}}/x_{\mathrm{RS}} = \sum^{L} y^{*}_{h\mathrm{R}} \Big/ \sum^{L} x^{*}_{h\mathrm{R}}$$

$$= \sum^{L} Z_h r_{1h} \Big/ \sum^{L} Z_h r_{2h} = \sum^{L} Z_h \bar{y}_h \Big/ \sum^{L} Z_h \bar{x}_h \tag{10.51}$$

A variance estimator of r' is

$$s^2_{r'} = (s^2_{y_{\mathrm{RS}}} + r'^2 s^2_{x_{\mathrm{RS}}} - 2r' s_{y_{\mathrm{RS}} x_{\mathrm{RS}}})/x^2_{\mathrm{RS}} \tag{10.52}$$

where $s^2_{y_{\mathrm{RS}}}$ and $s^2_{x_{\mathrm{RS}}}$ are given by estimating equations of the type (10.44), and the estimated covariance is

$$s_{y_{\mathrm{RS}} x_{\mathrm{RS}}} = \sum^{L} s_{y^{*}_{h\mathrm{R}} x^{*}_{h\mathrm{R}}} \tag{10.53}$$

where

$$s_{y^{*}_{h\mathrm{R}} x^{*}_{h\mathrm{R}}} = \sum^{n_h} \frac{(y^{*}_{hi} - r_{1h} z^{*}_{hi})(x^{*}_{hi} - r_{2h} z^{*}_{hi})}{n_h(n_h - 1)} \tag{10.54}$$

For an example, see exercise 3 of this chapter. •

• 10.6 Gain due to stratification

From the data of a stratified sample, it is possible to examine the gain, if any, due to stratification leading to a reduction in the variance of a sample estimator, as compared with an unstratified simple random sample. We will consider in the next section the question of estimating from the data of an unstratified srs, the gain had stratification been adopted.

The unbiased estimator of the variance of estimated total of the study variable y, obtained from a stratified simple random sample, is

$$s^2_y = \sum^{L} s^2_{y^{*}_{h0}} = \sum^{L} N^2_h \cdot \frac{\sum^{n_h} y^2_{hi} - \left(\sum^{n_h} y_{hi}\right)^2 / n_h}{n_h(n_h - 1)} \tag{10.16}$$

and an unbiased estimator of the variance of the estimator of the total, had the data come from an unstratified simple random sample, is

$$\frac{\left(N \sum^{L} N_h \sum^{n_h} y_{hi}^2/n_h - y^2 + s_y^2\right)}{n} \qquad (10.55)$$

Example 10.7

For the data of Example 10.2, estimate the gain in the efficiency of the sample estimate of the total population due to stratification, as compared to that for an unstratified srs.

The required computations are given in Table 10.15. The value of the expression (10.55) is

$(30 \times 297\,072.5 - 2961.5^2 + 9744.25)/6 = 25\,239.5.$

The efficiency of the estimate from the stratified sample as compared with that from the unstratified is thus

$25\,239.5/9744.25 = 2.59$ or 259% •

Table 10.15 Computation of the gain due to stratification: data of Example 10.2

Stratum	$\sum_{i=1}^{n_h} y_{hi}^2$	$\sum_{i=1}^{n_h} y_{hi}^2/n_h$	$N_h \sum_{i=1}^{n_h} y_{hi}^2/n_h$
I	14 468	7234.0	72 340.0
II	19 861	9930.5	109 235.5
III	25 666	12 833.0	115 497.0
Total			297 072.5 $\left(\sum_1^L N_h \sum_1^{n_h} y_{hi}^2/n_h\right)$

• 10.7 Stratification after sampling

For a stratified simple random sample, information on the total number and the listing of the units in each stratum for the 'stratification variable' is required prior to sample selection. Such information on the desirable variables suitable for stratification, such as sex, age, tribe etc. in a demographic survey, is often not available prior to sampling; it is collected for the selected sample units only in the course of the actual enumeration.

In such a situation, under suitable conditions, the sample data from an unstratified simple random sample may be classified into a number of strata (say L) and analysed in the same manner as a stratified simple random sample. Two cases may arise:

1. The total number of units in stratum h, N_h, is known. The sample data (obtained from an unstratified simple random sample) are classified into L strata (*post-strata*); let the number of sample units falling in the hth post-stratum and the mean of the values of the study variable there be denoted respectively by n_h and $\bar{y}_h$. Then for the estimator of the universe total Y, the same estimating equation (10.5(d)) for a stratified simple random sample is used, namely

$$y = \sum^{L} N_h \bar{y}_h \qquad (10.56)$$

and the variance estimator of y is given by equation (10.16). For an example, see exercise 4 of this chapter.

2. The total number of units in the post-stratum h, namely N_h, is not known, but the proportion is assumed to be known; e.g. if age is the desirable stratification variable, the age structure of the population can be assumed to be effectively the same as in the last population inquiry. The estimator y of equation (10.56) can be written in the form

$$y = \sum^{L} N_h \bar{y}_h$$
$$= N \sum^{L} (N_h/N) \bar{y}_h$$
$$= N \sum^{L} P'_h \bar{y}_h \tag{10.57}$$

where
$$P'_h = (N_h/N) \tag{10.58}$$

are the 'weights', i.e. the proportions of the total number of units in the different post-strata, the values of P'_h being known or taken from previous inquiries. This is a biased estimator.

The variance estimator of y is

$$s_y^2 = N^2 \sum^{L} P_h'^2 s_{\bar{y}_h}^2 \tag{10.59}$$

$s_{\bar{y}_h}^2$, the variance estimator of the sample mean in the hth post-stratum, being given, as usual, by

$$s_{\bar{y}_h}^2 = \mathrm{SS}y_{hi}/n_h(n_h - 1) \tag{10.60}$$

The estimator of the universe mean is obtained, as usual, on dividing N into the estimator of the universe total, and so also for its estimated standard error.

Notes

1. If the sample is reasonably large, say over twenty in each stratum, and the errors in the weights negligible, the method of post-stratification can give results almost as efficient as *proportional stratified sampling* (see Chapter 12). This is also obvious from the fact that for large samples, the sample units are likely to be distributed in proportion to N_h.
2. The method can also be used when the sample is already stratified according to some variable other than that desired for the study in hand. ●

Further reading

Cochran, sections 5.1–5.4, 5.6, 5.10, 5A.8, 5A.13, and 6.10–6.12; Hansen *et al.*, Vol. I, Chapters 5A, 5B, 5D, and Vol. II, sections 5.1–5.3, 5.5, 5.6, 5.12, and 5.13; Murthy, sections 7.1–7.3, 7.5–7.7, 7.14, and 10.6; Raj, sections 4.1, 4.2, and 5.18; Sukhatme and Sukhatme, sections 3.1–3.2, 3.5, 3.7, 3.8, and 4.11; Yates, sections 3.3–3.5, 6.5–6.7, 7.6, 7.9–7.11, and 8.3–8.5.

Exercises

1. In a sample survey designed to estimate the total number of cattle, the universe of 2072 farms was stratified into 5 strata by the total acreage of the farms. A simple random sample of farms (of size n_h) was taken from each stratum; the total number of cattle in the sample farms $\sum_{1}^{n_h} y_{hi}$ and the raw sum of squares $\sum_{i=1}^{n_h} y_{hi}^2$ are given in Table 10.16; y_{hi} denotes the number of cattle in the ith sample farm ($i = 1, 2, \ldots, n_h$; $h = 1, 2, \ldots, 5$). Estimate the total and average number of cattle per farm along wit[h] with their standard errors (United Nations *Manual*, Process 4, Example 4b, modified to 'with replacement' sampling).

2. For a socio-economic inquiry, the 23 671 villages in an area, including the uninhabited ones,

Table 10.16 Total number of cattle in the srs of farms and the raw sum of squares in the different strata

Stratum (acres)	*No. of farms in stratum* (N_h)	*No. of farms sampled* (n_h)	$\sum_{i=1}^{n_h} y_{hi}$	$\sum_{i=1}^{n_h} y_{hi}^2$
(1)	(2)	(3)	(4)	(5)
0–15	635	153	619	5579
16–30	570	138	1423	24 253
31–50	475	115	1758	34 082
51–75	303	73	1691	51 419
76–100	89	21	603	18 305
All strata	2072 (N)	500	6094	133 658

Table 10.17 Number of households in the srs of 10 villages in the different strata

Stratum	*Total number of villages*	*Total number of households in sample villages*									
		1	*2*	*3*	*4*	*5*	*6*	*7*	*8*	*9*	*10*
I	1411	43	84	98	0	10	44	0	124	13	0
II	4705	50	147	62	87	84	158	170	104	56	160
III	2558	228	262	110	232	139	178	334	0	63	220
IV	14 997	17	34	25	34	36	0	25	7	15	31

Table 10.18 Total acreage (a) and acreage in wheat (b) in the srs of farms in Hertfordshire, U.K., 1939

District 1 (15 farms)	(a)	188 60 192 48 44 79 14 465 197 163 198 78 6 35 168 **(1935)**
	(b)	16 0 0 0 0 33 0 92 0 0 0 0 0 0 0 **(141)**
District 2 (8 farms)	(a)	8 294 597 8 2 200 14 262 **(1385)**
	(b)	0 29 107 0 0 65 0 58 **(259)**
District 3 (40 farms)	(a)	370 26 369 212 153 287 28 14 4 17 2 3 7 6 335
		4 1 4 180 120 40 28 221 31 6 34 316 116 4 409
		6 115 19 274 3 144 3 482 156 302 **(4851)**
	(b)	67 0 58 45 20 44 0 0 0 0 0 0 0 0 82
		0 0 0 0 11 0 0 59 0 0 0 75 33 0 102
		0 0 0 6 0 0 0 62 28 71 **(763)**
District 4 (24 farms)	(a)	11 6 543 822 654 3 158 4 68 55 4 2 192 4 491 224 280
		90 3 3 6 4 161 246 **(4034)**
	(b)	0 0 80 265 112 0 50 0 27 12 0 0 24 0 24 28 75
		0 0 0 0 0 80 60 **(837)**
District 5 (4 farms)	(a)	4 312 8 11 **(335)**
	(b)	0 102 0 0 **(102)**
District 6 (24 farms)	(a)	8 87 6 44 4 614 192 10 24 2 9 3 2 120 58 20 30 197
		14 32 2 285 138 126 **(2027)**
	(b)	0 14 0 0 0 72 20 0 0 0 0 0 0 24 0 0 0 6
		3 6 0 29 0 0 **(174)**
District 7 (10 farms)	(a)	128 4 46 181 17 24 10 36 12 89 **(547)**
	(b)	5 0 0 20 0 0 0 0 0 0 **(25)**

Grand total: 125 farms
(a) 15 114
(b) 2301

were grouped into 4 strata on the basis of their altitude above sea-level and population density, and from each stratum a simple random sample of 10 villages was taken with replacement. The data on the number of households in each of the sample villages are given in Table 10.17.

(a) Estimate the total number of households with its CV;

(b) Examine whether there has been any gain due to stratification as compared to unstratified simple random sample with replacement (Murthy, Problem 7.1).

3. For the data of Example 10.5, estimate the average household size from the separate ratio estimates of numbers of persons and households, and its standard error.

4. A simple random sample of 125 farms out of a total of 2496 farms in Hertfordshire, 1939, gave the data on (a) total acreage and (b) acreage of wheat, classified by districts after selection (Table 10.18). Estimate the total area of wheat and the number of farms growing wheat (i) directly from the sample, and (ii) by stratification by size, given the total number of farms in the specific size-groups in cols. 2 in Table 2 to the answer (Yates, Examples 6.6 and 7.6.b, modified to 'with replacement' sampling).

5. In the Micro-Census Sample Survey of the Federal Republic of Germany, conducted on a continuing basis from October 1957, the sample design for the annual survey has been stratified single-stage since October 1962. In each stratum, formed according to the size of the communities, a one per cent systematic sample of area units (enumeration districts) is selected, and all dwellings, households, families, and persons in the sample areas are enumerated.

(a) The Federal Statistical Office uses the following estimators: the unbiased estimator of Y is $y = 100 \sum_h n_h \bar{y}_h$, and the unbiased variance estimator of y is $0.99 \times 10^4 \sum_h n_h \, SSy_{hi}/(n_h - 1)$.

Verify these, noting that the finite multiplier is not ignored for the variance estimators (Herberger,1971).

(b) also verify that

$$s_r^2 = 0.99 \sum_h \left[\frac{n_h}{n_h - 1} (SSy_{hi} + r^2 SSx_{hi} - 2r \, SPy_{hi}x_{hi}) \right] \Big/ \left(\sum_h n_h \bar{x}_h \right.$$

(*Hint:* (a) Use respectively equations (10.5.d) and (10.24), noting that $f_h = n_h/N_h = 0.01$.)

11 Stratified Varying Probability Sampling

11.1 Introduction

As for simple random sampling, so also for varying probability sampling the extension from unstratified to stratified sampling is straightforward and follows directly from the fundamental theorems of Chapter 9. The use of different schemes of selection – srs, pps etc. – in the different strata, the special cases of crop surveys, the ratio method of estimation, and the gain due to stratification will also be considered in the chapter.

11.2 Estimation of totals, means, ratios and their variances

11.2.1 Structure of a stratified varying probability sampling

Consider the total universe to be divided, on certain criteria, into L strata, and let the hth stratum ($h = 1, 2, \ldots, L$) contain a total of N_h units. The values of the two study variables for the ith unit ($i = 1, 2, \ldots, N_h$) in the hth stratum will be denoted, as in section 10.2, respectively by Y_{hi} and X_{hi}. The total number of units in the universe is, as before,

$$N = \sum_{h=1}^{L} N_h \tag{11.1}$$

In the hth stratum, a sample of n_h units is selected (with replacement) with varying probabilities, the probability of selection of the ith unit ($i = 1, 2, \ldots, N_h$) being π_{hi}. Let the values of the two study variables for the ith selected sample unit ($i = 1, 2, \ldots, n_h$) be denoted respectively by y_{hi} and x_{hi}. The total number of sample units is

$$n = \sum_{h=1}^{L} n_h \tag{11.2}$$

11.2.2 Estimators of totals, means, and ratio of two totals

An unbiased estimator of the stratum total Y_h, obtained from the ith sample unit in the hth stratum is, from equation (5.2),

$$y_{hi}^* = y_{hi}/\pi_{hi} \tag{11.3}$$

and the combined unbiased estimator of Y_h, obtained from the n_h sample units in the hth stratum, is, from equation (5.3) or (9.1),

$$y_{ho}^* = \sum_{i=1}^{n_h} y_{hi}^*/n_h \tag{11.4}$$

Thus, if the sample units are selected with probability proportional to size z_{hi}, i.e. $\pi_{hi} = z_{hi}/Z_h$, where $Z_h = \sum_{i=1}^{N_h} z_{hi}$, then

$$y^*_{hi} = y_{hi}/\pi_{hi} = Z_h\, y_{hi}/z_{hi} \qquad (11.5)$$

Some universe parameters and their sample estimator are defined in Table 11.1. These follow from the results of section 5.3 for any stratum and section 9.3 for all strata combined.

Table 11.1 Some universe parameters and their sample estimators for a stratified varying probability sample

	For the hth stratum		*For all strata combined*		
	Universe parameter (a)	*Sample estimator (b)*	*Universe parameter (c)*	*Sample estimator (d)*	
For a study variable:					
Total	$Y_h = \sum_{i=1}^{N_h} Y_{hi}$	$y^*_{ho} = \sum_{i=1}^{n_h} y^*_{hi}/n_h$	$Y = \sum_{h=1}^{L} y^*_{ho}$	$y = \sum_{h=1}^{L} y^*_{ho}$	(11.6)
Mean	$\bar{Y}_h = Y_h/N_h$	$\bar{y}^*_{ho} = y^*_{ho}/N_h$	$\bar{Y} = Y/N$	y/N	(11.7)
For two study variables:					
Ratio of totals	$R_h = Y_h/X_h$	$r_h = y^*_{ho}/x^*_{ho}$	$R = Y/X$	$r = y/x$	(11.8)

11.2.3 Sampling variances and their estimators for a stratum

The sampling variance of the sample estimator y^*_{ho} of the stratum total Y_h in ppswr is, from equation (5.4),

$$\sigma^2_{y^*_{ho}} = \frac{1}{n_h}\sum_{i=1}^{N_h}\left(\frac{Y_{hi}}{\pi_{hi}} - Y_h\right)^2 \pi_{hi} = \frac{1}{n_h}\left(\sum_{i=1}^{N_h}\frac{Y^2_{hi}}{\pi_{hi}} - Y^2_h\right) \qquad (11.9)$$

an unbiased estimator of which is (from equation (5.5) or (9.2))

$$s^2_{y^*_{ho}} = \sum_{i=1}^{n_h}(y^*_{hi} - y^*_{ho})^2/n_h(n_h - 1) = \mathrm{SS}y^*_{hi}/n_h(n_h - 1) \qquad (11.10)$$

The sampling variance of the sample estimator of the hth stratum mean is

$$\sigma^2_{\bar{y}^*_{ho}} = \sigma^2_{y^*_{ho}}/N^2_h \qquad (11.11)$$

which is estimated unbiasedly from equation (5.7) by

$$s^2_{\bar{y}^*_{ho}} = s^2_{y^*_{ho}}/N^2_h \qquad (11.12)$$

An unbiased estimator of the covariance of y^*_{ho} and x^*_{ho} (x^*_{ho} being defined similarly for another study variable) is, from equation (5.8) or (9.10),

$$s_{y^*_{ho}x^*_{ho}} = \sum_{i=1}^{n_h}(y^*_{hi} - y^*_{ho})(x^*_{hi} - x^*_{ho})/n_h(n_h - 1)$$

$$= \mathrm{SP}y^*_{hi}x^*_{hi}/n_h(n_h - 1)$$

An estimator of the variance of r_h is, from equation (5.10) or (9.14),

$$\begin{aligned} s_{r_h}^2 &= (s_{y_{ho}^*}^2 + r_h^2 s_{x_{ho}^*}^2 - 2r_h s_{x_{ho}^* y_{ho}^*})/x_{ho}^{*2} \\ &= (\text{SS}y_{hi}^* + r_h^2 \text{SS}y_{hi}^* - 2r_h \text{SP}y_{hi}^* x_{hi}^*)/n_h(n_h - 1)x_{ho}^{*2} \end{aligned} \tag{11.13}$$

11.2.4 Sampling variances and their estimators for all strata combined

The sampling variance of the estimator y of the universe total Y is

$$\sigma_y^2 = \sum^L \sigma_{y_{ho}^*}^2 = \sum_{h=1}^{L} \frac{1}{n_h} \sum_{i=1}^{N_h} \left(\frac{Y_{hi}}{\pi_{hi}} - Y_h\right)^2 \pi_{hi} \tag{11.14}$$

which is estimated by the sum of the unbiased variance estimators of y_{ho}^* from equation (9.6), by

$$s_y^2 = \sum^L s_{y_{ho}^*}^2 \tag{11.15}$$

The sampling variance of the estimator of the universe mean is

$$\sigma_y^2/N^2 \tag{11.16}$$

which is estimated unbiasedly by

$$s_y^2/N^2 \tag{11.17}$$

An unbiased estimator of the covariance of the estimators y and x is, from equation (9.12),

$$s_{yx} = \sum^L s_{y_{ho}^* x_{ho}^*} \tag{11.18}$$

A variance estimator of r is (from equation (9.16))

$$s_r^2 = (s_y^2 + r^2 s_x^2 - 2r s_{yx})/x^2 \tag{11.19}$$

For the notes relating to the estimation of ratios, see the notes to section 2.7.

Example 11.1

From each of the three zones (Appendix IV), select with probability proportional to the previous census population and with replacement a sample of two villages each, and on the basis of the data on the current number of households and of persons in these sample villages, estimate for the three zones separately and also for the three zones combined, the total numbers of households and of persons and average household size, with standard errors.

Here the zones are the strata. The sample villages are selected with probability proportional to the previous census population by adopting the procedure of cumulation of the previous census population in each zone (section 5.4). The information on the selected sample villages is given in Table 11.2; the Z_h values (the total previous census population in the zones) are 863, 1010, and 942 respectively. As two sample villages are selected from each zone, the simplified procedure of section 9.5 for estimation will be followed, as was done in Example 10.2 relating to a stratified srs. The required computations are shown in Tables 11.2–11.4 and the final results in Table 11.5.

Table 11.2 Population and number of households in each of the two sample villages selected with probability proportional to previous census population in each of the three zones (Appendix IV): Stratified pps sample of villages

Zone	*Sample village*	*Village serial no.*	*Size (previous census population)*	*Reciprocal of probability*	*Number of households*		*Population*	
					In sample village	*Stratum estimate*	*In sample village*	*Stratum estimate*
h	i		z_{hi}	$1/\pi_{hi} = Z_h/z_{hi}$	x_{hi}	$x^*_{hi} = x_{hi}/\pi_{hi}$	y_{hi}	$y^*_{hi} = y_{hi}/\pi_{hi}$
(1)	(2)	(3)	(4)	(5)	(6)	(7)	(8)	(9)
I	1	5	92	9.3804	24	225.1296	112	1050.6048
	2	7	72	11.9861	20	239.7220	88	1054.7768
					Total	464.8516		2105.3816
					Mean (= stratum estimate of total)	232.45 (x^*_{ho})		1052.6908 (y^*_{ho})
					Difference	−14.59 (d_{xh})		−4.1720 (d_{yh})
					½ |Difference| (= estimated standard error)	7.295 ($s_{x^*_{ho}}$)		2.0860 ($s_{y^*_{ho}}$)
II	1	3	73	13.8356	17	235.2052	80	1106.8480
	2	7	85	11.8824	19	225.7656	95	1128.8280
					Total	460.9708		2235.6760
					Mean (= stratum estimate of total)	230.49 (x^*_{ho})		1117.8380 (y^*_{ho})
					Difference	9.44 (d_{xh})		−21.9800 (d_{yh})
					½ |Difference| (= estimated standard error)	4.72 ($s_{x^*_{ho}}$)		10.9900 ($s_{y^*_{ho}}$)
III	1	2	112	8.4107	21	176.6247	121	1017.6947
	2	4	117	8.0513	21	169.0773	129	1038.6177
					Total	345.7020		2056.3124
					Mean (= stratum estimate of total)	172.85 (x^*_{ho})		1028.1562 (y^*_{ho})
					Difference	7.55 (d_{xh})		−20.9230 (d_{yh})
					½ |Difference| (= estimated standard error)	3.775 ($s_{x^*_{ho}}$)		10.4614 ($s_{y^*_{ho}}$)

Table 11.3 Computation of estimated total population and number of households and their standard errors: data of Table 11.2

Zone	*Number of households*			*Population*		
	Estimate	*Standard error*	*Variance*	*Estimate*	*Standard error*	*Variance*
h	x^*_{ho}	$s_{x^*_{ho}}$	$s^2_{x^*_{ho}}$	y^*_{ho}	$s_{y^*_{ho}}$	$s^2_{y^*_{ho}}$
I	232.43	7.295	53.2170	1052.69	2.086	4.3514
II	230.49	4.720	22.2784	1117.84	10.990	120.7801
III	172.85	3.775	14.2506	1028.16	10.462	109.4534
All zones combined	635.77 (x)	9.473* (s_x)	89.7460 (s^2_x)	3198.69 (x)	15.316* (s_x)	234.5849 (s^2_x)

* No additive.

Table 11.4 Computation of the estimated average household size and sampling variance: data of Tables 11.2 and 11.3

Zone *h*	*Average household size* $r_h = y^*_{ho}/x^*_{ho}$	$s_{y^*_{ho}x^*_{ho}} = \frac{1}{4}d_{yh}d_{xh}$	$s^2_{y^*_{ho}} + r^2_h s^2_{x^*_{ho}} - 2r_h s_{y^*_{ho}x^*_{ho}}$	$s^2_{r_h} =$ col. $(4)/x^{*2}_{ho}$
(1)	(2)	(3)	(4)	(5)
I	4.5291	15.0861	959.3228	0.0177574
II	4.8499	−51.8728	1147.9573	0.0216084
III	5.9482	−39.4918	1083.4655	0.0362654
All zones combined	5.0312 $(r = y/x)$	−76.2785 (s_{yx})	3273.8702 $(s^2_y + r^2 s^2_x - 2r s_{yx})$	0.00809956 (s^2_r)

Table 11.5 Estimates and standard errors, computed from the data of Table 11.2: Stratified pps sample of villages

Item	*Zone I*	*Zone II*	*Zone III*	*All zones combined*
1. *Total population*				
(a) Estimate	1052.7	1117.8	1028.2	3198.7
(b) Standard error	2.09	10.99	10.46	15.32
(c) Coefficient of variation (%)	0.20	0.98	1.02	0.48
2. *Total number of households*				
(a) Estimate	232.4	230.5	172.8	635.8
(b) Standard error	7.30	4.72	3.78	9.47
(c) Coefficient of variation (%)	3.14	2.05	2.18	1.49
3. *Average household size*				
(a) Estimate	4.53	4.85	5.95	5.03
(b) Standard error	0.1333	0.1470	0.1904	0.0900
(c) Coefficient of variation (%)	2.95	3.04	3.20	1.79

Example 11.2

For the Indian National Sample Survey (1957–8) in the rural sector, the total number of 2522 *tehsils* were grouped into a number of geographical strata. In each stratum, two villages were selected from the total number of villages with probability proportional to the 1951 Census population; in the sample villages thus selected, the existing households were listed for the purpose of sub-sampling of households for socio-economic inquiries, the sampling design for which was thus stratified two-stage (see section 1.9 and Chapter 14). However, at the time the list of households in the sample villages was constructed, information on the total number of births during the preceding 365 days in the premises occupied by the households was collected, along with the household size from all the households. The sample design for this inquiry on births was thus stratified single-stage pps. The estimating equations for the total population, births and the birth rate and their estimated variances will thus take the forms given in section 11.2.

Estimates of the birth rates and estimated standard errors for rural India as a whole as also for the five zones are given in Table 11.6. The zones comprised one or more states, each state containing several strata: for the zonal estimates, the same types of formulae as for the overall estimates for the country as a whole were used, excepting that only those strata constituting a zone were considered (for further details, see Som *et al.*, 1961). There were 924 villages, over 135 000 households, and over 680 000 persons in the sample.

By definition, births occurring in hospitals and

Table 11.6 Estimated birth rate (per 1000 persons): Indian National Sample Survey, rural sector, 1957–8

Zone	Estimate	Standard error	CV (%)
North	37.08	0.86	2.32
Central	42.36	0.89	2.10
East	32.96	1.25	3.79
South	28.07	0.75	2.67
West	32.81	0.52	2.80
India (rural)	35.16	0.45	1.28

other institutions were not recorded, nor, for households occupying the premises for less than a year, the births occurring in the previous residence. Even with the limited definition, it appears likely that a number of births were not reported because of the lack of emphasis and of probes and other associate items and of cross-checks on this item in the 'listing schedule'. The methods of obtaining adjusted estimates from demographic data are described briefly in section 25.5.

11.3 Different schemes of selection in different strata

As mentioned in section 9.2, it is possible to adopt different sampling schemes – srs, pps, or systematic – in different strata. The fundamental theorems of section 9.3 permit the estimation of the total of a study variable and its variance for all the strata combined by adding up the corresponding unbiased stratum estimators. The ratio of the totals of two study variables and its variance can also be estimated, as also can other related measures such as the correlation coefficient, from the results of section 9.3. The procedure will be illustrated with an example.

Example 11.3

A universe of 112 villages was divided into three strata with 51, 37, and 24 villages respectively. From the first stratum, a simple random sample of 6 villages was selected without replacement, from the second stratum a sample of 5 villages with probability proportional to the cultivated area (the total cultivated area in the stratum was 26 912 acres) and with replacement, and from the third stratum two linear systematic samples of 4 villages each. For each selected sample village, the total area under wheat was observed: this information is given in Table 11.7, along with the total areas of the sample villages of stratum II. Estimate the total area under wheat in each stratum separately and also for all strata combined, along with the standard errors (Murthy, Problem 7.3, adapted).

Table 11.7 Area under wheat (y_{hi}) for all the sample villages and cultivated area (x_{hi}) for the sample villages of stratum II

Sample village	Stratum (h)				
	Stratum I	Stratum II		Stratum III	
i	y_{1i}	x_{2i}	y_{2i}	y_{3i}	$y_{3'i}$
1	75	729	247	427	335
2	101	617	238	326	412
3	5	870	359	481	503
4	78	305	129	445	348
5	78	569	223		
6	45				

Table 11.8 Estimated wheat acreage, obtained from the data of Table 11.7

Area under wheat	*Stratum I*	*Stratum II*	*Stratum III*	*All strata combined*
1. Estimate (acres)	3247	10 507	9831	23 585
2. Estimated variance (acre)^2	435 540	153 506	59 049	648 095
3. Estimated standard error (acres) [square root of row (2)]	660	392	243	843
4. Estimated CV = row (3)/row (1)	20.3(%)	3.7(%)	2.5(%)	3.4(%)

For stratum I, the methods of Example 2.1 are followed. The estimated area under wheat is 3247 acres with estimated variance 435 540 (acres)^2.

For stratum II, we follow the methods of section 11.2. From equations (11.5.b) and (11.10), the estimated area under wheat is 10 507 acres with estimated variance 153 506 (acres)^2.

For stratum III, the methods of Example 4.1 are followed. The estimated area under wheat is 9831 acres with variance 59 049 (acres)^2.

The estimated total area under wheat in the three strata separately and combined are shown in Table 11.8, along with their estimated variance, standard errors, and coefficients of variation.

11.4 Special cases of crop surveys

11.4.1 Introduction

As for unstratified pps sampling (section 5.5), so also for stratified sampling the sampling of fields (or farms or plots) with *probability proportional to total (geographical) area* (ppa) simplifies the estimating procedures, in addition to possible improvements in the efficiency of the estimators.

11.4.2 Area surveys of crops

The results for any stratum follow from those in section 5.5.2. If in the hth stratum, n_h fields are selected with ppa

$$\pi_{hi} = a_{hi}\Big/\sum_{i=1}^{N_h} a_{hi} = a_{hi}/A_h$$

where a_{hi} is the area of the ith field ($i = 1, 2, \ldots, N_h$ for the universe, and $i = 1, 2, \ldots, n_h$ for the sample) and $\sum_{i=1}^{N_h} a_{hi} = A_h$, the total area of all the N_h fields in the stratum, and if for the ith sample field y_{hi} denotes the area under a particular crop, then an unbiased estimator of the average proportion under the crop in the stratum is, from equation (5.13),

$$\bar{p}_h = \sum_{i=1}^{n_h} p_{hi}/n_h \tag{11.20}$$

i.e. the simple (unweighted) average of the sample proportions $p_{hi} = y_{hi}/a_{hi}$ in the sample fields, with an unbiased variance estimator (from equation (5.14)) of

$$s^2_{\bar{p}_h} = \mathrm{SS}p_{hi}/n_h(n_h - 1) \tag{11.21}$$

An unbiased estimator of the total area under the crop in the hth stratum is, from (5.11),

$$y^*_{ho} = A_h \bar{p}_h \tag{11.22}$$

with an unbiased variance estimator (from equation (5.12))

$$s^2_{y^*_{ho}} = A_h^2 s^2_{\bar{p}_h} \tag{11.23}$$

For all the strata combined, the unbiased estimator of the total area under the crop Y is, from (11.6.d),

$$y = \sum^L y^*_{ho} = \sum^L A_h \bar{p}_h \tag{11.24}$$

with an unbiased variance estimator (from (11.15))

$$s_y^2 = \sum^L s^2_{y^*_{ho}} = \sum^L A_h^2 s^2_{\bar{p}_h} \tag{11.25}$$

An estimator of the average proportion under the crop in the universe is

$$\bar{p} = y/A = \sum^L A_h \bar{p}_h / A \tag{11.26}$$

where $A = \overset{L}{\Sigma} A_h$ is the total area of all the strata, and an unbiased variance estimator of $\bar{p}$ is

$$s^2_{\bar{p}} = s_y^2/A^2 \tag{11.27}$$

Note: If the crop is such that it either occupies the whole of a field or no part of it, or if the fields are small enough for this assumption to hold, i.e. the proportion of the total area under the crop p_{hi} is either 1 (whole) or 0 (none), then the estimator of the variance of the total area under the crop in equation (11.25) reduces to

$$s_y^2 = \overset{L}{\Sigma} A_h^2 p_h(1 - p_h)/(n_h - 1) \tag{11.28}$$

(see note to section 5.5.2).

11.4.3 Yield surveys of crops

Similar considerations apply for estimating the average yield of a crop per unit of area. The results for any stratum follow from those in section 5.5.3. If $r_{hi} = x_{hi}/y_{hi}$ is the yield per unit area in the ith sample field (obtained on harvesting the crop and measuring the yield x_{hi}) in the hth stratum, then an unbiased estimator of the average yield per unit area is

$$\bar{r}_h = \sum_{i=1}^{n_h} r_{hi}/n_h \tag{11.29}$$

with an unbiased variance estimator

$$s^2_{\bar{r}_h} = \mathrm{SS}r_{hi}/n_h(n_h - 1) \tag{11.30}$$

An unbiased estimator of the total yield X_h in the hth stratum is

$$x^*_{ho} = A_h \bar{r}_h \tag{11.31}$$

with an unbiased variance estimator

$$s^2_{x^*_{ho}} = A_h^2 s^2_{\bar{r}_h} \tag{11.32}$$

For all the strata combined, the unbiased estimator of the total yield is

$$x = \sum^L x^*_{ho} \tag{11.33}$$

with an unbiased variance estimator

$$s_x^2 = \sum^L s^2_{x^*_{ho}} \tag{11.34}$$

An unbiased estimator of the average yield per unit area in the universe is

$$r = x/A \tag{11.35}$$

with an unbiased variance estimator

$$s_r^2 = s_x^2/A^2 \tag{11.36}$$

A generally biased but consistent estimator of the average yield per unit of crop-area in the hth stratum is

$$r'_h = x^*_{ho}/y^*_{ho} = \bar{r}_h/\bar{p}_h \tag{11.37}$$

with a variance estimator

$$s^2_{r'_h} = (s^2_{x^*_{ho}} + r'^2_h s^2_{y^*_{ho}} - 2r'_h s_{y^*_{ho}x^*_{ho}})/y^{*2}_{ho} \tag{11.38}$$

where

$$s_{y^*_{ho}x^*_{ho}} = A_h^2 [\text{SP} p_{hi} r_{hi}/n_h(n_h - 1)] \tag{11.39}$$

For all the strata combined, a generally biased but consistent estimator of the average yield per unit of crop-area is

$$r' = x/y \tag{11.40}$$

with a variance estimator

$$s^2_{r'} = (s_x^2 + r'^2 s_y^2 - 2r' s_{yx})/y^2 \tag{11.41}$$

where s_{yx} is an unbiased estimator of the covariance of y and x, given by equation (11.18).

11.5 Ratio method of estimation: Combined and separate ratio estimators

As for unstratified pps sampling (section 5.6), so also for stratified pps sampling the ratio method of estimation may be used to improve the efficiency of estimators; and as for stratified srs (section 10.4), for the whole universe, 'combined' and 'separate' ratio estimators may be used.

If w is the ancillary variable used for the ratio estimation, then an unbiased estimator of the universe total W_h in the hth stratum from a pps (the 'size' variable being z) sample of n_h units is, from equation (11.6.b),

$$w^*_{ho} = \sum_{i=1}^{n_h} w^*_{hi}/n_h$$

$$= \sum_{i=1}^{n_h} w_{hi}/(\pi_{hi} n_h)$$

$$= Z_h \sum_{i=1}^{n_h} w_{hi}/(z_{hi} n_h) \tag{11.42}$$

where $Z_h = \sum_{i=1}^{n_h} z_{hi}$, the total of the values of the size variable in the hth stratum.

The ratio estimator of the total Y_h in the stratum is (from equation (5.27))

$$y_{hR}^* = W_h y_{ho}^* / w_{ho}^* = W_h r_{1h} \tag{11.43}$$

where

$$W_h = \sum_{i=1}^{N_h} w_{hi} \text{ and } r_{1h} = y_{ho}^* / w_{ho}^* \tag{11.44}$$

A consistent but generally biased estimator of the variance of y_{hR}^* is (from equation (5.29))

$$\begin{aligned} s_{y_{hR}^*}^2 &= W_h^2 s_{r_{1h}}^2 \\ &= s_{y_{ho}^*}^2 + r_{1h}^2 s_{w_{ho}^*}^2 - 2r_{1h} s_{y_{ho}^* w_{ho}^*} \end{aligned} \tag{11.45}$$

where

$$\begin{aligned} s_{y_{ho}^*}^2 &= \text{SS}y_{hi}^* / n_h(n_h - 1) \\ s_{w_{ho}^*}^2 &= \text{SS}w_{hi}^* / n_h(n_h - 1) \\ s_{y_{ho}^* x_{ho}^*} &= \text{SP}y_{hi}^* w_{hi}^* / n_h(n_h - 1) \end{aligned} \tag{11.46}$$

For the ratio estimator of the total of the study variable $Y = \sum^{L} Y_h$, two types of ratio estimators could be used.

1. *Separate ratio estimator.* This is obtained by summing up the stratum ratio estimators, namely

$$y_{RS} = \sum^{L} y_{hR}^* \tag{11.47}$$

the variance estimator of y_{RS} being provided by summing the stratum variance estimators, namely

$$s_{y_{RS}}^2 = \sum^{L} s_{y_{hR}^*}^2 \tag{11.48}$$

2. *Combined ratio estimator.* This is obtained by applying the ratio method of estimation at the overall level, namely,

$$y_{RC} = Wr = Wy/w \tag{11.49}$$

where y and w are the unbiased estimators of the universe totals Y and W respectively, defined by equations of the type (11.6.d), namely,

$$w = \sum^{L} w_{ho}^*$$

w_{ho}^* being given by equation (11.6.b), or more specifically by equation (11.42) and

$$r = y/w \tag{11.50}$$

The variance estimator of y_{RC} is (from equation (5.29))

$$s_{y_{RC}}^2 = W^2 s_r^2 = s_y^2 + r^2 s_w^2 - 2rs_{yw} \tag{11.51}$$

where

$$s_{yw} = \sum_{h=1}^{L} s_{y^*_{ho} w^*_{ho}}$$

$$= \sum_{h=1}^{L} \left[\sum_{i=1}^{n_h} y^*_{hi} w^*_{hi} - \left(\sum_{i=1}^{n_h} y^*_{hi} \right) \left(\sum_{i=1}^{n_h} w^*_{hi} \right) \Big/ n_h \right] \tag{11.52}$$

Note: For the ratio method of estimation to be efficient, the probability of selection should be appropriate for both y and w. See also the notes in section 10.4

• 11.6 Ratio of ratio estimators of totals of two study variables

For two or more study variables, the ratio method of estimation may be applied, using the same ancillary variable. However, in computing the ratio of the totals of two study variables, the ratio method is equivalent to using the unbiased estimators of the totals for the stratum levels, as also for all the strata combined if combined ratio estimators for totals are used.

If separate ratio estimators are used for the totals of the two study variables, namely

$$y_{RS} = \sum^{L} y^*_{hR} = \sum^{L} W_h\, y^*_{ho}/w^*_{ho} = \sum^{L} W_h r_{1h} \tag{11.47}$$

$$x_{RS} = \sum^{L} x^*_{hR} = \sum^{L} W_h\, x^*_{ho}/w^*_{ho} = \sum^{L} W_h r_{2h} \tag{11.53}$$

are respective estimators of the universe total Y and X, where x_{RS} is defined as for y_{RS} and

$$r_{2h} = x^*_{ho}/w^*_{ho} \tag{11.54}$$

An estimator of the universe ratio $R = Y/X$ is

$$r' = \frac{y_{RS}}{x_{RS}} = \frac{\sum^{L} y^*_{hR}}{\sum^{L} x^*_{hR}} = \frac{\sum^{L} W_h r_{1h}}{\sum^{L} W_h r_{2h}} \tag{11.55}$$

A variance estimator of r' is

$$s^2_{r'} = (s^2_{y_{RS}} + r'^2 s^2_{x_{RS}} - 2r' s_{y_{RS} x_{RS}})/x^2_{RS} \tag{11.56}$$

where $s^2_{y_{RS}}$ and $s^2_{x_{RS}}$ are given by estimating equations of the type (11.48), and the estimated covariance by

$$s_{y_{RS} x_{RS}} = \sum^{L} s_{y^*_{hR} x^*_{hR}} \tag{11.57}$$

where

$$s_{y^*_{hR} x^*_{hR}} = \frac{\sum^{n_h} [(y^*_{hi} - r_{1h} w^*_{hi})(x^*_{hi} - r_{2h} x^*_{hi})]}{n_h (n_h - 1)} \tag{11.58}$$ •

See notes to section 10.6.•

11.7 Gain due to stratification

As for stratified srs (section 10.6), so also for stratified pps samples it is possible to examine the gain, if any, due to stratification as compared to unstratified pps sampling.

We define

$$\pi_{ho} = \sum^{N_h} \pi_l = \sum^{N_h} z_{hi}/Z$$

as the sum of the probabilities of selection of the units of an unstratified pps sample, the summation being over the N_h units in the hth stratum.

The gain due to stratification in the variance estimator of the total of the study variable is

$$\sum^{L} (1/n\pi_{ho} - 1/n_h)\, s^2_{y^*_{ho}} + \frac{\left(\sum^{L} y^{*2}_{ho}/\pi_{ho} - y^2\right)}{n} - \frac{\sum^{L} (1/\pi_{ho} - 1) s^2_{y^*_{ho}}}{n} \tag{11.59}$$

When $\pi_{ho} = N_h/N$, the expression (11.59) takes the form

$$\frac{N \sum^{L} (n_h s^2_{y^*_{ho}}/N_h)}{n} - \sum^{L} s^2_{y^*_{ho}} + \frac{N \sum^{L} N_h (\bar{y}^*_{ho} - \bar{y})^2}{n} - \frac{\sum^{L} (N/N_h - 1)\, s^2_{y^*_{ho}}}{n} \tag{11.60}$$

Note: The conditions under which stratification would lead to a gain will be discussed in Chapter 12.●

Further reading

Murthy, sections 7.8, 7.8b, and 10.6; Raj, section 4.2; Sukhatme and Sukhatme, sections 3.16, 3.19, and 3.20; Yates, sections 3.10, 6.17, and 7.16.

Exercise

1. Table 11.9 gives the acreage of wheat (y_{hi}) in sample parishes, selected with probability proportional to size (the total acreage of crops and grass, x_{hi}) in four districts of Hertfordshire, U.K. Estimate the total acreage of wheat in these four districts taken together with its standard error, given the total sizes of the districts at the bottom of the Table (Yates, Examples 6.17 and 7.16, modified).

Table 11.9 Acreage of wheat (y_{hi}) in sample parishes selected with probability proportional to size (the total acreage of crops and grass, x_{hi}) in four districts of Hertfordshire, U.K.

District I (3 parishes)		*District II* (4 parishes)		*District III* (5 parishes)		*District IV* (2 parishes)	
y_{hi}	x_{hi}	y_{hi}	x_{hi}	y_{hi}	x_{hi}	y_{hi}	x_{hi}
766	3040	311	2370	558	2300	225	2520
701	3440	228	3330	775	4430	738	3740
503	2040	249	2290	495	2890		
		686	2930	565	2420		
				862	4160		
Total	43 591		57 263		73 946		34 437

12 Size of Sample and Allocation to Different Strata

12.1 Introduction

In this chapter we will consider first the problems of the allocation of the total sample size into different strata and of the formation of strata, and second, the determination of the total size of the sample.

12.2 Principle of stratification

Equation (10.10) shows that the sampling variance of y (unbiasedly estimating the universe total Y) in a stratified srs depends on the within-stratum variances σ_h^2; and from equation (11.14), we see that in a stratified pps sample, the sampling variance of y depends on the within-stratum variability of Y_{hi}/z_{hi}. For efficient stratification, therefore, the strata should be so formed as to be internally homogeneous with respect to Y_{hi} in srs and with respect to Y_{hi}/z_{hi} in pps sampling, so that within-stratum variability of Y_{hi} (in srs) or Y_{hi}/z_{hi} (in pps sampling) is minimized.

12.3 Allocation of sample size to different strata

12.3.1 Formulation of the problem

Given a total sample size n, its allocation to the different strata would be based on the same principle as in Chapter 7, namely, that for a specified total cost of surveying the sample, the sampling variance should be the minimum or *vice versa*.

The sampling variance of the sample estimator y of the universe total Y in srs and pps (both with replacement) takes the form

$$\sum_{h=1}^{L} N_h^2 V_h / n_h \tag{12.1}$$

where V_h is the variance per unit in the hth stratum, and n_h is the sample size in the hth stratum. Thus in stratified srs (with replacement),

$$V_h = \sigma_h^2 = \sum_{i=1}^{N_h} (Y_{hi} - \bar{Y}_h)^2 / N_h \tag{12.2}$$

(see also equation (10.8)) and in stratified srs (with replacement) for proportions

$$V_h = \sigma_h^2 = P_h(1 - P_h) \tag{12.3}$$

from equation (2.68).

In stratified pps (with replacement)

$$V_h = \sum_{i=1}^{N_h} (Y_{hi}/\pi_{hi} - Y_h)^2 \pi_{hi} / N_h^2 \tag{12.4}$$

where $\pi_{hi} = z_{hi}/Z_h$ is the probability of selection of the ith universe unit ($i = 1, 2, \ldots, N_h$) in the hth stratum, z_{hi} is the 'size' of the hith universe unit, and Z_h is the total 'size' of the hth stratum from equation (5.4).

A simple cost function in stratified sampling is

$$C = c_o + \sum_1^L n_h c_h \tag{12.5}$$

where c_o is the overhead cost, and c_h the average cost of taking a sample unit in the hth stratum, which may vary from stratum to stratum, depending on field conditions.

12.3.2 Optimum allocation

It can be shown that with the general variance function (12.1) and the cost function (12.5), the optimum value of n_h that minimizes the variance for a given total cost or *vice versa* is proportional to $N_h \sqrt{(V_h/c_h)}$, and is given by

$$\frac{n_h}{n} = \frac{N_h \sqrt{(V_h/c_h)}}{\Sigma N_h \sqrt{(V_h/c_h)}}$$

When the total cost is fixed at C',

$$n_h = (C' - c_0) \frac{N_h \sqrt{(V_h/c_h)}}{\Sigma N_h \sqrt{(V_h c_h)}} \tag{12.6}$$

For a fixed variance V', the optimum allocation is

$$n_h = \left[\sum N_h \sqrt{(V_h c_h)}\right] \left[N_h \sqrt{(V_h/c_h)}\right] / V' \tag{12.7}$$

These are the *optimum allocations* of the total sample size n into different strata and show that a large sample should be taken in a stratum which is large (N_h large), is more variable internally (V_h large), and sampling is inexpensive (c_h small).

Notes

1. Theoretical proofs are given in Appendix II, section A2.3.10.
2. The optimum n_h values would have been the same had we considered the estimation of the universe mean Y/N, rather than that of the total Y. In comparing expression (12.7) relating to the total with equivalent expressions for the mean in other textbooks, note that the variance for the total is N^2 times the variance for the mean.
3. For stratified srs without replacement, the optimum allocation is similarly obtained by taking the variance function (10.21) and the cost function (12.5). The optimum n_h is proportional to $N_h S_h/\sqrt{c_h}$ (where S_h is defined by equation (10.22)), and is given by the relation

$$\frac{n_h}{n} = \frac{N_h S_h/\sqrt{c_h}}{\Sigma N_h S_h/\sqrt{c_h}}$$

For a fixed total cost C', the optimum n_h is obtained on substituting S_h for $\sqrt{V_h}$ in equation (12.6). For a fixed variance V', the optimum n_h is obtained on substituting in the above relation the value of n given in note (4) to section 12.6.2.

Some other methods of allocation will now be considered.

12.3.3 Neyman allocation

If the cost per unit c_h is assumed to be the same ($= \bar{c}$) in all the strata, then the cost function (12.5) becomes

$$C = c_o + n\bar{c} \tag{12.8}$$

(see also equation (7.27)) which determines the total sample size, given the total cost C', thus

$$n = (C' - c_o)/\bar{c} \tag{12.9}$$

(see also equation (7.28)). The optimum allocation for a fixed total cost is then (from equation (12.6))

$$n_h = \frac{nN_h\sqrt{V_h}}{\Sigma N_h\sqrt{V_h}} \tag{12.10}$$

This is known as the *Neyman allocation.*

In both the optimum and the Neyman allocations, the values of V_h will not be known, and estimates would have to be obtained from a previous or a pilot survey relating to the study variable or from a survey relating to an ancillary variable, which has a high positive correlation with it. Lacking any information on the universe standard deviation, the range of the values of the study variable (R_h) may be substituted for $\sqrt{V_h}$ in the Neyman allocation (also see section 7.3), i.e.

$$n_h = \frac{nN_hR_h}{\Sigma N_hR_h} \tag{12.11}$$

Note: The Neyman allocation is sometimes called the optimum allocation. For theoretical proof, see Appendix II, section A2.3.10.

12.3.4 Proportional (Bowley) allocation

In the absence of any information on V_h, if it can be assumed to be the same in all the strata, then the Neyman allocation takes the simple form

$$n_h = nN_h/N \tag{12.12}$$

i.e. n_h is proportional to N_h, or that the sample is allocated to the different strata in proportion to the number of universe units N_h.

12.3.5 Allocation proportional to the stratum total

When the stratum coefficients of variation are assumed not to vary considerably, the Neyman allocation (12.10) becomes

$$n_h = nY_h/Y \tag{12.13}$$

i.e. n_h is proportional to the stratum total Y. The condition is likely to be met when stratification is adopted for administrative or operational conveniences. However, as the stratum totals Y_h will not be known, in practice the allocation will have to be made in proportion to the total of an ancillary variable.

For skew universes, such as industrial establishments, income of individuals, and agricultural farms, allocation proportional to the stratum total is more efficient than proportional allocation.

12.3.6 Allocation in srs

A stratified srs with the Neyman or proportional allocation is more efficient than an unstratified srs, but the efficiency decreases with departure from these allocations. While moderate departures do not affect the efficiency seriously, an extremely wide departure may lead to a loss due to stratification.

12.3.7 Allocation in pps sampling

In stratified pps sampling, the sample units are selected within each stratum with probability proportional to the size of an ancillary variable ($\pi_{hi} = z_{hi}/Z_h$). The formulae for the different types of allocation can be derived from equations (12.4)–(12.13); the difficulty in applying the optimum and the Neyman allocations lies, as in stratified srs, in the non-availability of advanced estimates of the stratum variances. A simple and appropriate allocation is, however, that proportional to the stratum total sizes Z_h (see section 12.3.5 above), i.e.

$$n_h = nZ_h/Z \tag{12.14}$$

In such a case, stratified pps sampling is always more efficient than unstratified pps sampling. The wider the variation of the stratum proportions Y_h/Z_h, the greater will be the efficiency.

12.3.8 Acreage surveys of crops

In section 11.4.2 was considered stratified sampling with probability proportional to total area in the different strata for crop area surveys: the unbiased estimator of the total area under a crop in such a plan is

$$y = \sum_{h=1}^{L} A_h \sum_{i=1}^{n_h} p_{hi}/n_h \tag{12.15}$$

where A_h is the total geographical area of the hth stratum, n_h the number of sample units (fields, farms, or plots) in the hth stratum, and p_{hi} is the proportion of the total area under the crop in the hith sample unit.

If the allocation of the sample units is made proportional to the total areas of the strata, i.e. if n_h is proportional to A_h or

$$n_h = nA_h/A \tag{12.16}$$

where $A = \overset{L}{\Sigma} A_h$ is the total area of the universe, then the design becomes self-weighting (section 13.3).

Notes

1. Similar considerations apply to yield surveys of crops if the sample units are selected with probability proportional to area (section 11.4.3).
2. A special case of the above is when the crop is such that it either occupies the whole of a plot ($p_{hi} = 1$) or not at all ($p_{hi} = 0$), when the sampling variance of the estimated total area under the crop is

$$\Sigma A_h^2 P_h(1 - P_h)/n_h \tag{12.17}$$

where P_h is the universe proportion of the area under the crop in the hth stratum.

With n_h proportional to A_h, the variance formula (12.17) becomes

$$A \sum A_h P_h (1 - P_h)/n \qquad (12.18)$$

The optimum allocation is

$$n_h = \frac{n A_h \sqrt{[P_h(1 - P_h)]}}{\Sigma A_h \sqrt{[P_h(1 - P_h)]}} \qquad (12.19)$$

If the universe proportion P_h could be estimated from the data of an earlier survey, an approximation to the optimum allocation could be obtained.

3. *Fractional values of* n_h. The allocations (n_h s) will mostly turn out to be fractional. They may be rounded off to whole numbers in the following three ways:
   (a) The usual rule of rounding off of the decimal system may be followed.
   (b) The smaller n_h s may be rounded off to higher integers at the cost of larger n_h s.
   (c) n_h may be rounded off to the neighbouring integers in a randomized manner such that its expected value remains the same. Thus let $n_h = k_h + d_h$, where k_h is an integer and d_h is a fraction; then n_h should be rounded off to k_h with probability $(1 - d_h)$ and to $(k_h + 1)$ with probability d_h. For example, suppose in one stratum $n_h = 5.3$ up to one decimal figure; then a one digit random number is chosen; if it falls between 1 and $(10 - 3 =)$ 7, then n_h is taken as 5, and if the random number is 8, 9, or 0, then n_h is taken as 6. This procedure may be necessary in self-weighting designs (Chapter 13), but otherwise procedure (b) may be more reasonable.

12.4 Formation of strata

12.4.1 Formulation of the problem

In large-scale surveys, the primary strata are usually compact administrative areas, within which ultimate strata are constituted. The problem is to determine the number of strata and to demarcate them. In stratified srs the strata should be so formed as to maximize principally the variation between stratum means and, to a lesser extent, the stratum standard deviations; in stratified pps sampling, on the other hand, the units within each stratum should be homogeneous with respect to $Y_{hi}/3_{hi}$; in practice, the values of the study variable relating to an earlier survey or those of a related ancillary variable would have to be used. The variable on basis of which stratification is made is called the *stratification variable.*

12.4.2 Demarcation of strata

There are several practical rules for the demarcation of the different strata.

(i) *Equalization of* $N_h \sqrt{V_h}$ *(Dalenius-Gurney rule).* The strata may be so formed as to equalize the values of $N_h \sqrt{V_h}$ ($= N_h \sigma_h$ in srswr) in the different strata, and then an equal allocation of the total sample to the different strata may be made. This requires, of course, a prior knowledge of $\sqrt{V_h}$ for different sets of stratification points.

(ii) *Equalization of* $N_h R_h$ *(Ekman rule).* In srs, lacking information on σ_h, the values of $N_h R_h$ may be equalized, where R_h is the range of the study variable in the hth stratum.

(iii) *Equalization of stratum totals* Y_h *(Mahalanobis-Hansen-Hurwitz-Madow rule).* The strata may be formed by equalizing the stratum totals Y_h (or the stratum totals of the stratification variable, in practice) with an equal allocation. This is a good rule when the stratum coefficients of variation are approximately the same (see also section 12.3.5) and would not change appreciably with changes in stratum boundaries; this situation is likely to be met when geographical stratification is made on administrative considerations. The rule is simple and also helps in achieving a self-weighting design.

For example, in the Indian National Sample Survey (1958–9), with the main emphasis on a demographic inquiry, the primary strata in the rural areas consisted of geographically compact areas formed by equalizing the 1951 Census population.

Note: For estimates of ratios (such as birth rate, death rate etc., in a vital rate survey, or average income per person in a household budget inquiry), the stratum sizes may be made equal with respect to a measure of the size that is highly correlated with the denominator of the ratios (total population in our examples).

(iv) *Equalization of Cumulatives of* $\sqrt{[f(y)]}$ *(Dalenius-Hodges rule).* If the number of strata is large and stratification effective, the distribution of the study variable can be assumed to be rectangular within the different strata. A very good rule in this case is to form the strata by equalizing the cumulatives of $\sqrt{[f(y)]}$ where $f(y)$ is the frequency distribution of the study variable. This makes $N_h \sigma_h$ approximately constant in srs, so that the Neyman allocation gives a constant sample size $n_0 = n/L$ in all strata. As the optimum is generally flat with respect to variations in n_h, the use of the Dalenius-Hodges rule is highly efficient.

12.4.3 Number of strata

Stratification with proportional allocation is always more efficient than unstratified sampling. But stratification may be carried up to a point where only one unit is selected from each stratum: the estimates of sampling variances cannot be computed then without resort to approximate methods (see section 24.8). One consideration is therefore that the number of strata should at most be $\frac{1}{2}n$ if n is even and $\frac{1}{2}(n - 1)$ if n is odd, so that at least two units are selected from each stratum. The other consideration is that although in general the efficiency increases with the number of strata, beyond a certain point the gain in efficiency will not be worth the efforts required to increase the number of strata.

12.5 Multiple stratification

If a universe is stratified according to two or more factors, it is said to be multiple stratified. In practice, this may be difficult to apply even when the required data are available, due to possible conflicts in the different stratification criteria. For example, for surveys on areas under different crops, the total geographical area may be a suitable stratification variable, but population may be more suitable for purposes of stratification in demographic or labour force surveys.

A compromise in such a case would be to stratify the universe first according to the most important stratification variable, and then within each stratum so formed to sub-stratify according to the next most important variable, and so on. This is known as *multiple stratification* or *deep stratification.*

12.6 Determination of the total sample size

12.6.1 Specified cost

For the optimum allocation, with the given total cost C', the total sample size is

$$n = \frac{(C' - c_o)\, \Sigma N_h \sqrt{(V_h/c_h)}}{\Sigma N_h \sqrt{(V_h c_h)}} \qquad (12.20)$$

For proof, see Appendix II, section A2.3.10.

For the Neyman allocation, from equation (12.8) the total sample size is determined as

$$n = (C' - c_o)/\bar{c} \qquad (12.9)$$

For the proportional allocation (equation (12.12)), the cost function (12.5) gives the total sample size as

$$n = (C' - c_o)\Big/\sum w_h c_h \qquad (12.21)$$

where for the proportional allocation $w_h = n_h/n = N_h/N$. If cost function (12.8) is taken, then the total sample size is given by equation (12.9) with proportional allocation also.

Note: For stratified srs without replacement, the value of n is obtained on replacing $\sqrt{V_h}$ by S_h in equation (12.20) for optimum allocation).

12.6.2 Specified variance: continuous data

For continuous data, if V is the desired variance of a sample estimator y of the universe total Y in a stratified sample, then the sampling variance of the estimator y (equation (10.15) could be written as

$$V = \sigma_y^2 = \sum (N_h^2 \sigma_h^2 / w_h)/n \qquad (12.22)$$

where $n_h = nw_h$, where the w_h values have been chosen already.

From equation (12.22) the required sample size n for a specified V is

$$n = \sum (N_h^2 \sigma_h^2 / w_h)/V \qquad (12.23)$$

For the Neyman allocation, i.e. when $w_h \propto N_h \sigma_h$, equation (12.23) becomes

$$n = \left(\sum N_h \sigma_h\right)^2 \Big/ V \qquad (12.24)$$

and for proportional allocation, i.e. when $w_h = N_h/N$, the required total sample size is

$$n = N \sum N_h^2 \sigma_h^2 / V \qquad (12.25)$$

Notes

1. As σ_h will not in general be known, it has to be estimated from a pilot inquiry or from other available information, such as the range.
2. Estimation of the universe mean Y/N requires the same sample size as that for the total Y.

3. If the margin of error d is specified (see section 7.2.2), then

$$V = (d/t)^2 \tag{12.26}$$

where t is the normal deviate corresponding to the permissible probability that the error will exceed the specified margin.

4. For stratified srs without replacement, the value of n in Neyman allocation is

$$n = \frac{(\Sigma N_h S_h / c_h)\ \Sigma N_h S_h / \sqrt{c_h}}{V + \Sigma N_h S_h^2}$$

12.6.3 Specified variance: estimation of proportions

If V is the desired variance of the proportion P for the whole universe, then the required sample size n for the Neyman allocation is

$$n = \frac{\Sigma N_h \sqrt{(P_h Q_h)}}{N^2 V} \tag{12.27}$$

and for proportional allocation

$$n = \sum (N_h P_h Q_h)/NV \tag{12.28}$$

where P_h is the proportion in the hth stratum, and $Q_h = 1 - P_h$.

Notes

1. In practice, an estimate p_h of P_h will have to be used in formulae (12.27) and (12.28).
2. Estimation of the total number of universe units in a certain category, namely, NP, requires the same sample size as for that for P.

12.7 Examples

Example 12.1

The data on the number of cattle obtained from a recent census are given in Table 12.1, in the 5 strata according to the total acreage of the farms, along with the present total number of farms in these strata. The problem is to estimate the present total number of cattle in the universe and its variance, by taking a sample of 500 farms.

(a) Determine the allocations of the sample in the different strata according to the following principles: (i) Neyman allocation, (ii) proportional allocation, and (iii) allocation proportional to the total number of cattle in the different strata;

(b) Also compute the expected variance for each of these (United Nations *Manual*, Process 4, Example 4, adapted).

Table 12.1 Number of cattle, obtained from a previous census, and the present total number of farms in each stratum

Stratum (acres)	*Previous census*			*Present total number of farms*
h	*Total number of farms* N'_h	*Average number of cattle* $\bar{y}'_h$	*Estimated s.d. per unit* s'_h	N_h
I: 0–15	625	3.91	4.5	635
II: 16–30	564	10.38	7.3	570
III: 31–50	476	14.72	9.6	475
IV: 51–75	304	21.99	12.2	303
V: 76–100	86	27.38	15.8	89
All strata combined	2055 (N')			2072 (N)

Table 12.2 Computation of the allocations to different strata of the sample 500 farms: data of Table 12.1

Stratum	$N'_h s'_h$	$Y'_h = N'_h \bar{y}_h$	*Allocations* (n_h)		
			*Neyman**	*Proportional*†	*Proportional to* Y'_h‡
I	2812.5	2444	84	153	50
II	4117.2	5854	125	138	120
III	4569.6	7007	138	115	144
IV	3708.8	6685	112	73	137
V	1358.8	2355	41	21	49
All strata combined	16 566.9 ($\Sigma N'_h s'_h$)	24 345 ($\Sigma Y'_h$)	500 (n)	500 (n)	500 (n)

* $n_h = \dfrac{500\, N'_h s'_h}{\Sigma N'_h s'_h}$ (12.10)

† $n_h = 500\, N_h/N$ (12.12)

‡ $n_h = 500\, Y'_h/\Sigma Y'_h$ (12.13)

Table 12.3 Expected variances of total number of cattle according to different allocations: data of Tables 12.1 and 12.2

Stratum *h*	*Estimated variance* in*		
	Neyman allocation	*Proportional allocation*	*Allocation proportional to* Y'_h
I	97 206	53 368	163 306
II	138 511	125 463	144 283
III	150 678	180 814	144 400
IV	122 008	187 190	99 743
V	48 229	94 162	40 355
All strata combined	566 632	640 997	592 087

* $N_h^2 s_h'^2/n_h$

(a) The required computations are shown in Table 12.2. Formula (12.10) is used for the Neyman allocation, and formula (12.12) for the proportional allocation. For allocation proportional to the stratum totals, the stratum total $Y'_h = N'_h \bar{y}_h$ for the previous census is used, and the formula (12.13), modified as $n_h = nY'_h/Y'$, is applied ($Y' = \Sigma Y'_h$).

• (b) The estimated variances are computed in Table 12.3. The relevant formula for any stratum is

$$s^2_{y^*_{ho}} = N_h^2 s_h'^2/n_h$$

where $s_h'^2$ are the variances per unit obtained from the previous census, and N_h and n_h respectively the total and sample number in the present inquiry.

Note that in this case, allocation proportional to the stratum totals will be almost as efficient as the Neyman allocation. The results of an actual sample, using proportional allocation, is given in exercise 1, Chapter 10: the actual computed variance of the estimated total number of cattle was 574 597, as compared with the expected value of 640 997.•

Example 12.2

An inquiry is to be designed to estimate the proportion of families with transistor radios in two cities in Ethiopia. Rough estimates of the total number of families, the proportion with radios, and the cost of surveying one family are given in Table 12.4. Treating the cities as strata, and assuming srs (with replacement) in each stratum, obtain the total optimum sample size and its allocation to the two cities, if the total cost (excluding overheads) is fixed at Eth. $20 000.

The required computations are shown in Table 12.5. With V_h given by equation (12.3), the

Table 12.4 Rough estimates of the total number of families, proportion with transistor radios, and the cost of enumerating one family in two cities in Ethiopia

City h	*Number of families* N_h	*Proportion with transistor radios* P_h	*Cost of surveying one family* c_h
I	140 000	0.10	Eth. \$2.25
II	30 000	0.25	1.00

Table 12.5 Computation of the optimum allocations from the data of Table 12.4: Stratified srs of families

City	V_h $P_h(1-P_h)$	$\sqrt{(V_h c_h)}$	$N_h\sqrt{(V_h c_h)}$	$\sqrt{(V_h/c_h)}$	$N_h\sqrt{(V_h/c_h)}$	$n_h = 20\,000 \times N_h\sqrt{(V_h/c_h)}/24\,525$
I	0.0900	0.1350	18 900	0.0600	8400	6850
II	0.1875	0.1875	5625	0.1875	5625	4587
Total			24 525		14 025	11 437 (n)

optimum allocations are given by equation (12.6). These are 6850 families in the first city and 4587 in the second, with a total of 11 437 sample families. (As a check on the computations, it can be verified that the total cost = 6850 x Eth. \$2.25 + 4587 x Eth. \$1.00 = \$19 999 or Eth. \$20 000).

Further reading

Cochran, Chapters 5 and 5A; Deming (1960), Chapter 20, and (1966), Chapter 6; Hansen *et al.*, Vols. I and II, Chapter 5; Kish, Chapter 3; Murthy, Chapter 7; Raj, Chapter 4; Sukhatme and Sukhatme, Chapter III.

Exercises

1. In a survey, using a stratified srs with five strata, rough estimates of the universe units (N_h), the standard deviations (σ_h) and the cost of surveying one unit (c_h in a certain unit) in the different strata are given in Table 12.6. The total cost is fixed at 10 000 and the overhead cost at 500. Determine the optimum total sample size and its allocations to different strata (Chakravarti *et al.*, Illustrative Example 4.1).

2. A survey is designed to estimate the proportion of illiterate persons in three communities. Rough estimates of the total number of persons and the proportion illiterate are given in Table 12.7. Assuming a stratified srs, with the communities as the strata, how would you allocate a total sample of 2000 persons in the strata so as to estimate the overall proportion of illiterates?

Table 12.6 Rough estimates of the total number of units, standard deviation, and cost of surveying one unit in the different strata

Stratum h	N_h	σ_h	c_h	*Optimum allocation*
I	37 800	28.5	3.50	587
II	52 600	18.6	2.75	601
III	82 000	27.6	2.25	1537
IV	41 600	21.2	3.00	519
V	28 800	16.8	2.50	312

Table 12.7 Rough estimates of the total number of persons, and proportion illiterate in three communities

Community h	*Total number of persons* N_h	*Proportion illiterate* P_h
I	60 000	0.4
II	10 000	0.2
III	30 000	0.6

13 Self-weighting Designs in Stratified Single-stage Sampling

13.1 Introduction

This chapter deals with the problems of making a stratified design self-weighting. We first consider stratified srs and then stratified varying probability sampling: the method of making any stratified design self-weighting at the tabulation stage will also be outlined.

13.2 Stratified simple random sampling

In stratified srs (as also in stratified circular systematic sampling with one sample), the unbiased estimator y of the universe total Y is

$$y = \sum_{h=1}^{L} N_h \sum_{i=1}^{n_h} y_{hi}/n_h \tag{13.1}$$

(see also (10.5.d)) where N_h and n_h are the total number of units and the number of sample units respectively in the hth stratum, and y_{hi} the value of the study variable for the ith selected sample unit in the hth stratum.

The weighting factor (or multiplier) for the hith sample unit is

$$w_{hi} = N_h/n_h \tag{13.2}$$

and for the design to be self-weighting with respect to y, this should be a constant, w_0. Then

$$w_0 = N_h/n_h = N/n \tag{13.3}$$

where $N(= \sum^{L} N_h)$ is the total number of units in the universe and $n(= \sum^{L} n_h)$ is the total number of units in the sample, i.e.

$$n_h = \frac{n}{N} N_h = \frac{N_h}{w_0} \tag{13.4}$$

This is the *proportional allocation* case (see section 10.2, note 2, and section 12.3.4).

Thus, a stratified srs will be self-weighting only with proportional allocation. Although the optimum and the Neyman allocations result in more efficient estimators, these require a prior knowledge of the variability in the different strata, which may not often be available. In such situations, there may be some advantage in the proportional allocation.

With a self-weighting stratified srs (i.e. a stratified srs with proportional allocations), some unbiased estimators defined in section 10.2 take the following form:

$$y_{hi}^{*} = N_h y_{hi} = w_0 n_h y_{hi};$$
$$y_{h0}^{*} = N_h \bar{y}_h = w_0 \sum^{n_h} y_{hi} \tag{13.5}$$

$$y = \sum^{L} y_{h0}^{*} = w_0 \sum^{L} \sum^{n_h} y_{hi} = w_0 \sum_{\text{sample}} y_{hi} \tag{13.6}$$

$$s^2_{y^*_{ho}} = N_h^2 \sum^{n_h} (y_{hi} - \bar{y}_h)^2/n_h(n_h - 1)$$
$$= w_0^2 n_h^2 \sum^{n_h} (y_{hi} - \bar{y}_h)^2/n_h(n_h - 1) \quad (13.7)$$

and
$$s_y^2 = \sum^{L} s^2_{y^*_{ho}} \quad (13.8)$$

The estimators of the ratio of the two universe totals (defined in estimating equations (10.7.b) and (10.7.d)) are

$$r_h = y^*_{h0}/x^*_{h0} = \sum^{n_h} y_{hi} \Big/ \sum^{n_h} x_{hi} \quad (13.9)$$

and

$$r = y/x = \sum^{L} \sum^{n_h} y_{hi} \Big/ \sum^{L} \sum^{n_h} x_{hi} \quad (13.10)$$

13.3 Stratified varying probability sampling

In stratified varying probability sampling, the unbiased estimator y of the universe total Y of a study variable is

$$y = \sum^{L} \frac{1}{n_h} \sum^{n_h} \frac{y_{hi}}{\pi_{hi}}$$
$$= \sum^{L} \frac{Z_h}{n_h} \sum^{n_h} \frac{y_{hi}}{z_{hi}} \quad \text{in pps sampling} \quad (13.11)$$

(see also (11.6.d)) where π_{hi} is the (initial) probability of selection of the ith universe unit ($i = 1, 2, \ldots, N_h$) in the hth stratum.

In pps sampling $\pi_{hi} = z_{hi}/Z_h$, where z_{hi} is the value of the 'size' variable of the hith universe unit, and $Z_h = \sum^{N_h} Z_{hi}$ is the total 'size' of the hth stratum.

The multiplier for the hith sample unit is

$$w_{hi} = 1/(n_h \pi_{hi}) = Z_h/(n_h z_{hi}) \quad (13.12)$$

If the ratio y_{hi}/z_{hi} can be observed and recorded easily in the field, then the design will be self-weighting when the ratios Z_h/n_h are the same in all the strata and equal to Z/n. This will be so if n_h is proportional to Z_h, i.e.

$$n_h = \frac{n}{Z} Z_h \quad (13.13)$$

A special case of this is when the strata are made equal with respect to the total 'sizes' and an equal number of sample units is selected with pps from each stratum (sections 12.3.7 and 12.4).

The above rule is especially helpful in acreage and yield surveys of crops, when a sample of fields (or farms or plots) is selected with probability proportional to total (geographical) area in each stratum. If the number of sample units allocated to a stratum is made proportional to the total geographical area of the stratum, the design becomes self-weighting as y_{hi}/z_{hi} (the proportion of the area under the crop or the crop-yield per unit area) can be observed easily (sections 11.4 and 12.3.8).

13.4 Fractional values of n_h

The values of n_h required to make a stratified srs or pps sample self-weighting will not in general be integers. In such cases, one of the following procedures may be adopted.

1. n_h may be rounded off to the neighbouring integers in a randomized manner such that its expected value remains the same (note 3 to section 12.3).
2. In stratified srs, the samples may be selected systematically with the same fractional interval N/n (see section 4.2), or the interval may be rounded off to N/n or $N/n + 1$ in a randomized manner so that the expected value of n_h remains the same; similarly, in stratified pps sampling the samples may be selected with the same fractional interval Z/n, or the intervals may be rounded off in the randomized manner.
3. Samples may be selected linear systematically with the same interval I (integer nearest to N/n) in each stratum, when the estimator y will become $$I \sum^{L} \sum^{n_h} y_{hi}.$$

• 13.5 Self-weighting designs at the tabulation stage

A design that is not originally self-weighting can be made so at the tabulating stage by selecting a sub-sample with probability proportional to the multipliers: the pps sampling can be with replacement or systematic. The procedure, will, however, lead to an increase in variance.

When a sub-sample of size $n'(< n)$ is taken out of the original $n(= \sum^{L} n_h)$ sample units in a stratified srs or pps sample with probability proportional to the multipliers (w_{hi}) and with replacement, then

$$y_0'^* = \frac{1}{n'} \sum_{j=1}^{n'} y_j'^* \tag{13.14}$$

gives an unbiased estimator of Y, where

$$y_j'^* = \left(\sum^{L} \sum^{n_h} w_{hi} \right) y_j' \tag{13.15}$$

y_j' being the value of the study variable of the jth unit in the sub-sample ($j = 1, 2, \ldots, n'$).

An unbiased estimator of the variance of $y_0'^*$ is

$$\sum^{n'} (y_j'^* - y_0'^*)^2 / n'(n' - 1) \tag{13.16}$$

Estimators of ratios and their variances estimators follow from the fundamental theorems of section 2.7.•

Further reading

Murthy, sections 12.2 and 12.5.

Part III :

Multi-stage Sampling

14 Multi-stage Sampling: Introduction

14.1 Introduction

In multi-stage sampling, the sample is selected in stages, the sampling units in each stage being sub-sampled from the (larger) units chosen at the previous stage, with appropriate methods of selection of the units – simple random sample (with or without replacement), systematic, probability proportional to size etc. – being adopted at each stage. In other words, the universe is divided into a number of first-stage (or primary sampling) units, which are sampled; then the selected first-stage units are sub-divided into a number of smaller second-stage (or secondary sampling) units, which are again sampled: the process is continued until the ultimate sampling units are reached.

This part of the book will deal with (unstratified) multi-stage sample designs. In this chapter, we shall consider the general theorems relating to multi-stage sampling; subsequent chapters will deal with two- and three-stage sampling with srs and pps sampling at different stages; the size and allocation of the sample; the estimation of the sampling variances at different stages; and self-weighting designs.

14.2 Reasons for multi-stage sampling

Multi-stage sampling is adopted in a number of situations:

1. Sampling frames may not be available for all the ultimate observational units in the universe, and it is extremely laborious and expensive to prepare such a complete frame. Here, multi-stage sampling is the only practical method. For example, in a rural household sample survey, conducted at intercensal periods, the households in rural areas could be reached after selecting a sample of villages (first-stage units), after which a list of households within the selected villages only is prepared, and then selecting a sample of households (second-stage units) in the selected villages. In a crop survey, villages may be first selected (first-stage units); and next a list of fields prepared within the selected villages and a sample of these fields taken (second-stage units); and finally, a list of plots prepared within the selected fields and a sample of these plots taken (third-stage units). In this way, great savings are achieved as sampling frames need be constructed only for the selected sampling units and not for all the sampling units in the universe. Moreover in multi-stage sampling, ancillary information collected on the sampling units while listing the units at each stage would help in improving the efficiency of sample designs, either by stratification of the units, or by selecting the sample with probability proportional to size when the ancillary information is available before sample selection at that stage, or by using the ratio or regression method of estimation.

2. Even when suitable sampling frames for the ultimate units are available for the universe, a multi-stage sampling plan may be more convenient than a single-stage sample of the ultimate units, as the cost of surveying and supervising such a sample in large-scale surveys can be very high due to travel, identification, contact, etc. This point is closely related to the consideration of

cluster sampling (Chapter 6). For instance, in a large-scale agricultural survey conducted in a developed country, although an up-to-date list of farms may be readily available from which a simple random sample of farms can be drawn, the cost of travelling and supervision of work on the widely scattered farms may be extremely high. Therefore, the procedure to be adopted would be to try to confine the sample of farms to certain area segments.

3. Multi-stage sampling can be a convenient means of reducing response errors and improving sampling efficiency by reducing the intra-class correlation coefficient observed in natural sampling units, such as households or villages. Thus, in opinion and marketing research, it becomes necessary to select only one individual in a sample household in order to avoid conditioned response, and also to spread the sample over a greater number of sample households because of the general homogeneity of responses of individuals in a household, even if the 'true' responses of all the members of the household could be obtained.

Note: Multi-stage sampling is cheaper and operationally easier than srs but not more than single-stage cluster sampling; considering sampling variability, however, multi-stage sampling is generally less efficient than srs but more efficient than cluster sampling: this is, of course, based on the assumption that the total sample size is fixed. Some of the lost efficiency may, however, be regained by using ancillary information.

14.3 Structure of a multi-stage design

Figure 14.1 illustrates the structure of a two-stage design for a household survey in rural areas for which there is no current list of households. Of the total N villages, a sample of n villages is selected in an appropriate way; all the extant

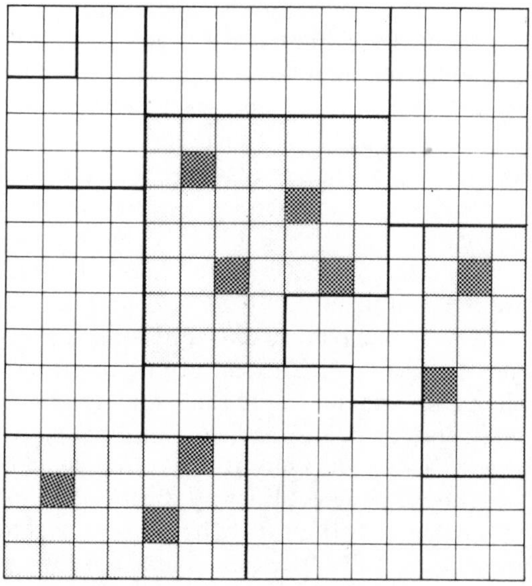

Figure 14.1 Schematic representation of a two-stage sample design.
The villages (first-stage units) are indicated by thick lines and the selected villages by hatchings; the households (second-stage units) are indicated by thin lines and the sampled households in the selected villages by cross-hatching.

households (M_i in number; $i = 1, 2, \ldots, n$) in these n sample villages are listed by the enumerators by actual field visit; and in the ith sample village, a sample of households (m_i in number) is finally selected for the survey and the required household characteristics recorded. This is a two-stage design, with villages as the first-stage and households as the second-stage units. The total number of households in all the villages is $\sum^{N} M_i$, which is generally not known; the sample number is $\sum^{n} m_i$.

Note: In such a plan, if all the N villages are included in the survey and a sample of households m_i ($i = 1, 2, \ldots, N$) selected in each village, the design becomes stratified single-stage (with total number of sample households, $\Sigma^{N} m_i$), and the methods of Part II of the book will apply. If in the sample of n villages, all the M_i households are included in the survey, the design becomes single-stage cluster sampling (with total number of sample households, $\Sigma^{n} M_i$), and the methods of Part I of the book will apply. Of course, if all the N villages are included in the survey and all the M_i households in all these villages surveyed, the survey is one of complete enumeration without any sampling (with total number of households, $\Sigma^{N} M_i$).

14.4 Fundamental theorems in multi-stage sampling

The fundamental theorems for the estimation of universe parameters in multi-stage sampling are analogous to those in section 2.7 (see note 1 to the section).

1. If n ($\geqslant 2$) units are selected out of the total N first-stage units with replacement (and with equal or varying probabilities), and t_i ($i = 1, 2, \ldots, n$) is an unbiased estimator of the universe parameter T, obtained from the ith selected first-stage unit, then a combined unbiased estimator of T is the arithmetic mean

$$\bar{t} = \sum^{n} t_i/n \tag{14.1}$$

and an unbiased estimator of the variance of $\bar{t}$ is

$$s_{\bar{t}}^2 = \sum^{n} (t_i - \bar{t})^2/n(n-1)$$
$$= \mathrm{SS}t_i/n(n-1) \tag{14.2}$$

We shall show later how to compute the t_i values for a multi-stage design.

If u_i ($i = 1, 2, \ldots, n$) are the n independent unbiased estimators, obtained also from the n first-stage sample units, of another universe parameter U, then a combined unbiased estimator of U is similarly

$$\bar{u} = \sum^{n} u_i/n \tag{14.3}$$

and an unbiased estimator of the variance of $\bar{u}$ is

$$s_{\bar{u}}^2 = \sum^{n} (u_i - \bar{u})^2/n(n-1)$$
$$= \mathrm{SS}u_i/n(n-1) \tag{14.4}$$

2. An unbiased estimator of the covariance of $\bar{t}$ and $\bar{u}$ is the estimator

$$s_{\bar{t}\bar{u}} = \sum^{n} (t_i - \bar{t})(u_i - \bar{u})/n(n-1)$$
$$= \mathrm{SP}t_i u_i/n(n-1) \qquad (14.5)$$

3. A consistent, but generally biased, estimator of the ratio of the universe parameters $R = T/U$ is the ratio of the sample estimators $\bar{t}$ and $\bar{u}$

$$r = \bar{t}/\bar{u} \qquad (14.6)$$

and an estimator of the variance of r is

$$s_r^2 = (s_{\bar{t}}^2 + r^2 s_{\bar{u}}^2 - 2rs_{\bar{t}\bar{u}}/\bar{u}^2 \qquad (14.7)$$

4. A consistent, but generally biased, estimator of the universe correlation coefficient ρ between the two study variables is the sample correlation coefficient

$$\hat{\rho}_{\bar{t}\bar{u}} = s_{\bar{t}\bar{u}}/s_{\bar{t}} s_{\bar{u}} \qquad (14.8)$$

• 5. For unbiased estimators of universe totals, obtained from the first-stage units, we change the notation and also deal with the totals of a study variable.
In a multi-stage sample design with u stages ($u > 1$), at the tth stage ($t = 1, 2, \ldots, u$)) let $n_{12\ldots(t-1)}$ sample units be selected with replacement out of the total $N_{12\ldots(t-1)}$ units and let the (initial) probability of selection of the $(12\ldots t)$th unit be $\pi_{12\ldots t}$. We define

$$f_t \equiv n_{12\ldots(t-1)} \pi_{12\ldots t} \qquad (14.9)$$

An unbiased estimator of the universe total Y of the study variable is

$$y = \sum_{\text{sample}} \left(\prod_{1}^{u} \frac{1}{f_t} \right) y_{12\ldots u} = \sum_{\text{sample}} w_{12\ldots u} y_{12\ldots u} \qquad (14.10)$$

where $y_{12\ldots u}$ is the value of the study variable in the $(12\ldots u)$th ultimate-stage sample units, and the factors

$$\prod_{1}^{u} \frac{1}{f_t} = w_{12\ldots u} \qquad (14.11)$$

are the weighting-factors (or the multipliers) for the $(12\ldots u)$th ultimate-stage sample units; the sum of the products of these multipliers and the values of the study variable for all the ultimate-stage sample units provides the unbiased linear estimator y of the universe total Y in equation (14.10).
An unbiased estimator of the variance of the sample estimator y in the general estimating equation (14.10) is

$$s_y^2 = \sum^{n} \left[n \sum_{2,3,\ldots,u} w_{12\ldots u} y_{12\ldots u} - \sum_{1,2,\ldots,u} w_{12\ldots u} y_{12\ldots u} \right]^2 /n(n-1) \qquad (14.12)$$

where n ($\geqslant 2$) is the number of first-stage units selected with replacement, the summation outside the square brackets being over these first-stage units.

The formulae can be put in simpler forms. Using a slightly different notation, an unbiased estimator of the universe total Y, obtained from the ith first-stage unit ($i = 1, 2, \ldots, n$), is

$$y_i^* = n \sum_{2, 3, \ldots, u} w_{12 \ldots u} y_{12 \ldots u} \qquad (14.13)$$

the summation being over the second- and subsequent-stage sampling units.

The combined unbiased estimator of the universe total Y from all the n sample first-stage units is, as we have seen from equation (14.1), the arithmetic mean

$$y_0^* = \sum^{n} y_i^*/n \qquad (14.14)$$

which is the same as the estimator y in equation (14.10), the summation in equation (14.14) being over the n sample first-stage units.

An unbiased estimator of the variance of y_0^* is, from equation (14.2),

$$s^2_{y_0^*} = \sum^{n} (y_i^* - y_0^*)^2/n(n-1) \qquad (14.15)$$

which is the same as equation (14.11).•

Notes

1. Estimating equation (14.10) applies to single- as well as multi-stage designs, sample selection being done with equal probability with or without replacement, or with varying probability with replacement or systematic sampling with equal or varying probability.
2. For srs, the fs are the sampling fractions at different stages.
3. The above formulae, which will be illustrated in later chapters for two- and three-stage designs, are adequate in obtaining estimates of universe totals, means, ratios, etc. of study variables, and also estimates of their variances.
4. The variance of a universe estimator is built up of the variances at different stages of a multi-stage design. The equations for estimating variances given above do not show these components separately and these are not required for estimating universe totals, means, ratios etc., with their respective standard errors. However, the decomposition of the total variance into the component stage-variances is essential in improving the design of subsequent sample surveys (Chapter 17).
5. For the estimation of the variance of y by equation (14.12), it is not necessary that the number of sample units at stages other than the first be two or more; they must, however, be so if estimates of stage-variances are required.

14.5 Setting of probability limits

The same results as for single-stage sampling given in section 2.8 apply also for multi-stage sampling. The $(100 - \alpha)$ per cent probability limits of the universe parameter T are set with the sample estimator $\bar{t}$ and its estimated standard error $s_{\bar{t}}$, thus

$$\bar{t} \pm t'_{\alpha,\, n-1} s_{\bar{t}} \qquad (14.16)$$

where $t'_{\alpha,\, n-1}$ is the 100α percentage point of the t-distribution for $(n-1)$ degrees of freedom. For large n, the percentage points of the normal distribution could be used in place of those for the t-distribution.

Further reading

Murthy, sections 9.1 and 9.2.

15 Multi-stage Simple Random Sampling

15.1 Introduction

In this chapter will be presented the estimating methods for totals, means, and ratios of study variables and their variances in a multi-stage sample with simple random sampling at each stage: the methods for two- and three-stage designs are illustrated in some detail and the general procedure for a multi-stage design indicated. The methods of estimation of proportion of units in a category and the use of ratio estimators will also be considered.

15.2 Two-stage srs: estimation of totals, means, ratios, and their variances

15.2.1 Universe totals and means

Let the number of first-stage units (fsu's) be N, and the number of second-stage units (ssu's) in the ith fsu ($i = 1, 2, \ldots, N$) be M_i; the total number of ssu's in the universe is $\sum^N M_i$ (Table 15.1). Denoting by Y_{ij} the value of the study

Table 15.1 Sampling plan for a two-stage simple random sample with replacement

Stage (t)	*Unit*	*No. in universe*	*No. in sample*	*Selection method*	*Selection probability*	f_t
1	First-stage	N	n	srswr	Equal = $1/N$	n/N
2	Second-stage	M_i	m_i	srswr	Equal = $1/M_i$	m_i/M_i

variable of the jth ssu ($j = 1, 2, \ldots, M_i$) in the ith fsu, we define the following universe parameters:

Total of the values of the study variable in the ith fsu:

$$Y_i = \sum_{j=1}^{M_i} Y_{ij} \tag{15.1}$$

Mean for the ith fsu:

$$\overline{Y}_i = Y_i/M_i \tag{15.2}$$

Grand total:

$$Y = \sum^N \sum^{M_i} Y_{ij} = \sum^N Y_i = \sum^N M_i \overline{Y}_i \tag{15.3}$$

Mean per fsu:

$$\overline{Y} = Y/N = \sum^N M_i \overline{Y}_i / N \tag{15.4}$$

Mean per ssu:

$$\bar{\bar{Y}} = Y\Big/\left(\sum^{N} M_i\right)$$
$$\equiv \sum^{N} M_i\bar{Y}_i\Big/\left(\sum^{N} M_i\right) \tag{15.5}$$

15.2.2 Structure of a two-stage srs

Out of the N fsu's, n are selected; and out of the M_i ssu's in the ith selected fsu ($i = 1, 2, \ldots, n$), m_i are selected; sampling at both stages is srs with replacement (also see section 14.3 and Figure 14.1). The total number of ssu's in the sample, i.e. the sample size, is $\sum^{n} m_i$.

Let y_{ij} denote the value of the study variable in the jth selected ssu ($j = 1, 2, \ldots, m_i$) in the ith selected fsu ($i = 1, 2, \ldots, n$).

15.2.3 Estimation of the universe total Y and the variance of the sample estimator

The problem is first to estimate the universe total Y and the variance of the sample estimator. This will be shown in three ways:

(a) Estimating downwards from the fsu's;
(b) estimating upwards from the ssu's;
(c) estimating from the general estimating equations given in section 14.4.

(a) *Estimating downwards from the first-stage units.* If all the M_i ssu's in each of the n sample fsu's are completely surveyed and y_i is the value of the study variable (i.e. the sum of the values for the M_i ssu's) in the ith sample fsu, then the estimating methods of Chapter 2 will apply. Thus, the unbiased estimator of the universe total Y from the ith selected fsu is, from equation (2.42),

$$y_i^* = Ny_i \tag{15.6}$$

The combined unbiased estimator of Y, obtained from all the n sample fsu's, is the simple arithmetic mean (from equation (2.43))

$$y_0^* = \frac{1}{n}\sum^{n} y_i^* = \frac{N}{n}\sum^{n} y_i \tag{15.7}$$

An unbiased estimator of the variance of y_0^* is, from equation (2.44),

$$s_{y_0^*}^2 = \sum^{n} (y_i^* - y_0^*)^2/n(n-1) \tag{15.8}$$

If, however, all the M_i ssu's in the ith sample fsu are not completely surveyed, but a sample of m_i ssu's is taken, then the value of the study variable in the sample fsu Y_i (or y_i above) has to be estimated on the basis of these sample m_i ssu's. Let the value of the study variable in the jth sample ssu ($j = 1, 2, \ldots, m_i$) in the ith sample fsu be y_{ij}. Then the unbiased estimator of the total Y_i (or y_i above) from the ijth sample ssu is, from equation (2.42),

$$y_{ij}^* = M_i y_{ij} \tag{15.9}$$

and the combined unbiased estimator of Y_i (from equation (2.43)) the simple arithmetic mean

$$y^*_{i0} = \frac{1}{m_i} \sum_{j=1}^{m_i} y^*_{ij} = \frac{M_i}{m_i} \sum_{j=1}^{m_i} y_{ij} = M_i \bar{y}_i \qquad (15.10)$$

where

$$\bar{y}_i = \sum_{j=1}^{m_i} y_{ij}/m_i \qquad (15.11)$$

is the unbiased estimator of the universe mean $\bar{Y}_i$ in the ith fsu.

This value of y^*_{i0} is substituted for y_i in equation (15.6) to provide the unbiased estimator of Y, from the ith sample fsu, namely

$$y^*_i = N y^*_{i0} = N \frac{M_i}{m_i} \sum_{j=1}^{m_i} y_{ij} \qquad (15.12)$$

and in equation (15.7) to provide the combined unbiased estimator of Y from all the n sample fsu's, namely,

$$y^*_0 = \frac{1}{n} \sum^{n} y^*_i = \frac{N}{n} \sum^{n} y^*_{i0} \equiv \frac{N}{n} \sum_{i=1}^{n} \frac{M_i}{m_i} \sum_{j=1}^{m_i} y_{ij} = \frac{N}{n} \sum^{n} M_i \bar{y}_i \qquad (15.13)$$

An unbiased estimator of the variance of y^*_0 is, from equation (14.2),

$$s^2_{y^*_0} = \sum^{n} (y^*_i - y^*_0)^2/n(n-1) \qquad (15.14)$$

(*b*) *Estimating upwards from the second-stage units.* If y_{ij} is the value of the study variable in the jth sample ssu of the ith sample fsu, the average value of the study variable in an srs of ssu's is obtained on summing up these y_{ij} values for the m_i sample ssu's and dividing by m_i, namely,

$$\bar{y}_i = \frac{1}{m_i} \sum_{j=1}^{m_i} y_{ij} \qquad (15.11)$$

To obtain the unbiased estimator of the total Y_i in the ith sample fsu in an srs, we multiply this average by the actual total number of ssu's M_i, to give

$$y^*_{i0} = M_i \bar{y}_i \qquad (15.10)$$

The unbiased estimator of the universe total Y obtained from the ith sample fsu is given in an srs on multiplication of y^*_{i0} by N, the total number of fsu's, namely,

$$y^*_i = N y^*_{i0} \qquad (15.12)$$

and the combined unbiased estimator of Y from all the n sample villages is the simple arithmetic mean

$$y_0^* = \frac{1}{n}\sum_{i=1}^{n} y_i^*$$

$$= \frac{N}{n}\sum_{i=1}^{n} y_{i0}^* \qquad (15.13)$$

$$= \frac{N}{n}\sum_{i=1}^{n} \frac{M_i}{m_i}\sum_{j=1}^{m_i} y_{ij} = \frac{N}{n}\sum^{n} M_i\bar{y}_i$$

an unbiased variance estimator of which is given by equation (15.14).

• (c) *From the general estimating equations for a multi-stage design.* We use the results of section 14.4(5). Noting that in simple random sampling, the (initial) probability of selection of an fsu is $\pi_i = 1/N$, and that of an ssu is $\pi_{ij} = 1/M_i$ (see Table 15.1), we find from equation (14.9),

$$\text{For the first-stage units:} \quad f_1 \equiv n\pi_i = n/N$$
$$\text{For the second-stage units:} \quad f_2 \equiv m_i\,\pi_{ij} = m_i/M_i \qquad (15.15)$$

These are shown in the last column of Table (15.1).

The unbiased estimator of the universe total Y is, from equation (14.10),

$$y = \sum_{\text{sample}} \frac{1}{f_1 \cdot f_2} y_{ij} = \sum_{i,j} w_{ij} y_{ij} \qquad (15.16)$$

where y_{ij} is the value of the study variable in the jth sample ssu in the ith sample fsu, the summation being over all the $\sum^{n} m_i$ sample second-stage units, and

$$w_{ij} = \frac{1}{f_1 \cdot f_2} = \frac{N}{n}\frac{M_i}{m_i} \qquad (15.17)$$

Putting this value of w_{ij} in equation (15.16), we get

$$y = \frac{N}{n}\sum_{i=1}^{n} \frac{M_i}{m_i}\sum_{j=1}^{m_i} y_{ij} \qquad (15.18)$$

which is the same as the estimating equation (15.13).

Also, from equation (14.13), the unbiased estimator of Y from the ith sample fsu is

$$y_i^* = n\sum_{j} w_{ij} y_{ij}$$

$$= \frac{nNM_i}{nm_i}\sum_{j=1}^{m_i} y_{ij} \qquad (15.19)$$

$$= \frac{NM_i}{m_i}\sum_{j=1}^{m_i} y_{ij}$$

which is the same as the estimating equation (15.12).

An unbiased estimator of the variance of y (or y_0^*) is given by equation (14.15).•

Notes:

1. If all the first-stage units are known to have the same total number of ssu's, i.e. if $M_i = M_0$ say, then the combined unbiased estimator y_0^* in equation (15.13) becomes

$$y_0^* = \frac{NM_0}{n} \sum_{i=1}^{n} \frac{1}{m_i} \sum_{j=1}^{m_i} y_{ij} = \frac{NM_0}{n} \sum^{n} \bar{y}_i \tag{15.20}$$

2. If, in addition, the sample number of second-stage units in each sample fsu is fixed, $m_i = m_0$ say, then the design becomes self-weighting and

$$y_0^* = \frac{NM_0}{nm_0} \sum_{i=j}^{n} \sum_{j=1}^{m_i} y_{ij} \tag{15.21}$$

3. In some textbooks, the number of ssu's M_i is referred to as the 'size' of the ith fsu. We shall, however, use the term 'size' to mean the value of the ancillary variable for pps selection (Chapters 5, 16, and 21); a special case of the size is M_i (sections 16.3.2 and 21.3.2).

15.2.4 Sampling variance of the estimator y_0^*

The sampling variance of y_0^* in srswr at both stages is

$$\sigma_{y_0^*}^2 = \frac{N^2 \sigma_b^2}{n} + \frac{N}{n} \sum^{N} \frac{M_i^2 \sigma_{wi}^2}{m_i} \tag{15.22}$$

where

$$\sigma_b^2 = \sum^{N} (Y_i - \bar{Y})^2/N \tag{15.23}$$

is the variance between the fsu's, and

$$\sigma_{wi}^2 = \sum^{M_i} (Y_{ij} - \bar{Y}_i)^2/M_i \tag{15.24}$$

is the variance between the ssu's within the ith fsu. $\sigma_{y_0^*}^2$ is estimated unbiasedly by $s_{y_0^*}^2$, defined by equation (15.14).

Notes

1. If sampling is simple random without replacement at both stages, y_0^*, defined in equation (15.13), still remains the (combined) unbiased estimator of Y. The sampling variance and its unbiased estimator are given in section 17.2.4.
2. If sampling is simple random with replacement at the first stage and simple random without replacement at the second stage, the estimator y_0^*, defined in equation (15.13), remains unbiased for Y. Its sampling variance is given in section 17.2.4, note 4, and the unbiased variance estimator is $s_{y_0^*}^2$, defined in equation (15.14).

15.2.5 Estimation of the ratio of the totals of two study variables

For another study variable, the unbiased estimator x_0^* of the universe total X can be obtained in the same manner as in section 15.2.3. Thus, the unbiased estimator of X, obtained from the ith sample fsu is, as in equation (15.12),

$$x_i^* = N \frac{M_i}{m_i} \sum_{j=1}^{m_i} x_{ij} \tag{15.25}$$

where x_{ij} is the value of the other study variable in the jth sample ssu in the ith sample fsu; and the combined unbiased estimator of X from all the n sample fsu's is, as in equation (15.13),

$$x_0^* = \frac{1}{n}\sum^n x_i^* \tag{15.26}$$

with an unbiased variance estimator $s^2_{x_0^*}$ defined as for $s^2_{y_0^*}$ in equation (15.14).

A consistent but generally biased estimator of the ratio of the two universe totals $R = Y/X$ is the ratio of the sample estimators y_0^* and x_0^* (from equation (14.6)),

$$r = y_0^*/x_0^* \tag{15.27}$$

and a variance estimator of r is, from equation (14.7),

$$s_r^2 = (s^2_{y_0^*} + r^2 s^2_{x_0^*} - 2rs_{y_0^* x_0^*})/x_0^{*2} \tag{15.28}$$

where

$$s_{y_0^* x_0^*} = \sum^n (y_i^* - y_0^*)(x_i^* - x_0^*)/n(n-1) \tag{15.29}$$

is the unbiased estimator of the covariance of y_0^* and x_0^* (from equation (14.5)).

Notes

1. If $M_i = M_0$, then from equations of the type (15.20), the estimator of the universe ratio $R = Y/X$ becomes

$$r = \frac{y_0^*}{x_0^*} = \frac{\left(\sum_{i=1}^{n}\sum_{j=1}^{m_i} y_{ij}/m_i\right)}{\left(\sum_{i=1}^{n}\sum_{j=1}^{m_i} x_{ij}/m_i\right)} = \frac{\sum_{i=1}^{n}\bar{y}_i}{\sum_{j=1}^{n}\bar{x}_i} \tag{15.30}$$

2. If in addition, the same number of sample units m_0 is selected in the sample fsu's, then the design becomes self-weighting and

$$r = y_0^*/x_0^* = \sum_{i=1}^{n}\sum_{j=1}^{m_0} y_{ij} \Big/ \sum_{i=1}^{n}\sum_{j=1}^{m_0} x_{ij} \tag{15.31}$$

i.e. the ratio of the sample totals.

15.2.6 Estimation of the means of a study variable

(a) *Estimation of the mean per first-stage unit.* In two-stage srs, the universe mean per first-stage unit $\bar{Y}$ (defined by equation (15.4)) is estimated by

$$\bar{y} = \frac{y_0^*}{N} = \frac{\sum^n M_i \bar{y}_i}{n} = \frac{\sum^n y_{io}^*}{n} \tag{15.32}$$

i.e. the average of the unbiased estimators of the fsu totals Y_i.

Note the similarity of the structures of the universe value and its estimator in srs.

An unbiased estimator of the variance of $\bar{y}$ is

$$s^2_{y^*_0}/N^2 \qquad (15.33)$$

where $s^2_{y^*_0}$ (defined by equation (15.14)) is the unbiased estimator of the variance of y^*_0.

(b) *Estimation of the mean per second-stage unit.* Two situations may arise: (i) the total number of ssu's M_i is known for all the N fsu's, the total being $\sum^N M_i$; (ii) the total number of ssu's is known only for the n sample fsu's, after these n first-stage units are enumerated for the list of ssu's.

(i) *Unbiased estimator.* When the total number of ssu's in the universe is known, namely, $\sum^N M_i$, then the unbiased estimator of the universe mean per second-stage unit $\bar{\bar{Y}}$ (defined by equation (15.5)) is obtained on dividing the unbiased estimator y^*_0 by $\sum^N M_i$, namely

$$\bar{y} = y^*_0 \Big/ \sum^N M_i \qquad (15.34)$$

and the unbiased estimator of the variance of this sample estimator $\bar{\bar{y}}$ is obtained on dividing $s^2_{y^*_0}$ (as given by equation (15.14)) by $\left(\sum^N M_i\right)^2$, i.e.

$$s^2_{\bar{\bar{y}}} = s^2_{y^*_0} \Big/ \left(\sum^N M_i\right)^2 \qquad (15.35)$$

Notes

1. If all the fsu's are known to have the same number of ssu's, i.e. if $M_i = M_0$, then $\bar{\bar{y}}$ becomes (from equations (15.20) and (15.34)),

$$\bar{\bar{y}} = y^*_0/NM_0 = \sum_{i=1}^{n} \sum_{j=1}^{m_i} y_{ij}/nm_i = \sum^{n} \bar{y}_i/n \qquad (15.36)$$

i.e. the simple mean of means (*see* (iii) of this section).

2. If, in addition, a fixed number m_0 of ssu's is selected in the sample ssu's, then from equations (15.21) and (15.34) the design becomes self-weighting and $\bar{\bar{y}}$ becomes

$$\bar{\bar{y}} = y^*_0/NM_0 = \sum_{i=1}^{n} \sum_{j=1}^{m_0} y_{ij}/nm_0 \qquad (15.37)$$

3. In crop surveys, if the design is two-stage srswr with villages as first-stage units, and fields as the second-stage units, and the total geographical area (A), and y_{ij} denotes the area under a particular crop in the ijth sample field, then the proportion of the area under the crop

in the universe is estimated unbiasedly by

$$y_0^*/A \tag{15.38}$$

and its sampling variance by

$$s_{y_0^*}^2/A^2 \tag{15.39}$$

Similarly, if x_{ij} denotes the yield of a crop in the ijth sample field, then the average yield per field is estimated unbiasedly by an equation of the type (15.34) and its variance by an equation of the type (15.35). Similar considerations apply when the total number of fields in the universe $\left(\sum^N\right) M_i$ is known.

(ii) *Ratio estimator.* When the total number of ssu's is known only for the n sample fsu's, the unbiased estimator of universe number of ssu's $\sum^N M_i$ is obtained from the sample. In two-stage srs, the average number of ssu's in the sample fsu's,

$$\sum^n M_i/n \tag{15.40}$$

is multiplied by the total number N of fsu's to provide the unbiased estimator, thus,

$$m_0^* = N\sum^n M_i/n \tag{15.41}$$

This can also be derived from the estimating equations of section 15.2.3 by putting $y_{ij} = 1$ for the selected sample units. Thus putting $y_{ij} = 1$ in equation (15.12), we get as the unbiased estimator of the total number of ssu's, obtained from the ith sample fsu,

$$m_i^* = NM_i m_i/m_i = NM_i \tag{15.42}$$

and the combined unbiased estimator (from equation (15.13))

$$m_0^* = \frac{1}{n}\sum^n m_i^* = \frac{N}{n}\sum^n M_i \tag{15.43}$$

An unbiased estimator of the variance of m_0^* is (from equation (15.14))

$$s_{m_0}^{2*} = \sum^n (m_i^* - m_0^*)^2/n(n-1)$$

$$= \frac{N^2 \sum^n \left[M_i - \left(\sum^n M_i/n\right)\right]^2}{n(n-1)} \tag{15.44}$$

A consistent but generally biased estimator of the universe mean $\overline{\overline{Y}}$, is from equation (15.27), the ratio

$$r = y_0^*/m_0^*$$
$$= \sum^n M_i\bar{y}_i \Big/ \sum^n M_i \tag{15.45}$$

A variance estimator of r is (from equation (15.28))

$$s_r^2 = (s_{y_0^*}^2 + r^2 s_{m_0^*}^2 - 2r s_{y_0^* m_0^*})/m_0^{*2} \tag{15.46}$$

where

$$s_{y_0^* m_0^*} = \sum^{n} (y_i^* - y_0^*)(m_i^* - m_0^*)/n(n-1) \tag{15.47}$$

is an unbiased estimator of the covariance of y_0^* and m_0^*.

Note: In crop surveys, even when the total geographical area or the total number of farms in the universe is known, in estimating the proportion of area under a particular crop, or the average yield per field, estimating equations of the ratio type, namely (15.27), and (15.45) respectively, may be used (see part (iv) of this section).

(iii) *Unweighted mean of means.* In estimating means, a third estimator – the simple (unweighted) mean of means – may be used. Denoting by $\bar{y}_i$ the mean $\sum_{j=1}^{m_i} y_{ij}/m_i$, this third estimator is

$$\sum^{n} \bar{y}_i/n = \sum_{i=1}^{n} \sum_{j=1}^{m_i} y_{ij}/nm_i \tag{15.48}$$

This is a biased estimator, unless all the M_is are the same (see note 1 to (i) of this section) or the design is self-weighting.

(iv) *Comparison of the estimators.* Of the three estimators, only the estimator given by equation (15.34) is unbiased. However, this estimator can be used only if the total number of ssu's in the universe is known, and even then its sampling variance is likely to be high. On the other hand, the ratio-type estimator, given by equation (15.45), although biased, is consistent. The third estimator, the simple mean of means, is neither unbiased nor consistent; however, if it is known that the number of ssu's (M_i) in the fsu's does not vary considerably and there is no appreciable correlation between M_i and the study variable, this estimator would be the most efficient. Crop surveys in India have shown the absence of any such correlation, but the specific situations should be considered for other studies.

15.3 Two-stage srs: estimation for sub-universes

As for single-stage srs (section 2.12), the unbiased estimator of the total of the values of the study variable in the sub-universe Y' is obtained from the estimating equation (15.13) by defining

$y'_{ij} = y_{ij}$ if the sample unit belongs to the sub-universe;
$y'_{ij} = 0$ otherwise.

An unbiased estimator of the variance of the estimator of the total for the sub-universe is similarly given by equation (15.14).

15.4 Two-stage srs: estimation of proportion of units

Let M'_i be the total number of ssu's possessing a certain attribute in the ith fsu ($i = 1, 2, \ldots, N$); the total number of such ssu's in all the N fsu's is

$$\sum^{N} M'_i \tag{15.49}$$

The proportion of such ssu's with the attribute in the ith fsu in the universe is

$$P_i = M'_i/M_i \tag{15.50}$$

and the universe proportion is

$$P = \sum^{N} M_i' \Big/ \sum^{N} M_i = \sum^{N} M_i P_i \Big/ \sum^{N} M_i \tag{15.51}$$

In a two-stage srs (as for single-stage srs, see section 2.13), the estimated total number of ssu's in the universe with the attribute and its estimated variance could be obtained respectively from estimating equations (15.13) and (15.14) by putting $y_{ij} = 1$ if the sample unit has the attribute, and 0 otherwise.

Thus in an srs if m_i' of the m_i sample ssu's in the selected ith fsu has the attribute, then the sample proportion

$$p_i = m_i'/m_i \tag{15.52}$$

is an unbiased estimator of P_i,

The unbiased estimator of M_i' is $M_i p_i$; and the unbiased estimator of $\sum^{N} M_i'$, the universe number of ssu's with the attribute, from the ith sample fsu, is (from equation (15.12))

$$m_i'^* = N M_i p_i \tag{15.53}$$

The combined unbiased estimator of $\sum^{N} M_i'$ from all the n sample fsu's is (from equation (15.13)) the arithmetic mean

$$m_0'^* = \sum^{n} m_i'^*/n = N \sum^{n} M_i p_i/n \tag{15.54}$$

An unbiased estimator of the variance of $m_0'^*$ is (from equation (15.14))

$$s^2_{m_0'^*} = \sum^{n} (m_i'^* - m_0'^*)^2/n(n-1) \tag{15.55}$$

If the M_i values are known for all the N fsu's, then the unbiased estimator of the universe proportion P is

$$m_0'^* \Big/ \left(\sum^{N} M_i \right) \tag{15.56}$$

and an unbiased estimator of the variance of this estimator is

$$s^2_{m_0'^*} \Big/ \left(\sum^{N} M_i \right)^2 \tag{15.57}$$

If $\sum^{N} M_i$ is not known, it is estimated by

$$m_0^* = N \sum^{n} M_i/n \tag{15.43}$$

an unbiased estimator of whose variance is given by equation (15.44).

A consistent but generally biased estimator of the universe proportion P is then

$$m_0'^*/m_0^* = \sum^{n} M_i p_i \Big/ \sum^{n} M_i \tag{15.58}$$

The estimated variance of this estimator is given by an equation of the type (15.46).

Notes

1. Even if $\sum^{N} M_i$ is known, the use of the estimator given by equation (15.58) may be preferred to that given in equation (15.56) as the former is likely to be more efficient.

2. If all the M_i values are known for the universe and are the same, M_0, then $\sum^{N} M_i = NM_0$, and the estimator in equation (15.56) becomes

$$m_0'^*/NM_0 = \sum^{n} p_i/n = \bar{p} \tag{15.59}$$

i.e. the simple (unweighted) arithmetic mean of the p_i sample values.

3. If, in addition, a fixed number m_0 of ssu's is selected in the sample fsu's, then the design is self-weighting and the estimator (15.59) becomes

$$\bar{p} = \sum^{n} p_i/n = \sum^{n} m_i'/nm_0 \tag{15.60}$$

i.e. the sample number of units possessing the attribute divided by the total number of sample units.

An unbiased estimator of the variance of $\bar{p}$ (equation (15.59) or (15.60)) is

$$s_{\bar{p}}^2 = \sum^{n} (p_i - \bar{p})^2/n(n-1) \tag{15.61}$$

In particular, the unbiased estimator of the variance of p in equation (15.60) is

$$s_{\bar{p}}^2 = \frac{\text{SS}m_i'/n(n-1)}{m_0^2} \tag{15.62}$$

For an example, see exercise 1(b) of this chapter.

15.5 Two-stage srs: ratio method of estimation

If information on an ancillary variable is available for all the second-stage units in the universe, then the universe total

$$Z = \sum_{i=1}^{N} \sum_{j=1}^{M_i} z_{ij} \tag{15.63}$$

is also known, where z_{ij} is the value of the ancillary variable of the ijth universe unit. The unbiased estimator of Z in a two-stage srs is (from equation (15.13))

$$\begin{aligned} z_0^* &= \frac{N}{n} \sum_{i=1}^{n} \frac{M_i}{m_i} \sum_{j=1}^{m_i} z_{ij} \\ &= \frac{N}{n} \sum^{n} z_i^* \end{aligned} \tag{15.64}$$

and its unbiased variance estimator is (from equation (15.14))

$$s_{z_0^*}^2 = \sum_{i=1}^{n} (z_i^* - z_0^*)^2/n(n-1) \tag{15.65}$$

where z_{ij} is the value of the ancillary variable in the ijth sample unit and

$$z_i^* = NM_i \sum^{m_i} z_{ij}/m_i.$$

The ratio estimator of the universe total Y, using the ancillary information, is (from equation (3.6))

$$y^*_R = Zy^*_0/z^*_0 = Zr_1 \quad (15.66)$$

where

$$r_1 = y^*_0/z^*_0 \quad (15.67)$$

A consistent but generally biased estimator of the variance of y^*_R is (from equations (3.10) and (3.11))

$$s^2_{y^*_R} = (s^2_{y^*_0} + r_1^2 s^2_{z^*_0} - 2r_1 s_{y^*_0 z^*_0})$$
$$= (\text{SS}y_i^* + r_1^2 \text{SS}z_i^* - 2r_1 \text{SP}y_i^* z_i^*)/n(n-1) \quad (15.68)$$

Notes

1. If x is another study variable, a ratio estimator x^*_R for the universe total X can be similarly defined as y^*_R. However, in estimating the ratio $R = Y/X$, the use of ratio estimators of the totals Y and X leads to the same result as those of the unbiased estimators, for $y^*_R/x^*_R = y^*_0/x^*_0$ (see note 5 to section 3.2).
2. The ratio estimator is likely to be very efficient when (i) the M_i values vary considerably, (ii) the ancillary variable is highly correlated with the study variable, and (iii) n is large. The ratio estimator of the universe mean $\overline{Y}$, namely, $\overline{Z}(y^*_0/z^*_0) = \overline{Z}r_1$ (where $\overline{Z}$, the universe mean of the ancillary variable is known) will then be more efficient than the other three estimators of the universe mean, considered in section 15.2.6.

Example 15.1

For the Rural Household Budget Survey in the Shoa province of Ethiopia, 1966–7, ten geographical strata were formed comprising a number of sub-divisions; in each stratum, three sub-divisions were selected with equal probability out of the total number of sub-divisions; in each selected sub-division, of the total number of households listed by the enumerators, twelve were selected with equal probability for the inquiry. Although the sample was not specifically designed to provide estimates of demographic parameters, the example shows the method of computation for estimating the average age of the household heads in one particular stratum and its standard error. Table 15.2 gives the required data for the stratum, which had a total of eighty sub-divisions.

The required computations are shown in Table 15.3, denoting by y_{ij} the age of the head of the ijth sample household ($i = 1, 2, 3$ for the sample sub-divisions; $j = 1, 2, \ldots, 12$ for the sample households); N denotes the total number of sub-divisions in the stratum = 80, and M_i and m_i (= 12) respectively the total and sample number of households in the ith sample sub-division.

From equation (15.13), the combined unbiased estimate of the total of the ages of the households heads in the stratum is

$$y^*_0 = \frac{N}{n} \sum_{i=1}^{n} \frac{M_i}{m_i} \sum_{j=1}^{m_i} y_{ij} = 80 \times 18\,472.89 \text{ years}$$

(shown at the bottom of column 5), and from equation (15.43), the combined unbiased estimate

Table 15.2 Total and sample number of households and total ages of the sample household heads in one stratum: Rural Household Budget Survey, Shoa Province, Ethiopia, 1966–7 (Source: Unpublished data obtained by courtesy of the Central Statistical Office, Ministry of Planning and Development, Imperial Ethiopian Government).

Sample sub-division (i)	*Total no. of households* (M_i)	*No. of households in sample* (m_i)	*Total ages (in years) of the sample household heads* $\left(\sum_{j=1}^{m_i} y_{ij}\right)$
1	222	12	474
2	42	12	503
3	913	12	590

Table 15.3 Computation of the average age of household heads and its sampling variance: data of Table 15.2

Sub-division i (1)	*Total no. of households* M_i (2)	*No. of households in sample* m_i (3)	*Total of ages of household heads* $\sum_{j=1}^{m_i} y_{ij}$ (4)	$y_i^* = NM_i \sum^{m_i} y_{ij}/m_i$ (5)	y_i^{*2} (6)
1	222	12	474	80 x 8769.00	80^2 x 76 895.361
2	42	12	503	1760.50	3099.360
3	913	12	590	44 889.17	2 015 040.277
			Total	80 x 55 418.67	80^2 x 2 095 035.998
			Mean	80 x 18 472.89 (y_0^*)	

Sub-division i (1)		For the number of households $m_i^* = NM_i$ (7)	m_i^{*2} (8)	For the sum of products $y_i^* m_i^*$ (9)
1		80 x 222	80^2 x 49 284	80^2 x 1 946 718
2		42	1764	73 941
3		913	833 569	40 983 809
	Total	80 x 1177	80^2 x 884 617	80^2 x 43 004 468
	Mean	80 x 392.33 (m_0^*)		

of the total number of households in the stratum is

$$m_0^* = N\sum^{n} M_i/n = 80 \times 392.33 = 31\ 387 \text{ households}$$

(shown at the bottom of column 7).

The estimated average age of household heads is (from equation (15.45))

$$r = y_0^*/m_0^* = 18\ 472.89/392.33 = 47.08 \text{ years}$$

An estimate of the variance of r is given by equation (15.46), namely,

$$s_r^2 = (s_{y_0^*}^2 + r^2 s_{m_0^*}^2 - 2r s_{y_0^* m_0^*})/m_0^{*2}$$

Since

$$SSy_i^* = \sum^{n} y_i^{*2} - \left(\sum^{n} y_i^*\right)^2 \Big/ n$$

$$= 80^2 \times [2\ 095\ 034\ 998 - 1\ 023\ 744\ 103]$$

$$= 80^2 \times 1\ 071\ 290\ 895$$

so, from equation (15.14),

$$s_{y_0^*}^2 = SSy_i^*/n(n-1) = 80^2 \times 1\ 071\ 290\ 895/6$$

Similarly from equation (15.44)

$$s_{m_0^*}^2 = SSy_i^*/n(n-1)$$

$$= 80^2 \times [884\ 617 - 461\ 776]/6$$

$$= 80^2 \times 422\ 841/6,$$

and from equation (15.47)

$$s_{y_0^* m_0^*} = \frac{\left[\sum^{n} y_i^* m_i^* - \left(\sum^{n} y_i^*\right)\left(\sum^{n} m_i^*\right)\Big/ n\right]}{n(n-1)}$$

$$= 80^2[43\ 004\ 468 - 21\ 742\ 590]/6$$

$$= 80^2 \times 21\ 262\ 878/6.$$

Noting that $r = 47.08$, $2r = 94.16$, and $r^2 = 2216.5264$, we get for s_r^2 (from equation (15.46)).

$$\frac{80^2[(1\ 071\ 290\ 895 + 937\ 238\ 070 - 2\ 002\ 018\ 432)/6]}{80^2 \times (392.33)^2}$$

$$= (6\ 510\ 533/6)/153\ 925.44 = 7.0494 \text{ years}^2$$

so that the standard error of estimate of r is $s_r = 2.66$ years, and the estimated CV of r is $2.66/47.08 = 0.0565$ or 5.65 per cent.

Example 15.2

In each of the simple random sample of 4 villages selected in Example 2.3 from the list of 30 villages (Appendix IV), select a simple random sample (with replacement) of 4 households from the total number of households, and on the basis of the collected data on the households, the total monthly income, and the number of adults (over 18 years of age) of the sample households, estimate for the 30 villages the total number of persons, the total monthly income, the per capita monthly income, and the total number and proportion of adults, along with their standard errors.

The sample data are shown in Table 15.4 and the required computations in Table 15.5, on page 170, denoting by y_{ij} the household size, by x_{ij} the monthly income, and by y'_{ij} the number of adults, in the ijth sample household ($i = 1, 2, 3, 4$ for sample villages numbered 15, 18, 19, and 24; $j = 1, 2, 3, 4$ for the sample households in the selected villages); the respective means in the ith sample village are denoted by $\bar{y}_i$, $\bar{x}_i$ and $\bar{y}'_i$ (defined by an equation of the type 15.11), $N = 30$ is the total number of villages, and M_i and m_i (= 4) denote respectively the total and the sample numbers of households in the ith sample village.

From equation (15.13), the combined unbiased estimate of the total number of persons in the 30 villages is

$$y_0^* = \frac{N}{n} \sum_{i=1}^{n} \frac{M_i}{m_i} \sum_{j=1}^{m_i} y_{ij}$$

$$= 30 \times 108.5 = 3255 \text{ persons;}$$

and from equation (15.14), an unbiased estimate of the variance of y_0^* is

$$s^2_{y_0^*} = SSy_i^*/n(n-1)$$
$$= 30^2 \times [47\,783.5 - 47\,089]/12$$
$$= 30^2 \times 694.5/12$$

so that the estimated standard error of y_0^* is

$$s_{y_0^*} = 30 \times 7.608 = 228 \text{ persons}$$

Similarly, the combined unbiased estimate of the total monthly income is

$$x_0^* = \sum^{n} x_i^*/n = 30 \times \$1129.9375 = \$33\,938$$

Table 15.4 Size, total monthly income, number of adults (18 years or over) and knowledge of development plans by the selected adult in sample households in each of the four sample villages of Example 2.3: two-stage simple random sample

Household serial no.	*Household size (y_{ij})*	*Total monthly income (x_{ij})*	*Number of adults (y'_{ij})*	*Knowledge of development plans by the selected adult (q'_{ij} = 1 for Yes; = 0 for No)*
Village serial no. 15; total number of households = 20				
1	8	$92	4	0
2	4	41	2	1
5	6	55	3	0
14	3	36	2	1
Village serial no. 18; total number of households = 23				
13	4	39	2	0
15	5	58	3	1
16	6	61	3	0
21	7	63	4	0
Village serial no. 19; total number of households = 25				
5	5	58	2	0
11	5	50	3	0
23	3	33	2	0
24	5	47	2	1
Village serial no. 24; total number of households = 18				
3	4	48	2	0
7	6	70	4	0
14	6	49	2	1
15	4	45	2	0

Table 15.5 Computation of the estimated total number of persons, total and *per capita* monthly income, number and proportion of adults for a two-stage srs of villages and households: data of Table 15.4.

Sample village i	Number of households: total M_i	Number of households: sample m_i	$\sum_{j=1}^{m_i} y_{ij}$	For the number of persons: $\bar{y}_i = \sum_{j=1}^{m_i} y_{ij}/m_i$	$y_i^* = NM_i\bar{y}_i$	y_i^{*2}
(1)	(2)	(3)	(4)	(5)	(6)	(7)
1	20	4	21	5.25	30 x 105.0	30^2 x 11 025.00
2	23	4	22	5.50	126.5	16 002.25
3	25	4	18	4.50	112.5	12.656.25
4	18	4	20	5.00	90.0	8 100.00
				Total	30 x 434.0	30^2 x 47 783.50
				Mean	30 x 108.5 (y_0^*)	—

Sample village	$\sum_{j=1}^{m_i} x_{ij}$	$\bar{x}_i = \sum_{j=1}^{m_i} x_{ij}/m_i$	For the total monthly income: $x_i^* = NM_i\bar{x}_i$	x_i^{*2}
(1)	(8)	(9)	(10)	(11)
1	\$224	\$56.00	\$30 x 1120.00	30^2 x 1 254 400.0000
2	221	55.25	1270.75	1 614 805.5625
3	188	47.00	1175.00	1 380 625.0000
4	212	53.00	954.00	910 116.0000
		Total	\$30 x 4519.75	30^2 x 5 159 946.5625
		Mean	\$30 x 1129.9375 (x_0^*)	

Sample village i	$\sum_{j=1}^{m_i} y'_{ij}$	$\bar{y}'_i = \sum_{j=1}^{m_i} y'_{ij}/m_i$	For the number of adults: $y_i'^* = NM_i\bar{y}'_i$	$y_i'^{*2}$
(1)	(12)	(13)	(14)	(15)
1	11	2.75	30 x 55.00	30^2 x 3025.0000
2	12	3.00	69.00	8728.5000
3	9	2.25	56.25	3164.0625
4	10	2.50	45.00	2025.0000
		Total	30 x 225.25	30^2 x 12 975.0625
		Mean	30 x 56.3125 ($y_i'^*$)	

Sample village i	For the sum of products: $y_i^* x_i^*$	$y_i^* y_i'^*$
(1)	(16)	(17)
1	30^2 x 117 600.000	30^2 x 5775.000
2	160 749.875	8728.500
3	132 187.500	6328.500
4	85 860.000	4050.000
Total	30^2 x 496 397.375	30^2 x 24 881.625

with estimated variance

$$s^2_{x_0^*} = 30^2[5\ 159\ 946.5625 - 5\ 107\ 035.0156]/12$$
$$= 30^2 \times 52\ 911.5469/12$$

so that the estimated standard error of x_0^* is

$$s_{x_0^*} = 30 \times \$66.4025 = \$1992$$

The combined unbiased estimate of the total number of adults is

$$y_0'^* = \sum^{n} y_i'^*/n = 30 \times 56.3125 = 1689$$

with estimated variance of

$$s^2_{y_0'^*} = 30^2 \times [12\ 975.0625 - 12\ 684.3906]/12$$
$$= 30^2 \times 290.6719/12$$

so that the estimated standard error of $y_0'^*$ is $s_{y_0'^*} = 148$.

The estimated monthly income per capita is equal to the estimated total monthly income divided by the estimated total number of persons, or

$$r = \frac{x_0^*}{y_0^*} = \frac{(30 \times \$1129.9375)}{(30 \times 108.50)} = \$10.414$$

As

$$s_{y_0^* x_0^*} = \text{SP}y_0^* x_0^*/n(n-1)$$
$$= 30^2 \times [496\ 397.3750 - 490\ 392.8750]/12$$
$$= 30^2 \times 6004.5/12$$

the estimated variance of r is (from equation (15.28))

$$s_r^2 = (s^2_{x_0^*} + r^2 s^2_{y_0^*} - 2r s_{y_0^* x_0^*})/y_0^{*2}$$

$$= \frac{30^2}{12}\left[\frac{52\ 911.5469 + (10.414)^2 \times 694.5 - 2 \times 10.414 \times 6004.5}{(30 \times 108.50)^2}\right]$$

$$= 3169.3182/(12 \times 11\ 772.25) = 0.02243495,$$

so that the estimated standard error of r is s_r = 0.1498.

The estimated proportion of adults is equal to the estimated total number of adults divided by the estimated total number of persons, or

$$p = \frac{y_0'^*}{y_0^*} = \frac{(30 \times 56.3125)}{(30 \times 108.50)} = 0.5190$$

As

$$s_{y_0^* y_0'^*} = 30^2 \times [24\ 881.625 - 24\ 439.625]/12$$
$$= 30^2 \times 442/12$$

the estimated variance of p is, from equation (15.28),

$$s_p^2 = (s^2_{y_0'^*} + p^2 s^2_{y_0^*} - 2p s_{y_0^* y_0'^*})/y_0^{*2}$$

$$= \frac{(30^2/12)[290.6719 + (0.519)^2 \times 694.5 - 2 \times 0.519 \times 442]}{(30 \times 108.50)^2}$$

$$= 0.0013411$$

so that the estimated standard error of p is $s_p = 0.0116$.

15.6 Three-stage srs: estimation of totals, means, ratios, and their variances

15.6.1 Universe totals and means

Extending the notations of section 15.2.1, let the number of third-stage units (tsu's) in the jth ssu ($j = 1, 2, \ldots, M_i$) of the ith fsu ($i = 1, 2, \ldots, N$) be Q_{ij}; the total number of tsu's in the universe is $\sum_{i=1}^{M_i} \sum_{j=1}^{Q_{ij}} Q_{ij}$. Denoting by Y_{ijk} the value of the study variable of the kth tsu ($k = 1, 2, \ldots, Q_{ij}$) in the jth ssu in the ith fsu, we define the following universe parameters:

Total of the values of the study variable in the jth ssu in the ith fsu:

$$Y_{ij} = \sum_{k=1}^{Q_{ij}} Y_{ijk} \qquad (15.69)$$

Mean for the jth ssu in the ith fsu:

$$\bar{Y}_{ij} = Y_{ij}/Q_{ij} \qquad (15.70)$$

Total for the ith fsu:

$$Y_i = \sum_{j=1}^{M_i} Y_{ij} \qquad (15.71)$$

Mean for the ith fsu per ssu:

$$\bar{Y}_i = Y_i/M_i \tag{15.72}$$

Grand total:

$$Y = \sum_{i=1}^{N} \sum_{j=1}^{M_i} \sum_{k=1}^{Q_{ij}} Y_{ijk} = \sum_{i=1}^{N} \sum_{j=1}^{M_i} Y_{ij} = \sum_{i=1}^{N} Y_i = \sum^{N} \sum^{M_i} Q_{ij} \bar{Y}_{ij} = \sum^{N} M_i \bar{Y}_i \tag{15.73}$$

Mean per fsu:

$$\bar{Y} = Y/N = \sum^{N} M_i \bar{Y}_i/N \tag{15.74}$$

Mean per ssu:

$$\bar{\bar{Y}} = Y\Big/\left(\sum^{N} M_i\right) = \sum^{N} M_i \bar{Y}_i\Big/\left(\sum^{N} M_i\right) \tag{15.75}$$

Mean per tsu:

$$\bar{\bar{\bar{Y}}} = Y\Big/\left(\sum^{N} \sum^{M_i} Q_{ij}\right) = \sum^{N} \sum^{M_i} Q_{ij} \bar{Y}_{ij}\Big/\left(\sum^{N} \sum^{M_i} Q_{ij}\right) \tag{15.76}$$

15.6.2 Structure of a three-stage srs

Extending the structure of a two-stage srs given in section 15.2.2, the design will be three-stage srswr if in the ijth sample ssu ($i = 1, 2, \ldots, n$; $j = 1, 2, \ldots, m_i$), q_{ij} sample tsu's are selected as an srs (and with replacement) out of the total Q_{ij} tsu's, the total number of tsu's in the sample, i.e. the total sample size is $\sum_{i=1}^{n} \sum_{j=1}^{m_i} q_{ij}$.

An example of a three-stage design for a crop survey has been given in section 14.2. Extending the example of a two-stage design for a household survey given in the same section, the design will be three-stage srs if a sample of persons (tsu's) is selected in the sample households (ssu's) in the sample villages (fsu's), sampling at the three stages being simple random. Or, to take another example, in an urban household inquiry, where all the towns constitute the urban area, an srs of towns (fsu's) may be first selected, then a sample of blocks (ssu's) in the selected towns, and finally a sample of households (tsu's) in the selected town-blocks. The sampling plan of a three-stage srs is shown in summary form in Table 15.6.

Table 15.6 Sampling plan for a three-stage simple random sample with replacement

Stage (t)	*Unit*	*No. in universe*	*No. in sample*	*Selection method*	*Selection probability*	f_t
1	First-stage	N	n	srswr	Equal = $1/N$	n/N
2	Second-stage	M_i	m_i	srswr	Equal = $1/M_i$	m_i/M_i
3	Third-stage	Q_{ij}	q_{ij}	srswr	Equal = $1/Q_{ij}$	q_{ij}/Q_{ij}

Let y_{ijk} denote the value of the study variable in the kth selected tsu ($k = 1, 2, \ldots, q_{ij}$) in the ijth sample ssu.

15.7.3 Estimation of the universe total Y and the variance of the sample estimator

Extending the methods of section 15.2.3 for a two-stage srs, it can be seen that an unbiased estimator of the universe total Y, obtained from the ith sample fsu ($i = 1, 2, \ldots, n$) is

$$y_i^* = \frac{NM_i}{m_i} \sum_{j=1}^{m_i} \frac{Q_{ij}}{q_{ij}} \sum_{k=1}^{q_{ij}} y_{ijk}$$

$$= \frac{NM_i}{m_i} \sum_{j=1}^{m_i} Q_{ij} \bar{y}_{ij} \tag{15.77}$$

where

$$\bar{y}_{ij} = \frac{1}{q_{ij}} \sum_{k=1}^{q_{ij}} y_{ijk} \tag{15.78}$$

is the simple arithmetic mean of the y_{ijk} values.

The combined unbiased estimator of Y from all the n fsu's is the simple arithmetic mean

$$y_0^* = \frac{1}{n} \sum^{n} y_i^* \tag{15.79}$$

an unbiased variance estimator of which is, from equation (14.15),

$$s_{y_0^*}^2 = \sum^{n} (y_i^* - y_0^*)^2 / n(n-1) \tag{15.80}$$

15.6.4 Estimation of the ratio of the totals of two study variables

The extension from the method in the two-stage srs is straightforward. The unbiased estimator of the universe total X of another study variable, obtained from the ith sample fsu, is (from equation 15.78))

$$x_i^* = \frac{NM_i}{m_i} \sum_{j=1}^{m_i} \frac{Q_{ij}}{q_{ij}} \sum_{k=1}^{q_{ij}} x_{ijk}$$

$$= \frac{NM_i}{m_i} \sum_{j=1}^{m_i} Q_{ij} \bar{x}_{ij} \tag{15.81}$$

where x_{ijk} is the value of the study variable in the ijkth sample third-stage unit, and

$$\bar{x}_{ij} = \sum_{k=1}^{q_{ij}} x_{ijk} / q_{ij} \tag{15.82}$$

The combined unbiased estimator of X, from all the n sample fsu's, is (from equation 15.79)

$$x_0^* = \frac{1}{n} \sum^{n} x_i^* \tag{15.83}$$

with an unbiased variance estimator $s^2_{x^*_0}$ defined as for $s^2_{y^*_0}$ in equation (15.80).

A consistent but generally biased estimator of the universe ratio $R = Y/X$ is the ratio of the sample estimators

$$r = y^*_0/x^*_0 \tag{15.84}$$

with a variance estimator

$$s^2_r = (s^2_{y^*_0} + r^2 s^2_{x^*_2} - 2rs_{y^*_0 x^*_0})/x^{*2}_0 \tag{15.85}$$

where

$$s_{y^*_0 x^*_0} = \sum^n (y^*_i - y^*_0)(x^*_i - x^*_0)/n(n-1) \tag{15.86}$$

is the unbiased estimator of the covariance of y^*_0 and x^*_0, from equation (14.5).

15.6.5 Estimation of the means of a study variable

For the estimation of the means of a study variable per fsu and ssu, see section 15.2.6(a) and (b): estimating equations of the same type, but relating to a three-stage srs, will apply. We consider here the estimating method for the mean per tsu $\bar{\bar{\bar{Y}}}$, as defined in equation (15.76).

The estimating procedure will depend on whether the total number of tsu's in the universe, namely $\sum_{i=1}^{N} \sum_{j=1}^{M_i} Q_{ij}$ is known or has to be estimated from the sample.

(a) *Unbiased estimator.* When the universe number of tsu's is known, then the universe mean per tsu $\bar{\bar{\bar{Y}}}$ is estimated unbiasedly by

$$\bar{\bar{\bar{y}}} = y^*_0 \Big/ \left(\sum_{i=1}^{N} \sum_{j=1}^{M_i} Q_{ij} \right) \tag{15.87}$$

with an unbiased variance estimator

$$s^2_{y^*_0} \Big/ \left(\sum_{i=1}^{N} \sum_{j=1}^{M_i} Q_{ij} \right)^2 \tag{15.88}$$

(b) *Ratio estimator.* When the universe number of third-stage units is not known, it is estimated from the sample by putting $y_{ijk} = 1$ in the estimating equations (15.77) and (15.79), thus from the ith sample fsu

$$q^*_i = \frac{NM_i}{m_i} \sum_{j=1}^{m_i} Q_{ij} \tag{15.89}$$

and from all the n sample fsu's

$$q^*_0 = \sum^n q^*_i / n \tag{15.90}$$

An unbiased variance estimator of q^*_0 is

$$s^2_{q^*_0} = \sum^n (q^*_i - q^*_0)^2 / n(n-1) \tag{15.91}$$

A consistent but generally biased estimator of the universe mean per third-stage unit is the ratio

$$r' = y_0^*/q_0^* \tag{15.92}$$

with a variance estimator

$$s_{r'}^2 = (s_{y_0^*}^2 + r'^2 s_{q_0^*}^2 - 2r' s_{y_0^* q_0^*})/q_0^{*2} \tag{15.93}$$

where

$$s_{y_0^* q_0^*} = \sum^{n} (y_i^* - y_0^*)(q_i^* - q_0^*)/n(n-1) \tag{15.94}$$

is an unbiased estimator of the covariance of y_0^* and q_0^*.

(c) *Unweighted mean of means.* A third estimator of the universe mean per third-stage unit is the unweighted mean of means

$$\sum_{i=1}^{n} \sum_{j=1}^{m_i} \bar{y}_{ij} \Big/ \sum_{i=1}^{n} \sum_{j=1}^{m_i} q_{ij} \tag{15.95}$$

where

$$\bar{y}_{ij} = \sum_{k=1}^{q_{ij}} y_{ijk}/q_{ij} \tag{15.96}$$

This estimator is both biased and inconsistent, unless the design is self-weighting (see note 3).

Notes

1. *Comparison of the estimators.* Although biased, the ratio-type estimator is likely to have a smaller sampling variance than the unbiased estimator: the latter is, of course, applicable only when the number of second- and third-stage units in the universe are known. The unweighted mean of means will be both biased and inconsistent, unless the design is self-weighting, but the observation made in section 15.2.6(b-iv) may also be noted.
2. In crop-surveys, when the total geographical area (A) or the total numbers of fields $(\sum^{N} M_i)$ and plots $(\sum_{i=1}^{N} \sum_{j=1}^{M_i} Q_{ij})$ are known in the universe, then in a three-stage srs with villages as the first-stage, fields as the second-stage, and plots as the third-stage units, the proportion of area under a particular crop in the universe is estimated unbiasedly by

$$y_0^*/A \tag{15.96}$$

where y_0^* is the unbiased estimator of the area under the crop in the universe, given by equation (15.79), y_{ijk} denoting the area under the crop in the ijkth sample plot. An unbiased variance estimator of y_0^*/A is

$$s_{y_0^*}^2/A^2 \tag{15.97}$$

$s_{y_0^*}^2$, the unbiased variance estimator of y_0^*, being defined in equation (15.80).

The average yield per unit area can be estimated similarly; and the average yield per plot is estimated unbiasedly by equation (15.87), and its unbiased variance estimator by equation (15.88), y_{ijk} denoting the yield of ijkth sample plot.

15.7 Three-stage srs: estimation for sub-universes

As for two-stage srs, estimators for sub-universes are obtained from the general results for the universe by defining

$y'_{ijk} = y_{ijk}$ if the sample unit belongs to the sub-universe; and
$y'_{ijk} = 0$ otherwise.

15.8 Three-stage srs: estimation of proportion of units

Let Q'_{ij} be the total number of tsu's in the ijth second-stage unit ($i = 1, 2, \ldots, N$; $j = 1, 2, \ldots, M_i$) possessing a certain attribute; the universe number of such tsu's is

$$\sum_{i=1}^{N} \sum_{j=1}^{M_i} Q'_{ij} \tag{15.98}$$

and the universe proportion of the tsu's possessing the attribute is

$$P = \sum_{i=1}^{N} \sum_{j=1}^{M_i} Q'_{ij} \Big/ \sum_{i=1}^{N} \sum_{j=1}^{M_i} Q_{ij} \tag{15.99}$$

As for two-stage srs, the estimated number of tsu's with the attribute could be obtained by putting $y_{ijk} = 1$, if the sample unit has the attribute, and 0 otherwise.

Thus if q'_{ij} of the q_{ij} sample third-stage units in the selected ijth ssu has the attribute, then the unbiased estimator of the total number of tsu's in the universe possessing the attribute ($\sum_{i=1}^{N} \sum_{j=1}^{M_i} Q'_{ij}$) from the ith sample fsu is

$$q_i'^{*} = \frac{NM_i}{m_i} \sum_{j=1}^{m_i} \frac{Q_{ij}}{q_{ij}} q'_{ij}$$

$$= \frac{NM_i}{m_i} \sum_{j=1}^{m_i} Q_{ij} p_{ij} \tag{15.100}$$

and from all the n sample fsu's

$$q_0'^{*} = \sum^{n} q_0'^{*}/n \tag{15.101}$$

where
$$p_{ij} = q'_{ij}/q_{ij} \tag{15.102}$$

is an unbiased estimator of the universe proportion $P_{ij} = Q'_{ij}/Q_{ij}$.

An unbiased variance estimator of $q_0'^{*}$ is

$$s^2_{q_0'^{*}} = \sum^{n} (y_i'^{*} - y_0'^{*})^2 / n(n-1) \tag{15.103}$$

A consistent but generally biased estimator of the universe proportion P is (from the general estimating equation (14.6))

$$p = q_0'^{*}/q_0^{*} \tag{15.104}$$

where q_0^{*} is defined by equation (15.90).

A variance estimator of p is given by the general estimating equation (14.7).

Note: If all the Q_{ij} values for the universe are known, then an unbiased estimator of the universe proportion P is

$$q_0'^{*} \Big/ \sum_{i=1}^{N} \sum_{j=1}^{M_i} Q_{ij} \tag{15.105}$$

However, this estimator is likely to have a much larger sampling variance than the estimator defined in estimating equation (15.104).

15.9 Three-stage srs: ratio method of estimation

As for two-stage srs (section 15.5), so also for three-stage srs the ratio method of estimation may be used to increase the efficiency of the estimators. The estimating procedures are the same as in section 15.5 for two-stage srs, except that for the unbiased estimators of the study and the ancillary variables and their estimated variances, formulae of the type (15.79) and (15.80) would be used.

Example 15.3

In example 15.2 are given the number of adults in each of the sample households in the srs of 4 villages out of the total 30 villages. In each of these sample households, one adult member was further selected at random from the total number of adults in the household, and asked about his knowledge of development plans in the country. This information is given also in Table 15.4. Estimate the total number of adults with knowledge of development plan and the proportion they constitute of the total number of adults.

Table 15.7 Computation of the estimated number and proportion of adults with knowledge of development plans: data of Tables 15.4 and 15.5

Sample village i	*Number of households* Total M_i	Sample m_i	*For the number of persons with knowledge of development plans and sum of products* $\sum_{j=1}^{m_i} Q_{ij}q'_{ij}$	$\dfrac{\text{Col.(4)}}{m_i}$	$q_i'^{*} = NM_i \times$ col.(5)	$q_i'^{*2}$	$q_i^{*}q_i'^{*}$
(1)	(2)	(3)	(4)	(5)	(6)	(7)	(8)
1	20	4	4	1.00	30 x 20.00	30^2 x 400.0000	30^2 x 1100.000
2	23	4	3	0.75	17.15	297.5625	1190.250
3	25	4	2	0.50	12.50	156.2500	703.125
4	18	4	2	0.50	9.00	81,0000	405.000
				Total	30 x 58.75	30^2 x 934.8125	30^2 x 3398.375
				Mean	30 x 14.6875 $(q_0'^{*})$		

Extending the notation of Example 15.1, let Q_{ij} ($=y'_{ij}$ in Example 15.1) denote the total number of adults, and q_{ij} ($=1$) the number selected for interview in the ijth sample household ($i = 1, 2, 3, 4$; $j = 1, 2, 3, 4$). As only one sample third-stage unit is selected, we can dispense with the subscript k in y_{ijk} by which we had denoted the value of the study variable for ijkth third-stage sample unit.

Following the method of section 15.8, we put $q'_{ij} = 1$, if the selected adult in the ijth sample household knows of development plans, and 0, otherwise. The required computations are shown in Table 15.7.

From estimating equation (15.101), the unbiased estimate of the total number of adults with knowledge of development plans is

$$q_0'^{*} = \sum^{n} q_i'^{*}/n = 30 \times 14.6875 = 441,$$

where (from equation 15.100)

$$q_i'^{*} = \frac{NM_i}{m_i}\sum_{j=1}^{m_i} Q_{ij}q'_{ij}$$

An unbiased estimate of the variance of $q_0'^{*}$ is (from equation (15.103))

$$\begin{aligned} s^2_{q_0'^{*}} &= SSq_i'^{*}/n(n-1) \\ &= 30^2[934.8125 - 862.8906]/12 \\ &= 30^2 \times 71.9219/12 \end{aligned}$$

so that the estimated standard error of $q_0'^{*}$ is 30 x 2.448 = 73, and the estimated CV 73/441 = 16.55 per cent.

The estimated proportion of adults with knowledge of development plans is equal to the estimated number of adults with such knowledge divided by the estimated total number of adults, or

$$\begin{aligned} p' &= q_0'^{*}/q_0^{*} \\ &= (30 \times 14.6875)/(30 \times 56.3125) \\ &= 0.2608 \text{ or } 26.08 \text{ per cent} \end{aligned}$$

As

$$s_{q_0^* q_0'^*} = \mathrm{SP} q_i^* q_i^* / n(n-1)$$
$$= 30^2 \times 90.0156/12$$

the estimated variance of p' is

$$s_{p'}^2 = (s_{q_0^*}^2 + p'^2 s_{q_0'^*}^2 - 2p' s_{q_0^* q_0'^*})/q_0^{*2}$$

$$= \frac{30^2}{12} \frac{71.9212 + (0.2608)^2 \times 290.6719 - 2 \times 0.2608 \times 90.0156}{(30 \times 56.3125)^2}$$

$$= \frac{54.1636}{12 \times 3171.0977} = 0.0014234$$

so that the estimated standard error of p' is $s_{p'} = 0.0377$, with estimated CV of 14.46 per cent.

Further reading

Cochran, sections 10.1–10.5, 10.8, 11.1–11.2, 11.5–11.7, and 11.12–11.13; Deming (1966), Chapter 5a; Hansen *et al.*, Vols. I and II, Chapters 6, 8, 9, and 10; Kish, section 5.3; Murthy, sections 9.3 and 9.10; Raj, section 6.3; Sukhatme and Sukhatme, sections 7.1–7.3, 7.5–7.8, 7.10–7.12, and 7.14; Yates, Chapters 3.8, 6.18–6.19, and 7.17.

Exercises

1. A hospital has received 1000 bottles of 100 tablets each. A simple random sample of 6 bottles is taken and from each sample bottle a sample of 20 tablets is taken at random. Given the data in Table 15.8, estimate (a) the average weight per tablet, (b) the proportion of sub-standard tablets, and (c) the ratio between the composition of two active substances A and B in the tablet, with their standard errors (Weber, Example 4.4).

2. Of 53 communes in an area, 14 were selected at random; from each of the selected communes, of the total number of farms, a sample of farms (1 in 4) was taken also at random. Table 15.9 gives the required information on the sample, including the total number of cattle in the sample farms in the selected communes.

(a) Estimate the total number of cattle in the area;
(b) estimate the average number of cattle per farm, by (i) using the ratio of the unbiased estimators of the total number of cattle and the total number of farms; (ii) using the unweighted mean of means; and (iii) using the additional information that the average number of farms per commune in the universe is 39.09.

Estimate also the standard errors of these estimates, assuming sampling with replacement. (United Nations *Manual*, Processes 20, 21, and 22, modified to 'with replacement' sampling.)

3. To estimate the total number of beetles in a square plot of land, 250 ft by 250 ft, the plot was divided into 625 square cells, 10 ft by 10 ft. A random sample of 10 cells was taken, and for each cell two independent estimates of the number of beetles were obtained by actual survey of two randomly located circles within the cell. The data are given in Table 15.10. (Chakravarti *et al.*, Vol. II, Example 3.4).

Table 15.8 Total weight, number of sub-standard tablets, and the percentage composition of two active substances in 20 sample tablets selected from 100 tablets in 6 out of 1000 bottles received by a hospital

Bottle number	*Total weight (gr) of the sample 20 tablets*	*Number of sub-standard tablets*	*Composition in percentage for the sample of 20 tablets*	
			Substance A	*Substance B*
1	21.8	1	5.42	31.2
2	19.3	0	5.01	30.3
3	22.8	1	5.43	30.5
4	23.1	3	5.43	32.4
5	22.9	1	5.20	31.4
6	19.7	1	5.56	32.9
Total	129.6	7	32.05	188.7
Mean per tablet	1.08	0.058	0.267	1.572
Raw sum of squares	2813.68	13	171.3999	5939.91
Raw sum of products			1008.658	

Table 15.9 Number of cattle in sample farms

Sample commune i	*Number of farms*		*Number of cattle*		
	Total M_i	*Sample* m_i	*Total in sample* $\sum_{j=1}^{m_i}$	*Average per farm* $\bar{y}_i = \sum_{j=1}^{m_i} y_{ij}/m_i$	$M_i\bar{y}_i$
1	46	11	88	8.0000	368.0000
2	39	10	114	11.4000	444.6000
3	25	6	96	16.0000	400.0000
4	23	5	82	16.4000	377.2000
5	32	8	83	10.3750	332.0000
6	31	8	207	25.8750	802.1250
7	60	15	208	13.8667	832.0020
8	28	7	73	10.4286	292.0008
9	59	14	195	13.9286	821.7874
10	24	6	73	12.1667	292.0008
11	84	21	191	9.0952	763.9968
12	30	7	79	11.2857	338.5710
13	64	16	226	14.1250	904.0000
14	66	17	166	9.7647	644.4702
Total	611	151	1881	182.7112	7 612.7540

For checking and facility of computation, the following are given:

$$\sum_{i=1}^{n} (M_i\bar{y}_i)^2 = 4\,857\,967.5243; \quad m_0^* = 2313; \quad \sum_{i=1}^{n} M_i^2 = 31\,625; \quad s_{m_0^*}^2 = 53^2 \times 4959.2143/182;$$

$$\sum_{i=1}^{n} (M_i\bar{y}_i)M_i = 376\,746.97; \quad s_{y_0^* m_0^*} = 53^2 \times 44\,504.6347/182; \quad \sum_{i=1}^{n} \bar{y}_i^2 = 2644.3481.$$

Table 15.10 Estimated number of beetles in cells of 10 ft by 10 ft obtained from two randomly located circles within each cell

Cell	1	2	3	4	5	6	7	8	9	10
Estimate 1	61	38	25	0	71	95	32	50	10	0
Estimate 2	29	81	32	45	46	69	26	39	24	5
Total	90	119	57	45	117	164	58	89	43	5

16 Multi-stage Varying Probability Sampling

16.1 Introduction

As for single-stage, so also for multi-stage sampling, sampling with varying probabilities – more specifically with probability proportional to size – would, under favourable conditions, lead to an improvement in the efficiency of estimators if ancillary information is available for such selection. Although pps sampling can be resorted to in all stages of sampling, in general it is restricted to the stages that can be covered in offices; the selection of sample units at other stages, generally the ultimate or the penultimate stages or both which can readily be sampled either systematically or by simple random sampling, are left to the enumerators in the field.

The general estimating equations for a multi-stage pps design are given in section 14.4. In this chapter we will consider first the specific cases of two-stage and three-stage designs with pps either at all stages or only at the first stage, and then the special cases of crop surveys.

16.2 General estimating equations: two- and three-stage pps designs

16.2.1 Two-stage pps design

The structure of a two-stage pps is given in Table 16.1. In the first stage, a sample of n units is selected with pps (and replacement) from N units, the

Table 16.1 Sampling plan for a two-stage pps sample design with replacement

Stage (t)	*Unit*	*No. in universe*	*No. in sample*	*Selection method*	*Selection probability*	f_t
1	First-stage	N	n	ppswr	$\pi_i = z_i/Z$	$n\pi_i$
2	Second-stage	M_i	m_i	ppswr	$\pi_{ij} = w_{ij}/W_i$	$m_i\pi_{ij}$

initial probability of selection of the ith first-stage unit ($i = 1, 2, \ldots, N$) being

$$\pi_i = z_i/Z \tag{16.1}$$

where z_i is the value ('size') of an ancillary variable, $Z = \sum^{N} z_i$ being known. In each of the n first-stage sample units (fsu's) thus selected, a sample of m_i second-stage units (ssu's) is selected with pps (and replacement) from the M_i units, the (initial) probability of selection of the jth ssu ($j = 1, 2, \ldots, M_i$) being

$$\pi_{ij} = w_{ij}/W_i \tag{16.2}$$

where w_{ij} is the value of another ancillary variable and $W_i = \sum_{j=1}^{M_i} w_{ij}$ is known.

The universe totals and means are defined as for a two-stage srs in section 15.2.1.

If y_{ij} is the value of the study variable in the jth selected ssu ($j = 1, 2, \ldots, m_i$) in the ith selected fsu ($i = 1, 2, \ldots, n$), then from the general estimating equation (14.13), or an extension from single-stage pps to two-stage pps in the same manner as was done for srs in section 15.2.3(a), it will be seen that an unbiased estimator of the universe total Y of the study variable, obtained from the ith sample fsu, is

$$y_i^* = n \sum_j \frac{1}{f_1 f_2} y_{ij}$$

$$= \frac{1}{\pi_i m_i} \sum_{j=1}^{m_i} \frac{y_{ij}}{\pi_{ij}}$$

$$= \frac{ZW_i}{z_i m_i} \sum_{j=1}^{m_i} \frac{y_{ij}}{w_{ij}} \qquad (16.3)$$

and the combined unbiased estimator of the universe total Y from all the n sample fsu's is

$$y_0^* = \frac{1}{n} \sum^{n} y_i^* \qquad (16.4)$$

An unbiased estimator of the variance of y_0^* is (from equation 14.15))

$$s_{y_0^*}^2 = \sum^{n} (y_i^* - y_0^*)^2 / n(n-1) \qquad (16.5)$$

Estimators of the means of study variables, and the ratio of two totals follow directly from the fundamental theorems of section 14.4.

Notes

1. *Sampling variance of the estimator* y_0^*. The sampling variance of y_0^* in ppswr at both stages is

$$\sigma_{y_0^*}^2 = \frac{1}{n}\left(\sum_{i=1}^{N} \frac{Y_i^2}{\pi_i} - Y^2\right) + \frac{1}{n} \sum_{i=1}^{N} \frac{1}{\pi_i m_i} \left(\sum_{j=1}^{M_i} \frac{Y_{ij}^2}{\pi_{ij}} - Y_i^2\right) \qquad (16.6)$$

which is estimated unbiasedly by $s_{y_0^*}^2$, defined by equation (16.5).

2. If the same ancillary 'size' variable is used for pps in both the stages, i.e. if $\pi_i = z_i/Z$, and also $\pi_{ij} = z_{ij}/Z_i$ (note that $z_i = Z_i$ in our notation), then estimating equation (16.3) simplifies to

$$y_i^* = \frac{Z}{m_i} \sum^{m_i} \frac{y_{ij}}{z_{ij}} \qquad (16.7)$$

and estimating equation (16.4) becomes

$$y_0^* = \frac{Z}{n} \sum^{n} \frac{1}{m_i} \sum^{m_i} \frac{y_{ij}}{z_{ij}} \qquad (16.8)$$

3. If, in the above case, a fixed number m_0 of ssu's is selected in each sample fsu, then estimating equation (16.7) becomes

$$y_i^* = \frac{Z}{m_0} \sum^{m_0} \frac{y_{ij}}{z_{ij}} \qquad (16.9)$$

and estimating equation (16.8) becomes

$$y_i^* = \frac{Z}{nm_0} \sum^{n} \sum^{m_0} \frac{y_{ij}}{z_{ij}} \tag{16.10}$$

The ratio Y/Z is then estimated unbiasedly by

$$\frac{y_0^*}{Z} = \frac{1}{nm_0} \sum^{n} \sum^{m_0} \frac{y_{ij}}{z_{ij}} \tag{16.11}$$

i.e. by the simple (unweighted) mean of the ratios y_{ij}/z_{ij}. This result is particularly useful in crop surveys (see section 16.5).

16.2.2 Three-stage pps design

The sampling plan for a three-stage pps sample design is given in Table 16.2. The universe totals and means are defined as for a three-stage srs in section 15.6.1. If

Table 16.2 Sampling plan for a three-stage pps sample design with replacement

Stage (t)	*Unit*	*No. in universe*	*No. in sample*	*Selection method*	*Selection probability*	f_t
1	First-stage	N	n	ppswr	$\pi_i = z_i/Z$	$n\pi_i$
2	Second-stage	M_i	m_i	ppswr	$\pi_{ij} = w_{ij}/W_i$	$m_i\pi_{ij}$
3	Third-stage	Q_{ij}	q_{ij}	ppswr	$\pi_{ijk} = v_{ijk}/V_{ij}$	$q_{ij}\pi_{ijk}$

y_{ijk} is the value of the study variable in the kth selected third-stage unit (tsu) of the jth selected ssu of the ith selected fsu ($i = 1, 2, \ldots, n$; $j = 1, 2, \ldots, m_i$; $k = 1, 2, \ldots, q_{ij}$) in a three-stage pps sample, then from the general estimating equation (14.13) or an extension from the case of the two-stage pps design, it will be seen that an unbiased estimator of the universe total Y of the study variable, obtained from the ith selected fsu, is

$$y_i^* = n \sum_{j,k} \frac{1}{f_1 f_2 f_3} y_{ijk}$$

$$= \frac{1}{\pi_i m_i} \sum_{j=1}^{m_i} \frac{1}{\pi_{ij} q_{ij}} \sum_{k=1}^{q_{ij}} \frac{y_{ijk}}{\pi_{ijk}}$$

$$= \frac{ZW_i}{z_i m_i} \sum^{m_i} \frac{V_{ij}}{w_{ij} q_{ij}} \sum^{q_{ij}} \frac{y_{ijk}}{v_{ijk}} \tag{16.12}$$

where $Z = \sum^{N} z_i$; $W_i = \sum^{M_i} W_i$, and $V_{ij} = \sum^{Q_{ij}} v_{ijk}$

The combined unbiased estimator of the universe total Y from all the n sample fsu's is

$$y_0^* = \frac{1}{n} \sum^{n} y_i^* \tag{16.13}$$

with an unbiased variance estimator

$$s_{y_0^*}^2 = \sum^{n} (y_i^* - y_0^*)^2 / n(n-1) \tag{16.14}$$

Estimators of the means of study variables and the ratio of two totals follow directly from the fundamental theorems of section 14.4.

Note: As for two-stage pps sampling, so also for three-stage sampling, great simplifications result if the same ancillary ('size') variable is used for selection in all three stages, i.e. if

$$\pi_i = z_i/Z \qquad \pi_{ij} = z_{ij}/Z_i \qquad \pi_{ijk} = z_{ijk}/Z_{ij} \tag{16.15}$$

(In our notation, $z_i = Z_i$ and $z_{ij} = Z_{ij}$). This is particularly useful in crop surveys, as will be seen in section 16.5.

16.3 Two-stage design with pps and srs

16.3.1 General case

As mentioned in section 16.1, from operational and technical considerations in a two-stage design, the first-stage units may be selected with pps, and the second-stage with srs (or systematically). The sampling plan for such a design is given in Table 16.3.

Table 16.3 Sampling plan for a two-stage sample design with pps sampling at the first-stage and srs at the second-stage

Stage (t)	*Unit*	*No. in universe*	*No. in sample*	*Selection method*	*Selection probability*	f_t
1	First-stage	N	n	ppswr	$\pi_i = z_i/Z$	$n\pi_i$
2	Second-stage	M_i	m_i	srswr	Equal = $1/M_i$	m_i/M_i

An unbiased estimator of the universe total Y of a study variable, obtained from the ith selected fsu, is (from the general estimating equation 14.13 or from the combination of sections 16.2.2 and 15.2.3)

$$y_i^* = n \sum_j \frac{1}{f_1 f_2} y_{ij}$$

$$= \frac{M_i}{\pi_i m_i} \sum_{j=1}^{m_i} y_{ij} = \frac{M_i}{\pi_i} \bar{y}_i \tag{16.16}$$

where

$$\bar{y}_i = \sum_{j=1}^{m_i} y_{ij}/m_i \tag{16.17}$$

is the unbiased estimator of the mean $\bar{Y}_i = Y_i/M_i$ in the ith sample fsu (as sampling is simple random), and $\pi_i = z_i/Z$ is the (initial) probability of selection of the ith fsu.

The combined unbiased estimator of Y from all the n sample fsu's is the arithmetic mean

$$y_0^* = \frac{1}{n} \sum^{n} y_i^* \tag{16.18}$$

with an unbiased estimator of the variance of y_0^*

$$s_{y_0^*}^2 = \sum^{n} (y_i^* - y_0^*)^2/n(n-1) \tag{16.19}$$

Estimators of the ratio of the totals of two study variables, their means etc. as also their variances follow from the fundamental theorems in section 14.4, and will be illustrated by an example.

Example 16.1

In each of the four sample villages selected out of 30 villages with probability proportional to the previous census population in Example 5.2, select a simple random sample of 4 households each from the total number of households in the villages, and on the basis of the collected data on household size, the total monthly income and the number of adults (over 18 years of age) of these sample households (given in Table 16.4), estimate for the 30 villages the total number of persons, the total and per capita monthly income, and the proportion of adults, along with their standard errors.

The required computations are shown in Table 16.5 on page 185, denoting by the y_{ij} the household size, by x_{ij} the monthly income, and by y'_{ij} the number of adults in the jth sample household of the ith sample village ($i = 1, 2, 3, 4$ for sample villages numbered 5, 6, 11, and 18 respectively; $j = 1, 2, 3, 4$). The means in the ith sample villages are denoted by $\bar{y}_i$, $\bar{x}_i$, and $\bar{y}'_i$ respectively, defined by equations of the type (16.17).

$N = 30$ is the total number of villages, and M_i and m_i (= 4) denote respectively the total and the sample numbers of households in the ith sample village.

From equation (16.18), the combined unbiased estimate of the total number of persons in the 30 villages is $y_0^* = \sum^n y_i^*/n = 3\,349$ (shown at the bottom of column 7 of Table 16.5). An unbiased variance estimate of y_0^* is (from equation 16.19)

$$s^2_{y_0^*} = (45\,056\,019.02 - 44\,870\,538.61)/12$$
$$= 15\,456.70$$

Similarly, the combined unbiased estimate of the total monthly income is $x_0^* = \sum^n x_i^*/n = \$33\,548$ (shown at the bottom of column 11 of Table 16.5), an unbiased variance estimate of which is 596 340.5; and the combined unbiased estimate of the total number of adults is

$$y_0'^* = \sum^n y_i'^*/n = 1930$$

(shown at the bottom of column 15 of Table 16.5) with an unbiased variance estimate 11 916.32.

Table 16.4 Size, total monthly income, number of adults (18 years or over) and knowledge of development plans by the selected adult in the srs of the sample of households in each of the four sample villages of Example 5.2 selected with probability proportional to previous census population

Household serial no.	*Household size* y_{ij}	*Total monthly income* x_{ij}	*Number of adults* y'_{ij}	*Knowledge of development plans by the selected adult* (q'_{ij} = 1 for Yes; = 0 for No)
Village serial No. 5				
10	4	\$37	3	0
14	5	53	3	0
20	4	40	2	1
23	5	48	3	0
Village serial No. 6				
4	6	52	3	0
5	5	42	2	1
15	3	35	2	0
17	6	60	4	0
Village serial No. 11				
2	4	45	2	1
4	6	69	3	0
10	5	46	3	0
13	6	57	3	0
Village serial No. 18				
3	5	60	3	0
12	5	43	3	1
18	5	52	3	0
20	6	65	4	0

Table 16.5 Computations of the estimated total number of persons, total and per capita monthly income, number and proportion of adults for a two-stage design: data of Table 16.4

Sample village i	Reciprocal of probability $1/\pi_i$	Number of households M_i	M_i/π_i *	For the number of persons $\sum_{j=1}^{m_i} y_{ij}$	$\bar{y}_i = \sum_{j=1}^{m_i} y_{ij}/m_i$	$y_i^* = (M_i/\pi_i)\bar{y}_i$	y_i^{*2}
(1)	(2)	(3)	(4)	(5)	(6)	(7)	(8)
1	30.5978	24	734.35	18	4.50	3304.58	10 920 248.98
2	43.3077	17	736.23	20	5.00	3681.15	13 550 865.32
3	39.0972	15	586.46	21	5.25	3078.92	9 479 748.37
4	27.5980	23	634.75	21	5.25	3332.44	11 105 156.35
					Total	13 397.09	45 056 019.02
					Mean	3349.27 (y_0^*)	—

Sample village i	For the total monthly income $\sum_{j=1}^{m_i} x_{ij}$	$\bar{x}_i = \sum_{j=1}^{m_i} x_{ij}/m_i$	$x_i^* = (M_i/\pi_i)\bar{x}_i$	x_i^{*2}
(1)	(9)	(10)	(11)	(12)
1	$178	$44.50	$32 678.58	1 067 890 244
2	189	47.25	34 786.87	1 210 127 368
3	217	54.25	31 815.46	1 012 224 768
4	220	55.00	34 911.25	1 218 793 631
		Total	$134 192.16	4 509 036 011
		Mean	$33 548.04 ($x_0^*$)	

Sample village i	For the number of adults $\sum_{j=1}^{m_i} y'_{ij}$	$\bar{y}'_i = \sum_{j=1}^{m_i} y'_{ij}/m_i$	$y_i'^* = (M_i/\pi_i)\bar{y}'_i$	$y_i'^{*2}$
(1)	(13)	(14)	(15)	(16)
1	11	2.75	2019.19	4 077 128.26
2	11	2.75	2024.63	4 099 126.64
3	11	2.75	1612.76	2 600 994.82
4	13	3.25	2062.94	4 255 721.44
		Total	7719.32	15 032 971.16
		Mean	1929.88 ($y_0'^*$)	

Sample village i	For the sum of products $y_i^* x_i^*$	$y_i^* y_i'^*$
(1)	(17)	(18)
1	108 315.768	6 672 574.89
2	128 055.687	7 452 966.72
3	97 957.256	4 965 559.02
4	116 339.646	6 874 623.77
Total	450 668.357	25 965 724.40

* Also the estimated number of households (h_i^*) in Table 5.7.

The estimated monthly income per capita is, from equation (14.6), the estimated total monthly income divided by the estimated total number of persons, or

$$r = x_0^*/y_0^* = \$10.02$$

As

$$s_{x_0^* y_0^*} = (450\ 668\ 357 - 449\ 446\ 581)/12$$
$$= 101\ 814.67$$

the estimated variance of r is (from equation (14.7))

$$s_r^2 = (s_{x_0^*}^2 + r^2 s_{y_0^*}^2 - 2r s_{y_0^* x_0^*})/y_0^{*2} = 0.00961286$$

Similarly, the estimated proportion of adults is the estimated total number of adults divided by the estimated total number of persons, or

$$p = y_0'^*/y_0^* = 0.5762$$

and the estimated variance of p is

$$s_p^2 = (s_{y_0'^*}^2 + p^2 s_{y_0^*}^2 - 2p s_{y_0^* y_0'^*})/y_0^{*2}$$
$$= 0.0005643645$$

as

$$s_{y_0^* y_0'^*} = (25\ 965\ 724.40 - 25\ 854\ 125.48)/12$$
$$= 9299.91$$

16.3.2 First-stage units selected with probability proportional to the number of second-stage units

In a two-stage sample design, with pps sampling at the first-stage and srs at the second, the estimating procedure becomes simpler if for the pps sampling at the first stage, the 'size' is the number of second-stage units, i.e. if the fsu's are selected with probability proportional to the number of ssu's contained in each fsu: this assumes, of course, that the information is for all the fsu's in the universe.

In this case, the (initial) probability of selection of the ith fsu ($i = 1, 2, \ldots, N$) is

$$\pi_i = M_i/M$$

where $M = \sum^N M_i$ is the total number of second-stage units in the universe.

Estimating equation (16.16) then becomes

$$y_i^* = M(M_i/M_i)\,\bar{y}_i = M\bar{y}_i \qquad (16.20)$$

and the combined unbiased estimator given by equation (16.18) becomes

$$y_0^* = \frac{1}{n}\sum^n y_i^*/n$$
$$= \frac{M}{n}\sum^n \bar{y}_i = M\bar{y} \qquad (16.21)$$

where

$$\bar{y} = \sum^n \bar{y}_i/n \qquad (16.22)$$

is the simple (unweighted) mean of the $\bar{y}_i$ values.

An unbiased estimator of the mean $\bar{\bar{Y}} = Y/M$ is

$$M\bar{y}/M = \bar{y} = \sum^n \bar{y}_i/n \qquad (16.23)$$

which is the simple (unweighted) mean of the $\bar{y}_i$ values, and an unbiased estimator of the variance of $\bar{y}$ is

$$s_{\bar{y}}^2 = \sum^n (\bar{y}_i - \bar{y})^2/n(n-1) \qquad (16.24)$$

Example 16.2

In an area there are 315 schools with a total of 27 215 students. Eight schools were first selected with probability proportional to the number of students, and in each selected school, 50 students were selected at random. Table 16.6 gives for each school the number of students with trachoma and the number with multiple scars. Estimate (a) the proportion of students with trachoma, and (b) among those with trachoma, the proportion with multiple scars (Weber, Example 4.5).

Here $N = 315$ schools, and $M = 27\ 215$ students. As the sample of 8 schools is selected with probability proportional to the number of students (M_i) and in each selected school, a sample of students is taken at random, the method of section 16.3.2 will apply here. The estimating equations are further simplified as a fixed number $m_0 = 50$ of students are selected in the sample schools.

An unbiased estimate of the variance of $\bar{y}$ is (from equation (16.24))

$$\begin{aligned} s_{\bar{y}}^2 &= \sum^{n} (y_i - \bar{y})^2/n(n-1) \\ &= (\mathrm{SS}y_i/m_0^2)/n(n-1) \\ &= (13\ 421 - 13.203.125)/(50^2 \times 8 \times 7) \\ &= 0.00155625, \end{aligned}$$

so that the estimated standard error of $\bar{y}$ is $s_{\bar{y}} = 0.03945$ with CV 4.86 per cent.

(b) Similarly, defining $y'_{ij} = 1$, if the ijth sample student has trachoma with multiple scars, and $y'_{ij} = 0$, otherwise, an unbiased estimate of the total number of students with multiple scars is (from equation (16.23))

$$\begin{aligned} \bar{y}'_i/n &= \sum^{n} y'_i/nm_0 \\ &= 55/(8 \times 50) = 0.1375. \end{aligned}$$

Table 16.6 Number of students with trachoma and number with multiple scars among samples of 50 students each selected at random from each of the 8 schools selected with probability proportional to the total number of students

School no. i	*Number of students with trachoma*	
	Total number y_i	*Number with multiple scars* y'_i
1	40	3
2	31	0
3	47	16
4	41	8
5	43	8
6	36	5
7	39	2
8	48	13
Total	325	55
Raw sum of squares	13 421	591
Raw sum of products	2426	

(a) Defining $y_{ij} = 1$, if the jth sample student in the ith sample school has trachoma; and $y_{ij} = 0$, otherwise ($i = 1, 2, \ldots, 8; j = 1, 2, \ldots, 50$), an unbiased estimate of the proportion of students with trachoma is (from equation (16.23))

$$\begin{aligned} \bar{y} &= \sum^{n} \bar{y}_i/n = \sum^{n} y_i/nm_0 \\ &= 325/(8 \times 50) = 0.8125 \text{ or } 81.25 \text{ per cent.} \end{aligned}$$

where $y_i = \sum_{j=1}^{m_0} y_{ij}$.

where $y'_i = \sum_{j=1}^{m_0} y'_{ij}$.

An unbiased estimate of the variance of $\bar{y}'$ is (from equation (16.24))

$$\begin{aligned} s_{\bar{y}'}^2 &= \sum^{n} (y'_i - \bar{y}')^2/n(n-1) \\ &= (\mathrm{SS}y'_i/m_0^2)/n(n-1) \\ &= (591 - 378.125)/(50^2 \times 8 \times 7) \\ &= 0.00152054. \end{aligned}$$

The estimated proportion of students with multiple scars among those with trachoma is $r' = \bar{y}'/\bar{y} = 55/325 = 0.1692$ or 16.92 per cent.

An estimate of the variance of r' is, from the general estimating equation (14.7),

$$s_{r'}^2 = (s_{\bar{y}'}^2 + r'^2 s_{\bar{y}}^2 - 2r' s_{\bar{y}\bar{y}'})/\bar{y}^2$$

Since

$$s_{\bar{y}\bar{y}'} = \sum^{n} (y_i - \bar{y})(y_i' - \bar{y}')/n(n-1)$$
$$= (\mathrm{SP}y_i y_i'/m_0^2)/n(n-1)$$
$$= (2426 - 2232.125)/(50^2 \times 8 \times 7)$$
$$= 193.875/(50^2 \times 8 \times 7)$$

therefore

$$s_{r'}^2 = \frac{212.875 + (0.1692)^2 \times 217.875 - 2 \times 0.1692 \times 193.875)/(50^2 \times 8 \times 7)}{[325/(8 \times 50)]^2}$$
$$= \frac{(212.875 + 6.237543 - 65.607300)}{325^2 \times 7/8}$$
$$= 0.0016609$$

so that the estimated standard error of r' is 0.04075, with CV of 24.08 per cent.

16.4 Three-stage design with pps, srs, and srs

In this section will be extended the method of section 16.3.1 to a three-stage sample design with pps sampling at the first stage and srs at the second and the third stages. The sampling plan is shown in Table 16.7.

Table 16.7 Sampling plan for a three-stage sample design with pps sampling at the first-stage and srs at the second- and third-stages

Stage (*t*)	*Unit*	*No. in universe*	*No. in sample*	*Selection method*	*Selection probability*	f_t
1	First-stage	N	n	ppswr	$\pi_i = z_i/Z$	$n\pi_i$
2	Second-stage	M_i	m_i	srs	Equal = $1/M_i$	m_i/M_i
3	Third-stage	Q_{ij}	q_{ij}	srs	Equal = $1/Q_{ij}$	q_{ij}/Q_{ij}

An unbiased estimator of the universe total Y of a study variable, obtained from the ith selected fsu, is (from the general estimating equation 16.12)

$$y_i^* = n \sum_{j=1}^{m_i} \sum_{k=1}^{q_{ij}} \frac{1}{f_1 f_2 f_3} y_{ijk}$$

$$= \frac{M_i}{\pi_i m_i} \sum_{j=1}^{m_i} \sum_{k=1}^{q_{ij}} y_{ijk} \qquad (16.25)$$

where y_{ijk} is the value of the study variable in the kth sample tsu in the jth sample ssu in the ith sample fsu ($i = 1, 2, \ldots, n$; $j = 1, 2, \ldots, m_i$; $k = 1, 2, \ldots, q_{ij}$) and π_i is the (initial) probability of selection of the ith first-stage unit in the universe ($i = 1, 2, \ldots, N$).

The combined unbiased estimator of Y from all the n sample fsu's is the arithmetic mean

$$y_0^* = \frac{1}{n} \sum^{n} y_i^* \qquad (16.26)$$

with an unbiased variance estimator

$$s_{y_0^*}^2 = \sum^{n} (y_i^* - y_0^*)^2/n(n-1) \qquad (16.27)$$

Estimators of the ratio of the totals of the values of two study variables, their means etc. as also their variances follow from the fundamental theorems in section 14.4 and will be illustrated with an example.

Example 16.3

In Example 16.1 are given the data on the number of adults in the srs of 4 households in each of the 4 sample villages, selected with probability proportional to the previous census population. In each of these sample households, one adult member was further selected at random from the total number of adults in the household, and asked about his knowledge of development plans in the country. This information is given also in Table 16.4. Estimate the total number of adults with knowledge of development plans and the proportion they constitute of the total number of adults.

Extending the notation of Example 16.1, let Q_{ij} ($= y'_{ij}$ in Example 16.1) denote the total number of adults, and q_{ij} ($= 1$) the number selected for interview in the ijth sample household ($i = 1, 2, 3, 4$; $j = 1, 2, 3, 4$). As only one sample third-stage unit is selected, we can dispense with the subscript k in y_{ijk} by which we had denoted the value of the study variable for the ijkth third-stage sample unit.

We define $q'_{ij} = 1$, if the selected adult in the ijth sample household knows of development plans, and 0 otherwise.

The computations are shown in Table 16.8.

From equation (16.26), the unbiased estimate of the total number of adults with knowledge of development plans is

$$q_0'^* = \sum^{n} q_i'^*/n = 376$$

(shown at the bottom of column 5 of Table 16.9) where

$$q_i'^* = \frac{M_i}{\pi_i m_i} \sum^{m_i} Q_{ij} q'_{ij} \qquad \text{from equation (16.25)}$$

An unbiased variance estimate of $q_0'^*$ is (from equation (16.27))

$$s^2_{q_0'^*} = (582\,950.43 - 565\,947.77)/12 = 1416.8883$$

so that the estimated standard error of $q_0'^*$ is 37.64, and the estimated CV is 10.00 per cent.

The estimated proportion of adults with knowledge of development plans equal the estimated total number of adults with such knowledge divided by the estimated total number of adults, or

$$p' = q_0'^*/q_0^* = 376.15/1929.88 = 0.1949$$

since

$$s_{q_0^* q_0'^*} = (2\,914\,705.81 - 2\,903\,602.92)/12 = 11\,102.89$$

the estimated variance of p' is

$$s^2_{p'} = (s^2_{q_0'^*} + p'^2 s^2_{q_0^*} - 2p' s_{q_0^* q_0'^*})/q_0^{*2} = 0.0004051344,$$

so that the estimated standard error of p' is 0.02013, with estimated CV of 10.33 per cent.

Table 16.8 Computation of the estimated number and proportion of adults with knowledge of development plans: data of Tables 16.4 and 16.5

Sample village i	$\sum_{j=1}^{m_i} Q_{ij} q'_{ij}$	$\sum_{j=1}^{m_i} Q_{ij} q'_{ij}/m_i$	$q_i'^* = (M_i/\pi_i) \times$ col.(3)	$q_i'^{*2}$	$q_i^* q_i'^*$ ★
(1)	(2)	(3)	(4)	(5)	(6)
1	2	0.50	367.18	134 821.15	714 406.18
2	2	0.50	368.12	135 512.33	745 306.80
3	2	0.50	293.23	85 983.83	472 909.61
4	3	0.75	476.06	226 633.12	982 083.22
Total	—	—	1 505.59	582 950.43	2 914 705.81
Mean	—	—	376.15 ($q_0'^*$)	—	—

★ q_i^* ($= y_i'^*$ in Table 16.5) denotes the estimated total number of adults.

16.5 Special cases of crop surveys

16.5.1 Introduction

As observed in the notes to sections 16.2.2 and 16.2.3, great simplifications result if the same size variable is used in all the stages of a multi-stage pps sample. This is particularly useful in crop surveys when ancillary information is available on each of the universe units.

16.5.2 Area surveys of crops

When the areas of all the fields in the universe are known, let the sample design for estimating the total area under a crop be pps two-stage with villages as the fsu's, selected with probability proportional to total (geographical) area, and fields as the ssu's, selected with probability proportional to their areas: the fields may be selected from a list of fields showing the areas or from a cadastral map of the fields, following the procedure of section 5.4. We further assume that the same number m_0 of fields is selected in each sample village.

If a_{ij} is the total area and y_{ij} is the area under the crop in the ijth sample field ($i = 1, 2, \ldots, n$; $j = 1, 2, \ldots, m_0$), then an unbiased estimator of the total area under the crop, from the ith sample village, is (from equation (16.9))

$$y_i^* = \frac{A}{m_0} \sum^{m_0} p_{ij} = A\bar{p}_i \qquad (16.28)$$

and from all the n sample villages is (from equation (16.10))

$$y_0^* = \sum^{n} y_i^*/n = A \sum^{n} \bar{p}_i/n = A\bar{p} \qquad (16.29)$$

where $p_{ij} = y_{ij}/a_{ij}$ is the proportion of the area under the crop in the ijth sample field and A the total area in the universe,

$$\bar{p}_i = \sum^{m_0} p_{ij}/m_0 \qquad (16.30)$$

is an unbiased estimator of the proportion of the area under the crop in the ith sample village; and

$$\bar{p} = \sum^{n} \bar{p}_i/n = \sum^{n} \sum^{m_0} p_{ij}/nm_0 \qquad (16.31)$$

is the simple (unweighted) average of the sample p_{ij} values and an unbiased estimator of the proportion of the area under the crop in the universe.

An unbiased variance estimator of $\bar{p}$ is

$$s_{\bar{p}}^2 = \sum^{n} (\bar{p}_i - \bar{p})^2/n(n-1) \qquad (16.32)$$

and an unbiased variance estimator of $y_0^* = A\bar{p}$ is

$$s_{y_0^*}^2 = A^2 s_{\bar{p}}^2 \qquad (16.33)$$

16.5.3 Yield surveys of crops

Similar considerations apply for estimating the yield of a crop. If $r_{ij} = x_{ij}/a_{ij}$ is the yield per unit area in the ijth sample field, then an unbiased estimator of the total yield in the universe is

$$x_0^* = A\bar{r} \qquad (16.34)$$

where

$$\bar{r} = \sum^{n} \sum^{m_0} r_{ij}/nm_0 \qquad (16.35)$$

is the simple average of the sample r_{ij} values and an unbiased estimator of the average yield per unit area in the universe. Unbiased variance estimators of x_0^*

and $\bar{r}$ are given respectively by estimating equations of the types (16.33) and (16.32).

A consistent but generally biased estimator of the yield rate per unit of crop area is the ratio

$$x_0^*/y_0^* = \sum^{n}\sum^{m_0} r_{ij} \Big/ \sum^{n}\sum^{m_0} p_{ij} \qquad (16.36)$$

a variance estimator of which is given by the general estimating equation (14.7).

Notes

1. If one crop cut of prescribed size and shape is located in the selected fields at random and the yield rate obtained from it, then although the design becomes three-stage, the same estimating procedure as above will apply.
2. In the far less common situation when the crop areas of all the fields in the universe are known, the villages and fields may be selected with probability proportional to their respective crop areas, and the total crop yield in the universe will be estimated by

   $$x_0^* = Y\bar{r}' \qquad (16.37)$$

   where

   $$\bar{r}' = \sum^{n}\sum^{m_0} r'_{ij}/nm_0 \qquad (16.38)$$

   is an unbiased estimator of the average yield per unit of crop area; $r'_{ij} = x_{ij}/y_{ij}$ is the yield per unit of crop area in the *ij*th sample field; and Y is the total crop area in the universe. Unbiased variance estimators of x_0^* and $\bar{r}'$ are given respectively by equations of the types (16.33) and (16.32).
3. As the area of a crop is likely to have a high positive correlation with the total area, sampling with probability proportional to total area (ppa) is likely to be more efficient than srs. However, yield rates may not be highly correlated with the area of the field, and in a crop survey with ppa sampling, estimators of crop areas will generally be more efficient than an srs, but not necessarily so for estimators of crop yields. In that case, for estimating the yield rate, a biased estimator of the unweighted mean of means may be used, after examining from a pilot survey that the bias is not appreciably high; some Indian surveys have shown that it was safe to use such biased estimators there.

Further reading

Cochran, sections 11.8–11.11, and 11.19; Hansen *et al.*, Vol. I, sections 8.6, and 8.14, Vol. II, section 8.1; Murthy, sections 9.4, 9.5, 9.8c, 9.9, and 9.10; Raj, sections 6.4–6.6, and 6.8; Sukhatme and Sukhatme, sections 8.1–8.3, 8.6, 8.9–8.10, and 8.12–8.13; Yates, sections 6.19, 7.17, and 8.11.

Exercises

1. Of 53 communes in an area, 14 were selected with probability proportional to the area of the communes (and with replacement), and in each selected commune, farms were randomly selected using a sampling fraction of $\frac{1}{4}$. For each selected commune, the probability of its selection, the total and the sample number of farms, and the number of cattle in the sample farms are given in Table 16.9 on page 192. Estimate the total number of cattle in the area with its standard error (United Nations *Manual*, Process 24).

2. Of the 53 communes in an area, 14 were selected with probability proportional to the number of farms in the communes (and with replacement), and in each selected commune, farms were randomly selected using a sampling fraction of $\frac{1}{4}$. For each selected commune the total number of farms (M_i), the number of sample farms (m_i), and the number of cattle in the sample farms are given in Table 16.10. on page 192. Estimate the total number of cattle in the area with its standard error; the total number of farms in the area is $M = \sum^{N} M_i = 2072$ (United Nations *Manual*, Process 23).

3. For estimating the total number of cultivators (Y), a sample n villages is selected from the N villages in the universe with ppswr, the 'size' being the current number of households (M_i), and from each sample village m_0 households are selected circular systematically. The number of cultivators in each of the nm_0 sample households is determined. Suggest an unbiased estimator of Y and obtain an unbiased variance estimator of it (Murthy, Problem 9.8). (*Hint:* See section 16.3.2).

Table 16.9 Number of cattle in sample farms selected at random from sample communes, selected with probability proportional to area

Sample commune i	Probability of selection π_i	Total number of farms M_i	No. of farm samples m_i	Number of cattle in sample farms $\sum_{j=1}^{m_i} y_{ij}$
1	0.0026	19	5	14
2	0.0098	23	5	82
3	0.0146	31	8	207
4	0.0167	40	10	124
5	0.0187	54	13	113
6	0.0187	54	13	113
7	0.0220	39	10	114
8	0.0249	55	14	242
9	0.0258	46	12	203
10	0.0298	83	20	256
11	0.0362	74	19	272
12	0.0370	70	17	131
13	0.0465	60	15	208
14	0.0465	60	15	208

Table 16.10 Number of cattle in sample farms selected at random from sample communes, selected with probability proportional to the number of farms

Sample commune i	Total no. of farms M_i	No. of farm samples m_i	Number of cattle in sample farms $\sum_{j=1}^{m_i} y_{ij}$
1	13	3	30
2	15	3	58
3	19	5	14
4	28	7	73
5	39	10	162
6	41	11	88
7	46	12	102
8	46	12	102
9	48	12	203
10	51	13	134
11	59	14	195
12	74	19	272
13	83	20	242
14	83	20	242

17 Size of Sample and Allocation to Different Stages

17.1 Introduction

In a multi-stage design, in addition to the determination of the total sample size there arises the problem of its allocation to the different stages: this requires a knowledge of the variability of the data and the unit costs at the different stages. These will be considered in this chapter, especially relating to a two-stage design; those for a three-stage design will also be indicated.

For a two-stage design, the problem and its treatment are similar to those for single-stage cluster sampling dealt with in Chapter 6 and section 7.6. As noted in section 6.1, single-stage cluster sampling may be considered to be a special case of two-stage sampling with all the second-stage units being surveyed in the selected first-stage units.

17.2 Two-stage design

17.2.1 Variance function

For simplicity, we shall consider a two-stage design with n ($\geqslant 2$) of fsu's being selected (with replacement) out of the N total fsu's with equal or varying probabilities, and a fixed number m_0 ($\geqslant 2$) of ssu's being selected (with replacement) out of the M_i total ssu's in the ith selected fsu with equal or varying probabilities.

We have seen in sections 15.2.3 (for a two-stage srs) and 16.2.1 (for a two-stage pps design) that if y_i^* is an unbiased estimator from the ith sample fsu ($i = 1, 2, \ldots, n$) of the universe total Y, then the combined unbiased estimator of Y from all the n sample fsu's is

$$y_0^* = \frac{1}{n}\sum^{n} y_i^* \qquad (17.1)$$

with an unbiased variance estimator

$$s_{y_0^*}^2 = \sum^{n} (y_i^* - y_0^*)^2/n(n-1) \qquad (17.2)$$

It is also seen from equations (15.22) and (16.6) that the sampling variance of y_0^* in both srs and pps sampling (with replacement at both stages) is of the form

$$V = \frac{V_1}{n} + \frac{V_2}{nm_0} \qquad (17.3)$$

The value of the study variable of the second-stage unit is considered to be the sum of two independent parts. One term, associated with the fsu's, has the same value for all the ssu's in an fsu, and varies from one fsu to another with variance

$$V_1 = N^2 \sigma_b^2$$

$$= N^2 \frac{\sum^N (Y_i - \bar{Y})^2}{N} \quad \text{in srs} \qquad (17.4)$$

$$= \sum^N \frac{Y_i^2}{\pi_i} - Y^2 \quad \text{in pps} \qquad (17.5)$$

where

$$Y_i = \sum_{j=1}^{M_i} Y_{ij} \qquad (17.6)$$

is the universe total for the ith fsu,

$$\bar{Y} = Y/N = \sum^N Y_i/N \qquad (17.7)$$

is the universe mean per fsu and σ_b^2 the variance between the fsu's.

The second term, which serves to measure the differences between the ssu's, varies independently from one unit to another with variance

$$V_2 = \sum^N M_i^2 \sigma_{wi}^2$$

$$= N \sum^N M_i^2 \frac{\sum_{j=1}^{M_i} (Y_{ij} - \bar{Y}_i)^2}{M_i} \quad \text{in srs} \qquad (17.8)$$

$$= \sum^N \frac{1}{\pi_i} \left(\sum_{j=1}^{M_i} \frac{Y_{ij}^2}{\pi_{ij}} - Y_i^2 \right) \quad \text{in pps} \qquad (17.9)$$

where σ_{wi}^2 is the variance between the ssu's within the ith fsu, and $\bar{Y}_i$ the mean for the ith fsu.

Thus, the sample as a whole consists on n independent values of the first term, and nm_0 independent values of the second term.

Note: The variance estimator of y_0^* in estimating equation (17.2) does not explicitly take account of the stage-variances; the estimation of the latter is shown in section 17.2.4.

17.2.2 Cost function

We consider a simple cost function

$$C = nc_1 + nm_0 c_2 \qquad (17.10)$$

where c_1 is the average cost per fsu and c_2 that per ssu. The overhead cost, not shown in the above cost function, may later be added to C. In general, the cost per fsu (c_1) will be greater than that per ssu (c_2).

17.2.3 Optimum size and allocation

With the above cost and variance functions, the optimum values of n and m_0 are those that minimize the variance V for a given total cost C or *vice versa*, and are given by

$$m_0 = \frac{\sqrt{(V_2 c_1)}}{\sqrt{(V_1 c_2)}} \qquad (17.11)$$

and

$$n = \frac{C\sqrt{(V_1/c_1)}}{\sqrt{(V_1c_1)}+\sqrt{(V_2c_2)}} \tag{17.12}$$

if the cost is preassigned (from equation (17.10)); or

$$n = \frac{\sqrt{(V_1c_1)}+\sqrt{(V_2c_2)}}{V\sqrt{(c_1/V_1)}} \tag{17.13}$$

if the variance is preassigned (from equation (17.3)).

The optimum number of ssu's can also be expressed in terms of the intra-class correlation coefficient ρ between the second-stage units

$$m_0 = \sqrt{\left[\frac{c_1}{c_2}\,\frac{1-\rho}{\rho}\right]} \tag{17.14}$$

Note that this is of the same form as that of (7.37) for the optimal cluster size.

Equations (17.11) and (17.14) show that a larger sub-sample should be taken when sub-sampling is relatively inexpensive (c_1 large in relation to c_2) and the variability between the ssu's within the fsu's (V_2) is larger than that between the fsu's (V_1) i.e. the fsu's are internally heterogeneous.

This equation often leads to the same value of m_0 for a wide range of ratios c_1/c_2. Thus a choice of the sub-sampling fraction can often be made even when information on relative costs is not too definite.

17.2.4 Estimation of stage-variances

We have seen from the fundamental theorems of section 14.4 that if sampling is with replacement at least at the first stage, then the sampling variance is estimated properly from the estimators obtained from the first-stage units by estimating equation (14.2) the variability of the data at the different stages does not enter explicitly in this equation. In estimating the stage-variances, to avoid the consistent appearance of negative variances (as it would happen with the 'with replacement' sampling when the variability between the fsu's is smaller than that between the ssu's), we assume that sampling is without replacement at both stages; for simplicity, we also assume that sampling is simple random, m_i ssu's being selected in the ith selected fsu's. The unbiased estimator y_0^* (estimating equation (17.1)) of the universe total Y remains the same, but its sampling variance is given by

$$V(y_0^*) = N^2\bar{M}^2(1-f_1)\frac{S_b^2}{n} + \frac{N}{n}\sum^{n} M_i^2(1-f_{2i})\frac{S_{wi}^2}{m_i} \tag{17.15}$$

where $\bar{M}$ is the average number of ssu's per fsu; $f_1 = n/N$; $f_{2i} = m_i/M_i$; S_b^2 is the variance between the fsu's and S_{wi}^2 that between the ssu's within the ith fsu:

$$S_b^2 = \frac{1}{N-1}\sum^{N}\left(\frac{M_i}{\bar{M}}\bar{Y}_i - \bar{Y}\right)^2 \tag{17.16}$$

$$S_{wi}^2 = \frac{1}{M_i-1}\sum_{j=1}^{M_i}(Y_{ij}-\bar{Y}_i)^2 \tag{17.17}$$

Y_{ij} is the value of the study variable of the ijth universe unit ($i = 1, 2, \ldots, N$; $j = 1, 2, \ldots, M_i$):

$$\bar{Y}_i = \sum_{j=1}^{M_i} Y_{ij}/M_i; \quad \bar{Y} = Y/N\bar{M} = \sum^{N} M_i \bar{Y}_i/N\bar{M}$$

An unbiased estimator of S_{wi}^2 is

$$s_{wi}^2 = \frac{1}{m_i - 1} \sum_{j=1}^{m_i} (y_{ij} - \bar{y}_i)^2 \tag{17.18}$$

where y_{ij} is the value of the study variable of the ijth sample unit ($i = 1, 2, \ldots, n$; $j = 1, 2, \ldots, m_i$) and $\bar{y}_i = \sum_{j=1}^{m_i} y_{ij}/m_i$.

Denoting by

$$s_b^2 = \frac{1}{n-1} \sum^{n} \left(\frac{M_i}{\bar{M}} \bar{y}_i - \bar{\bar{y}} \right)^2 \tag{17.19}$$

where $\bar{\bar{y}} = y_0^*/N\bar{M}$, an unbiased variance estimator of S_b^2 is

$$s_b^2 - \frac{1}{n} \sum^{n} \left(\frac{M_i}{\bar{M}} \right)^2 (1 - f_{2i}) \frac{s_{wi}^2}{m_i} \tag{17.20}$$

An unbiased estimator of $V(y_0^*)$ in (17.15) is

$$N^2 \bar{M}^2 (1 - f_1) \frac{s_b^2}{n} + \frac{N}{n} \sum^{n} M_i^2 (1 - f_{2i}) \frac{s_{wi}^2}{m_i} \tag{17.21}$$

$$= \frac{(1 - f_1)}{n(n-1)} \sum^{n} (y_i^* - y_0^*)^2 + \frac{N}{n} \sum^{N} M_i^2 (1 - f_{2i}) \frac{s_{wi}^2}{m_i} \tag{17.22}$$

where $y_i^* = NM_i\bar{y}_i$ is the unbiased estimator of Y from the ith sample fsu (estimating equation 15.12).

Notes

1. For the estimator of the mean, the stage-variances are obtained by dividing the components in equations (17.15), (17.21), and (17.22) by $N^2\bar{M}^2$.
2. When $M_i = M_0$, a constant, and $m_i = m_0$ is fixed, the unbiased variance estimator (equation (17.21)) for the total becomes

$$N^2 M_0^2 (1 - f_1) \frac{s_b^2}{n} + \frac{N}{n} \frac{M_0^2}{m_0} (1 - f_2) s_w^2 \tag{17.23}$$

where $f_2 = m_0/M_0$; and

$$s_w^2 = \sum^{n} s_{wi}^2 \tag{17.24}$$

For the estimator of the mean, equation (17.23) divided by $N^2M_0^2$ becomes

$$(1 - f_1) \frac{s_b^2}{n} + \frac{1}{N} (1 - f_2) \frac{s_w^2}{nm_0} \tag{17.25}$$

3. When $M_i = M_0$, and $m_i = m_0$, unbiased estimators of the stage-variances can be obtained readily from an analysis of variance of the sample observations; assuming sampling from infinite universes at both stages. In that case, the universe variance of y_0^* is of the form (17.3) and is estimated unbiasedly by

$$N^2 M_0^2 \text{ (mean square between fsu's)}/nm_0 = N^2 M_0^2 s_b^2/n \tag{17.26}$$

where

$$\text{mean square between fsu's} = \frac{1}{n(n-1)} \sum^{n} (\bar{y}_i - \bar{\bar{y}})^2$$

$$\bar{y}_i = \sum_{j=1}^{m_0} y_{ij}/m_0$$

$$\bar{\bar{y}} = \sum^{n} \bar{y}_i/n$$

Unbiased estimators of V_1 and V_2 in equation (17.3) are respectively

$$\hat{V}_1 = N^2 M_0^2 \text{ (mean square between fsu's} - \text{mean square within fsu's})/m_0 \tag{17.27}$$

$$\hat{V}_2 = N^2 M_0^2 \text{ (mean square within fsu's)} \tag{17.28}$$

where

$$\text{mean square within fsu's} = \frac{1}{n(m_0 - 1)} \sum^{n} \sum^{m_0} (y_{ij} - \bar{y}_i)^2$$

For the mean, its sampling variance is also of the form (17.3), and is estimated unbiasedly by

$$\text{(mean square between fsu's)}/nm_0 \tag{17.29}$$

and V_1 and V_2 are estimated unbiasedly by

$$\hat{V}_1 = \text{(mean square between fsu's} - \text{mean square within fsu's})/m_0 \tag{17.30}$$

$$\hat{V}_2 = \text{mean square with fsu's} \tag{17.31}$$

4. When sampling is simple random with replacement at the first stage and simple random without replacement at the second stage, the sampling variance of y_0^*, the (combined) unbiased estimator of Y, is

$$\sigma^2_{y_0^*} = N^2 \bar{M}^2 \frac{S_b^2}{n} + \frac{N}{n} \sum^{N} M_i^2 (1 - f_{2i}) \frac{S_{wi}^2}{m_i} \tag{17.32}$$

S_b^2 and S_{wi}^2 being defined by equations (17.15) and (17.16). $\sigma^2_{y_0^*}$ is estimated unbiasedly by $s^2_{y_0^*}$, defined in equation (17.2).

Example 17.1

Table 17.1 gives the analysis of variance of a sample of $n = 40$ from $N = 1000$ fsu's and $m_0 = 5$ sample ssu's in each selected fsu ($M_0 = 50$), sampling at both stages being simple random.

(a) Assuming that sampling has been from infinite universes at both the stages, estimate unbiasedly the variance of the mean and its components (Hendricks, pp. 190–191).

(b) Given the total cost (excluding overheads) of a survey, $C = £600$; the cost of including an fsu in the sample, $c_1 = £5$; and the cost of getting the data from an ssu within each selected fsu, $c_2 = £1$, obtain the optimum sample numbers of fsu's and ssu's within each selected fsu, and the expected variance of the mean.

(a) From equation (17.29), the unbiased variance estimate of the mean is

(mean square between fsu's)/nm_0 = 634/200 = 3.1700.

Table 17.1 Analysis of variance

Source of variation	*Degrees of freedom*	*Mean square*
Between first-stage units	39	634
Between second-stage units within first-stage units	160	409

From equations (17.30) and (17.31),

$\hat{V}_1$ = (mean square between fsu's – mean square within fsu's)/m_0

= (634 – 409)/5 = 45

$\hat{V}_2$ = mean square within fsu's = 409

so that the components of the variance of the mean are (from equation (17.3))

$$\hat{V}_1/n = 45/40 = 1.1250$$
$$\hat{V}_2/nm_0 = 409/200 = 2.0450$$

totalling 3.1700, as before.

(b) From equation (17.11), the optimum number of ssu's per sample fsu is

$$m_0 = \frac{\sqrt{(V_2c_1)}}{\sqrt{(V_1c_2)}} = \frac{\sqrt{(409 \times 5)}}{\sqrt{(45 \times 1)}} = \sqrt{(45.44)} = 6.7$$

or, to the nearest whole number, 7.

From equation (17.10),

$$n = C/(c_1 + m_0c_2) = 600/(5 + 7 \times 1) = 50$$

so that the total sample size is $nm_0 = 50 \times 7 =$ 350 ssu's.

The estimated (expected) variance of the mean with this sample is

$$\hat{V} = \frac{\hat{V}_1}{n} + \frac{\hat{V}_2}{nm_0}$$
$$= \frac{45}{50} + \frac{409}{350}$$
$$= 0.9 + 1.1686 = 2.0686$$

Example 17.2

For Example 15.2 relating to a two-stage srs, and assuming sampling without replacement, estimate the variance of the estimated total number of persons in the form of expression (17.22). Also obtain an unbiased estimate of S_b^2.

The required computations are shown in summary form in Table 17.2. Taking Expression (17.22), the first term is, from the data of Example 15.2,

$$\frac{(1 - f_1)}{n(n-1)} \sum^{n} (y_i^* - y_0^*)^2 = \frac{(1 - 4/30)}{4 \times 3} \times 625\ 050$$
$$= 45\ 142.67$$

and the second term is

$$\frac{N}{n} \sum^{n} M_i^2 (1 - f_{2i}) \frac{s_{wi}^2}{m_i} = \frac{30}{4} \times 790.67 = 5930.02,$$

Table 17.2 Computation of the stage-variances of the estimated total number of persons in Example 15.2, assuming sampling without replacement

Sample village (fsu) i	SSy_{ij}	$s_{wi}^2 = \frac{SSy_{ij}}{(m_i - 1)}$	$f_{2i} = m_i/M_i$	$(1 - f_{2i}) \times s_{wi}^2/m_i$	M_i^2	$M_i^2(1 - f_{2i}) \times s_{wi}^2/m_i$
(1)	(2)	(3)	(4)	(5)	(6)	(7)
1	14.75	4.916667	0.2000	0.983333	400	393.33
2	5.00	1.666667	0.1739	0.344209	529	182.09
3	3.00	1.000000	0.1600	0.210000	625	131.25
4	4.00	1.333333	0.2222	0.259266	324	84.00
					Total	790.67

Sample village (fsu) i	$M_i/\bar{M}$*	$(M_i/\bar{M})^2$	$(M_i/\bar{M})^2 (1 - f_{2i}) s_{wi}^2/m_i$
(1)	(8)	(9)	(10)
1	0.9302	0.865272	0.850851
2	1.0698	1.144472	0.393938
3	1.1628	1.352104	0.283942
4	0.8372	0.700904	0.181721
		Total	1.710452

* $\bar{M} = \frac{1}{4}(20 + 23 + 25 + 18) = 21.5$ households per village.

so that the estimated variance of the estimated total number of persons (y_0^*) is 45 142.67 + 5930.02 = 51 072.69. The difference between the fsu's (villages) accounts for 45 142.67/51 072.69 = 88 per cent of this variance and the variability within the fsu's between the ssu's (households) for 12 per cent.

The estimated standard error is 226.0, and the estimated CV is 6.94 per cent.

Note that the variance estimated on the basis of a 'without replacement' scheme, is, as known, less than that for a 'with replacement' scheme in Example 15.2, namely,

$$s_{y_0^*}^2 = \sum^n (y_i^* - y_0^*)^2/n(n-1)$$
$$= 625\ 050/12 = 52\ 087.5$$

the estimated standard error was 228.2, and the estimated CV 7.01 per cent.

From the first terms of (17.21) and (17.22), we find

$$s_b^2 = \frac{\sum^n (y_i^* - y_0^*)^2}{N^2\bar{M}^2(n-1)} = \frac{625\ 050}{416\ 025 \times 3} = 0.500811$$

so that an unbiased estimate of s_b^2 is (from equation (17.20))

$$s_b^2 - \frac{1}{n}\sum^n (M_i^2/\bar{M})^2\ (1 - f_{2i})\ s_{wi}^2/m_i$$
$$= 0.500811 - 0.427613 = 0.073198$$

17.4 Three-stage designs

The above procedures can be extended to sample designs with three and higher stages. Those for a three-stage design are given below: the procedures for higher-stage designs follow from symmetry.

Assume that the sample at the three stages is drawn with varying probabilities and with replacement, the respective numbers being n, m_0 and q_0. Then the universe variance of the estimator y_0^* (estimating equation 16.13) is

$$V = \frac{1}{n}\left(\sum_{i=1}^{N} \frac{Y_i^2}{\pi_i} - Y^2\right) + \frac{1}{nm_0}\sum^{N}\frac{1}{\pi_i}\left(\sum_{j=1}^{M_i}\frac{Y_{ij}^2}{\pi_{ij}} - Y_i^2\right)$$
$$+ \frac{1}{nm_0q_0}\sum^{N}\frac{1}{\pi_i}\sum_{j=1}^{M_i}\frac{1}{\pi_{ij}}\left(\sum_{k=1}^{Q_{ij}}\frac{Y_{ijk}^2}{\pi_{ijk}} - Y_{ij}^2\right) \tag{17.33}$$

where $Y_{ij} = \sum^{Q_{ij}} Y_{ijk}$; $Y_i = \sum^{M_i} Y_{ij}$; and $Y = \sum^{N} Y_i$ is the universe total. This is of the form

$$V = \frac{V_1}{n} + \frac{V_2}{nm_0} + \frac{V_3}{nm_0q_0} \tag{17.34}$$

where V_1, V_2 and V_3 are respectively the variations (1) between the fsu's, (2) between the ssu's within the fsu's, and (3) between the third-stage units within the ssu's, for an estimator based on any of the nm_0q_0 possible third-stage sample units.

Assuming the simple cost function (neglecting the over-head costs)

$$C = nc_1 + nm_0c_2 + nm_0q_0c_3 \tag{17.35}$$

where c_1 is the average cost per fsu, c_2 that per ssu, and c_3 that per third-stage unit, the optimum values of m_0 and q_0 are obtained as

$$m_0 = \frac{\sqrt{(V_2c_1)}}{\sqrt{(V_1c_2)}} \qquad q_0 = \frac{\sqrt{(V_3c_2)}}{\sqrt{(V_2c_3)}} \tag{17.36}$$

If the cost is preassigned, the optimum value of n is

$$n = \frac{C\sqrt{(V_1/c_1)}}{\sqrt{(V_1 c_1)} + \sqrt{(V_2 c_2)} + \sqrt{(V_3 c_3)}} \tag{17.37}$$

If the variance is preassigned, the optimum value of n is

$$n = \frac{\sqrt{(V_1 c_1)} + \sqrt{(V_2 c_2)} + \sqrt{(V_3 c_3)}}{V\sqrt{(c_1/V_1)}} \tag{17.38}$$

Further reading

Cochran, sections 10.6, 11.15, and 11.16; Deming (1966), Chapter 5A; Hansen *et al.*, Vols. I and II, Chapter 6; Hendricks, Chapter VII; Kish, section 5.6; Murthy, sections 9.3–9.10; Raj, section 6.9; Sukhatme and Sukhatme, Chapters VII and VIII; Yates, sections 7.17 and 8.11.

Exercises

1. In a sample survey for the study of the economic conditions of the population, it is proposed to take a two-stage sample with villages as the first-stage units and m_0 households per village as second-stage units, sampling at both the stage units to be with equal probability and with replacement. The variance of an estimate is $V = \sigma_b^2/n + \sigma_w^2/nm_0$, when $\sigma_b^2 = 5.2$ and $\sigma_w^2 = 11.2$. The cost of the survey is given by $C = c_0 + nc_1 + nm_0c_2$, with c_0 = Rs 600, c_1 = Rs 16, and c_2 = Rs 4. Determine the optimum values of n and m_0 for C = Rs 10.000 (Chakravarti *et al.*, Illustrative Example 4.2).

2. For Example 15.2, obtain estimates of the stage-variances for the total income, assuming sampling without replacement.

3. To develop the sampling technique for the determination of the sugar percentage in field experiments on sugar beets, ten beets were chosen from each of 100 plots in a uniformity trial, the plots being the first-stage units. The sugar percentage was obtained separately for each beet. Table 17.3 gives the analysis of variance between plots and between beets within plots.

(a) Estimate directly the estimated variance of the mean, and also separately its components.

(b) How accurate are the treatment means with 6 replications and 5 beets per plot? (Snedecor and Cochran, pp. 529–31).

Table 17.3 Analysis of variance of sugar percentage of beets (on a single-beet basis)

Source of variation	*Degrees of freedom*	*Mean square*
Between plots (first-stage units)	80	2.9254
Between beets (second-stage units) within plots	900	2.1374

18 Self-weighting Designs in Multi-stage Sampling

18.1 Introduction

In this chapter we will discuss the problem of how to make a multi-stage design self-weighting. The general case will first be covered, followed by the specific cases of two-stage designs, and sampling for opinion surveys and marketing research along with the situation where it is desirable to select a fixed number of ultimate stage units in each of the selected penultimate stage units.

18.2 General case

Taking the multi-stage design of section 14.4, an unbiased estimator of the universe total Y of a study variable is

$$y = \sum_{\text{sample}} \left(\prod_{1}^{u} \frac{1}{f_t} y_{12\ldots u} \right)$$

$$= \sum_{\text{sample}} w_{12\ldots u}\, y_{12\ldots u} \tag{18.1}$$

(see also equation (14.13)) where $y_{12\ldots u}$ is the value of the study variable in the $(12\ldots u)$th ultimate stage sample unit, the factor

$$\prod_{1}^{u} \frac{1}{f_t} \equiv w_{12\ldots u} \tag{18.2}$$

(see also (14.14)) is the multiplier (or the weighting-factor) for the $(12\ldots u)$th ultimate-stage sample unit; and $f_t \equiv n_{12\ldots(t-1)}\pi_{12\ldots t}$, there being $n_{12\ldots(t-1)}$ sample units each selected with probability $\pi_{12\ldots t}$ out of the total $N_{12\ldots(t-1)}$ units at the tth stage.

The design will be self-weighting with respect to y when the multiplier $w_{12\ldots u}$ defined above is a pre-determined constant, w_0, for all the ultimate-stage sample units. It will be so if for the ultimate stage sampling, the factor f_u could be made proportional to the product of the reciprocals of the corresponding factors for the preceding stages, i.e.

$$f_u \propto \prod_{1}^{u-1} \frac{1}{f_t} \tag{18.3}$$

For example, in a two-stage srs, where $f_1 = n/N$ and $f_{2i} = m_i/M_i$, the multiplier for the sample ssu's is

$$w_{ij} = \frac{1}{f_1 \cdot f_{2i}} = \frac{N}{n} \cdot \frac{M_i}{m_i}$$

and this will be a constant w_0, when

$$m_i = \frac{NM_i}{w_0 n}$$

In this case, the sampling fraction at the second stage should be the same ($= N/w_0 n$) in all the n sample fsu's.

In general, a multi-stage srs will be self-weighting if, in each stage, the sampling fraction is a constant for selecting the next-stage sample units. That is, $n_{12\ldots t}/N_{12\ldots t}$ should be a constant; e.g. the second-stage sampling fraction N_{12}/N_{12} should be a constant, the third-stage sampling fraction n_{123}/N_{123} another constant, and so on.

We shall further consider the requirement of having a fixed number, say n_0, of the ultimate stage sample units in each of the selected penultimate stage units. If these n_0 units are selected with equal probability, then

$$\pi_{12\ldots(u-1)} \propto N_{12\ldots(u-1)} \tag{18.4}$$

i.e. the penultimate stage units should be selected with probability proportional to the total number of ultimate stage units contained in each, and the sample number of penultimate stage units is given by

$$n_{12\ldots(u-2)} = \frac{N_{12\ldots(u-2)}}{w_0 n_0}\left(\prod_{1}^{u-2} \frac{1}{f_t}\right) \tag{18.5}$$

If the n_0 units are selected with probability proportional to 'size', i.e.

$$\pi_{12\ldots u} = \frac{z_{12\ldots u}}{z_{12\ldots(u-1)}} \tag{18.6}$$

where the numerator refers to the 'size' of the particular units to be selected and the denominator to the total size of all the units, and the design will be self-weighting when:

1. The variate value is the ratio of the characteristic under study $y_{12\ldots u}$ to the size in the ultimate stage units $z_{12\ldots u}$; and
2. the penultimate stage units are selected with probability proportional to total 'size', and the number of penultimate stage units is given by

$$\pi_{12\ldots(u-2)} = \frac{1}{w_0 n_0}\left(\prod_{1}^{u-2} \frac{1}{f_t}\right) z_{12\ldots u} \tag{18.7}$$

Notes

1. We have already seen in section 16.5 how two- and three-stage pps designs for crop surveys can be made self-weighting by using the same size variable, such as area, in all stages of sample selection.
2. For fractional values of the sample sizes at different stages, see section 13.4.
3. For making a multi-stage design self-weighting at the tabulating stage, the same principles as for a stratified sample, given in section 13.5, will apply, namely, selecting a sub-sample with probability proportional to the multipliers.

18.3 Two-stage sample design with pps and srs

Let us consider the sampling plan of section 16.3, with n fsu's selected with pps (π_i) out of a total N, and m_i ssu's selected with equal probability out of a total

M_i units in each selected fsu. Here $f_1 = n\pi_i$ and $f_2 = m_i/M_i$. The multiplier for the ijth sample ssu ($i = 1, 2, \ldots, n$; $j = 1, 2, \ldots, m_i$) is

$$w_{ij} = \frac{1}{f_1 f_2} = \frac{M_i}{n\pi_i m_i} \tag{18.8}$$

and the design will be self-weighting with a constant multiplier w_0 when

$$m_i = M_i/(w_0 n \pi_i) \tag{18.9}$$

Suppose the above sample design refers to a household inquiry with villages as the fsu's and households the ssu's. The difficulty is that the ratios M_i/π_i will vary from village to village and so will the sample number of households m_i, given by equation (18.9) required for making the design self-weighting; i.e. the work load per enumerator in each of the sample villages will vary greatly: this is apart from the fact that the total number of households (M_i) will not be known beforehand, unless the inquiry is conducted simultaneously with a census. The problem is to make the work load approximately equal, retaining a self-weighting design.

If a fixed number of m_0 of households is to be sampled in each village, then with a self-weighting design

$$w_0 = M_i/(n\pi_i m_0) \tag{18.10}$$

or

$$\pi_i = M_i/(w_0 n m_0) = M_i/M \tag{18.11}$$

where $M = \sum^{N} M_i$ is the total number of households in the universe, i.e. the villages are to be selected with probability proportional to the number of households. It can be seen that

$$\begin{aligned} w_0 &= M/(nm_0) \\ &= \text{total number of households/number of sample households} \\ &= 1/\text{sampling fraction} \end{aligned} \tag{18.12}$$

In practice, however, the number of households (M) in the universe is not known and the number of households existing at the time of enumeration in the ith sample village (M_i) will be known only after the listing of households when the village is reached and the households listed. In such a situation, the values which are known from the latest census may be used for selecting the sample villages, i.e. $\pi_i = M_i'/M'$, where M_i' is the number of census households in the ith village and $M' = \sum^{N} M_i'$ the total number of census households in the universe. The overall constant multiplier can be fixed on the basis of the expected number of households (M'') in the universe and the total sample size (nm_0); the value of M'' can, for example, be obtained by projecting the number of total households in the censuses. By using these values it is seen that

$$\begin{aligned} w_0 &= \text{estimated total number of households/number of sample households} \\ &= M''/(nm_0) = M'M_i/nM_i'm_i \end{aligned} \tag{18.13}$$

so that

$$m_i = m_0\, M'M_i/M''M_i' = m_0 I_i/I \tag{18.14}$$

where $I = M''/M'$ = estimated increase in the total number of households,
$I_i = M_i/M_i'$ = actual increase in the number of households in the ith village from the latest census.

This procedure is expected to restrict the variation of the number of sample households in the ith village within a short range. The operational steps will be:

1. Fix n and m_0 from cost and error considerations.
2. Find the overall sampling fraction which is the ratio of the number of households to be sampled to the expected total number of households in the universe. The reciprocal of this will be the constant multiplier w_0.
3. Select villages with probability proportional to the number of households as in the previous census.
4. Compute the estimated increase in the number of households in the universe from the census, $I = M''/M'$; and compute $Mm'_0 = m_0/I$.
5. Give to the enumerators the values m'_0 and M'_i (the number of households in the village at the time of the census) with instruction to select at random (or systematically) $m_i = m'_0 M''_i / M'_i$ households out of the M_i households listed in the ith village.

Note: The expected number of sample households = $nm_0 M/M''$. For fractional m_i, see section 13.4.

18.4 Opinion and marketing research

In opinion and market research, to obtain a valid, independent response it is important to interview only one individual in a selected household so that the responses of other household members, which are likely to be conditioned, are not included in the survey. Even if the 'true' responses of all the household members could be obtained, a situation not usually realized in practice, the sample may be more profitably spread over a greater number of households, because of the homogeneity of responses in a household often observed in such surveys. It is easy to see that with the selection of one individual from a household, the sampling design can be made self-weighting if the households are selected with probability proportional to the respective sizes.

To illustrate, consider a three-stage sample design, as in Examples 15.3 and 16.2, for the type of opinion and marketing surveys usually conducted in cities and towns with small first-stage units – city or town-blocks. Suppose that n sample blocks are selected with probability π_i out of N blocks; in the ith sample block ($i = 1, 2, \ldots, n$), m_i sample households are selected with probability π_{ij} ($j = 1, 2, \ldots, m_i$) out of the total M_i households, and finally, one adult is selected with equal probability out of the total Q_{ij} adults in the ijth sample household: sampling is with replacement at the first two stages. Here $f_1 = n\pi_i$; $f_2 = m_i \pi_{ij}$; and $f_3 = 1/Q_{ij}$; so that the multiplier for the selected adult in the ijth sample household is

$$1/f_1 f_2 f_3 = Q_{ij}/(n\pi_i m_i \pi_{ij}) \tag{18.15}$$

If the sample households are selected with probability proportional to the number of adults, i.e. if $\pi_{ij} = Q_{ij}/Q_i$, where $Q_i = \sum^{M_i} Q_{ij}$ is the total number of adults in the ith block, the multiplier becomes $Q_i/n\pi_i m_i$. If this is to be a constant, w_0, then

$$m_i = Q_i/(w_0 n\pi_i) \tag{18.16}$$

In practice, the factors $1/(w_0 n\pi_i)$ will be supplied to the enumerators; on reaching any block, the enumerators will list the households and the adults in these households. The total number of adults (Q_i) in the ith block, when multiplied by the factors $1/(w_0 n\pi_i)$, will determine the sample number of households, which is also the sample number of adults, in the block.

The blocks may be selected with probability proportional to size (the most recent census population, for example), or with equal probability. In the latter case, the factor $1/(w_0 n\pi_i) = N/(w_0 n)$ will be a constant for all the blocks.

Further reading

Hansen, *et al.*, Vol. I, sections 7.12 and 9.11;
Murthy, section 12.3; Som, 1958–59, 1959;
Sukhatme and Sukhatme, section 8.14.

Part IV :

Stratified Multi-stage Sampling

19 Stratified Multi-stage Sampling: Introduction

19.1 Introduction

In this part, we deal with stratified multi-stage sample designs – a synthesis of what has gone earlier. In such a design, the universe is divided into a number of sub-universes (strata) and sampling is conducted in stages separately within each stratum.

The general theorems relating to stratified multi-stage sampling will be considered in this chapter; subsequent chapters will deal with different types and aspects of stratified multi-stage sampling: simple random sample; varying probability sample; size of sample and allocation to different strata and stages; and self-weighting designs.

19.2 Reasons for stratified multi-stage sampling

Stratified multi-stage designs are the commonest of all types of sample designs. This is because they combine, as regards costs and efficiency, the advantages of both stratification and multi-stage sampling.

In stratified sampling, ideally the strata should be formed so as to be internally homogeneous and heterogeneous with respect to one another (with respect to the values of study variables in srs or with respect to the ratios of the values of study to the size-variable in pps sampling): this, as has been observed before, results in increased efficiency. On the other hand, in a multi-stage design the first-stage units should be internally heterogeneous and homogeneous with respect to one another. In a stratified multi-stage design, all the strata have of course to be covered, but when the strata are made internally homogeneous and the first-stage units internally heterogeneous, a small number of first-stage units need be selected from each stratum in order to provide an efficient sample.

Almost all nation-wide sample surveys, including those mentioned in section 1.1, use stratified multi-stage designs. For example, for the National Market Research Survey in Great Britain (1952), where the universe was the adult civilian population aged sixteen and over, the sample was stratified by factors such as geographical region, urban/rural areas, industrialization index and zoning, and was spread over four stages: administrative districts, polling districts, households, and individuals (Moser and Kalton, Chapter 8). In the monthly Current Population Survey in the U.S.A., the sampling design is stratified three-stage; heterogeneous first-stage units are defined, comprising individual counties (or sometimes two or more adjoining counties), and grouped into a number of strata; the first-stage units are sampled with varying probabilities; relatively small areal units of about two hundred households are the second-stage units; and the third-stage units are clusters of approximately six households, called 'segments': the sample contains 449 first-stage units, about 900 segments and about 50 000 households (Hansen *et al.*, Vol. I, Chapter 12; Hansen and Tepping, 1969). In the Indian National Sample Survey (1964–5), in the rural sector there were 353 strata, comprising administrative units, and each covering an average of 1.2 million persons; for the socioeconomic inquiries, villages and households were the two stages of sampling, and for crop yield surveys, the four stages of sampling were respectively villages, clusters of plots, crop-plots, and circular cuts of radius 2′ 3″ or 4′ (see Appendix V). In the Sample Survey on Goods Traffic Movement in Zambia (1967), the design was stratified two-stage, with proprietors of transport vehicles constituting the first-stage and the vehicles the second-stage units.

19.3 Fundamental theorems in stratified multi-stage sampling

The fundamental theorems for the estimation of universe parameters in stratified multi-stage sampling follow from those for a stratified single-stage sample (section 9.3) and for an unstratified multi-stage sample (section 14.4) taken in combination.

1. The universe is sub-divided into L mutually exclusive and exhaustive strata. In the hth stratum ($h = 1, 2, \ldots, L$), of the total N_h first-stage units, n_h ($\geqslant 2$) are selected with replacement (and with equal or varying probabilities); if t_{hi} ($h = 1, 2, \ldots, L$; $i = 1, 2, \ldots, n_h$) are the n_h independent and unbiased estimators of the universe parameter T_h, obtained from the ith selected first-stage unit in the hth stratum, then the combined unbiased estimator of T_h is the arithmetic mean

$$\bar{t}_h = \sum^{n_h} t_{hi}/n_h \tag{19.1}$$

An unbiased estimator of the variance of $\bar{t}_h$ is

$$\begin{aligned} s^2_{\bar{t}_h} &= \sum^{n_h} (t_{hi} - \bar{t}_h)^2/n_h(n_h - 1) \\ &= \mathrm{SS}t_{hi}/n_h(n_h - 1) \end{aligned} \tag{19.2}$$

An unbiased estimator of the universe parameter for all the strata combined

$$T = \sum_{h=1}^{L} T_h \tag{19.3}$$

is obtained on summing up the L unbiased stratum estimators $\bar{t}_h$ (given in equation (19.1)), namely

$$t = \sum^{L} \bar{t}_h \tag{19.4}$$

An unbiased estimator of the variance of t is obtained on summing the L unbiased estimators of the stratum variances $s^2_{\bar{t}_h}$ (sampling in one stratum is independent of sampling in another, so that the stratum estimators $\bar{t}_h$ are mutually independent),

$$s^2_t = \sum^{L} s^2_{\bar{t}_h} \tag{19.5}$$

We shall see later in this section how to compute the t_{hi} values for a stratified multi-stage design.

For another study variable, similarly defined, an unbiased estimator of the universe parameter

$$U = \sum^{L} U_h \tag{19.6}$$

is the sum of the L unbiased stratum estimators $\bar{u}_h$ of U_h

$$u = \sum^{L} \bar{u}_h \tag{19.7}$$

and an unbiased estimator of the variance of u is the sum of L unbiased estimators of the stratum variances $s^2_{\bar{u}_h}$,

$$s_u^2 = \sum^{L} s^2_{\bar{u}_h} \tag{19.8}$$

2. An unbiased estimator of the covariance of $\bar{t}_h$ and $\bar{u}_h$ for the hth stratum is

$$s_{\bar{t}_h \bar{u}_h} = \sum^{n_h} (t_{hi} - \bar{t}_h)(u_{hi} - \bar{u}_h)$$
$$= \text{SP}t_{hi}u_{hi}/n_h(n_h - 1) \tag{19.9}$$

For all the strata combined, an unbiased estimator of the covariance between t and u is the sum of the stratum estimators $s_{\bar{t}_h \bar{u}_h}$, i.e.

$$s_{tu} = \sum^{L} s_{\bar{t}_h \bar{u}_h} \tag{19.10}$$

3. A consistent but generally biased estimator of the ratio of two universe parameters in the hth stratum $R_h = T_h/U_h$ is the ratio of the sample estimators $\bar{t}_h$ and $\bar{u}_h$ in the stratum,

$$r_h = \bar{t}_h/\bar{u}_h \tag{19.11}$$

with a variance estimator

$$s^2_{r_h} = (s^2_{\bar{t}_h} + r_h^2 s^2_{\bar{u}_h} - 2r_h s_{\bar{t}_h \bar{u}_h})/\bar{u}_h^2 \tag{19.12}$$

For all the strata combined, a consistent but generally biased estimator of the ratio of two universe parameters $R = T/U$ is the ratio of the respective sample estimators t and u,

$$r = t/u \tag{19.13}$$

with a variance estimator

$$s_r^2 = (s_t^2 + r^2 s_u^2 - 2rs_{tu})/u^2 \tag{19.14}$$

4. A consistent but generally biased estimator of the correlation coefficient ρ_h between the two study variables for the hth stratum is

$$\hat{\rho}_h = s_{\bar{t}_h \bar{u}_h}/s_{\bar{t}_h} s_{\bar{u}_h} \tag{19.15}$$

For all the strata combined, an estimator of the correlation coefficient ρ is

$$\hat{\rho} = s_{tu}/s_t s_u \tag{19.16}$$

• 5. *Unbiased estimators of stratum universe totals obtained from the sample units.* Here we change the notations and also deal with the totals of a study variable (see section 14.4.5).

In a stratified multi-stage design with L strata and u stages ($u > 1$), in the hth stratum ($h = 1, 2, \ldots, L$), and tth stage ($t = 1, 2, \ldots, u$), let $n_{n12\ldots(t-1)}$ sample units be selected with replacement out of the total $N_{12\ldots(t-1)}$ units and let the (initial) probability of selection of the ($h12 \ldots t$)th unit be $\pi_{h12\ldots t}$. We define

$$f_{ht} \equiv n_{h12\ldots(t-1)}\,\pi_{h12\ldots t} \tag{19.17}$$

An unbiased estimator of the universe total Y_h of the study variable in the hth stratum is

$$y_h = \sum_{\substack{\text{stratum}\\\text{sample}}} \left(\prod_{t=1}^{u} \frac{1}{f_{ht}}\right) y_{h12\ldots u}$$

$$= \sum_{\substack{\text{stratum}\\\text{sample}}} w_{h12\ldots u}\, y_{12\ldots u} \tag{19.18}$$

where $y_{h12\ldots u}$ is the value of the study variable in the ($h12 \ldots u$)th ultimate-stage sample units and the factor

$$\prod_{t=1}^{u} \frac{1}{f_{ht}} = w_{h12\ldots u} \tag{19.19}$$

is the multiplier (or the weighting-factor) for the ($h12 \ldots u$)th ultimate-stage sample units; the sum of products of these multipliers and the values of the study variable for all the ultimate-stage sample units in the stratum provides the unbiased linear estimator of the universe total Y_h in equation (19.18).

An unbiased estimator of the variance of the sample estimator y_h in the general estimating equation (19.18) is

$$s_{y_h}^2 = \frac{\sum^{n_h} \left[n_h \sum_{2, 3, \ldots, u} w_{h12\ldots u}\, y_{h12\ldots u} - \sum_{1, 2, \ldots, u} w_{h12\ldots u}\, y_{h12\ldots u} \right]^2}{n_h(n_h - 1)} \tag{19.20}$$

where n_h ($\geqslant 2$) is the number of first-stage units selected with replacement in the hth stratum, the summation outside the square brackets being over these first-stage units.

The formulae can be put in simpler forms. Using a slightly different notation, an unbiased estimator of the universe total Y_h of the study variable obtained from the ith first-stage unit ($i = 1, 2, \ldots, n_h$) in the hth stratum is

$$y_{hi}^* = n_h \sum_{2, 3, \ldots, u} w_{h12\ldots u}\, y_{h12\ldots u} \tag{19.21}$$

the summation being over the second- and subsequent-stage sample units.

The combined unbiased estimator of the universe total Y_h from all the n_h sample first-stage units in the hth stratum is the arithmetic mean

$$y_{h0}^* = \sum^{n_h} y_{hi}^* / n_h \tag{19.22}$$

which is the same as the estimator y_h in equation (19.18), the summation in equation (19.22) being over the n_h sample first-stage units in the hth stratum.

An unbiased estimator of the variance of y_{ho}^* is, from equation (19.2),

$$s_{y_{ho}^*}^2 = \sum^{n_h} (y_{hi}^* - y_{ho}^*)^2 / n_h(n_h - 1) \qquad (19.23)$$

For all the strata combined, an unbiased estimator of the universe total $Y = \sum^{L} Y_h$ is the sum of the stratum estimators

$$y = \sum^{L} y_{ho}^* \qquad (19.24)$$

An unbiased estimator of the variance of y is, from equation (19.5),

$$s_y^2 = \sum^{L} s_{y_{ho}^*}^2 \qquad (19.25)\bullet$$

Notes

1. *Setting of probability limits.* The results for stratified single-stage sampling, given in section 9.4, also hold for stratified multi-stage sampling.
2. *Selection of two first-stage units from each stratum.* As for stratified single-stage sampling (section 9.5), so also for stratified multi-stage sampling, the computation of estimates is simplified considerably when in each stratum two first-stage units are selected with replacement out of the total number of units. The results of section 9.5 also hold here, t_{h1} and t_{h2} denoting the unbiased estimators of the universe parameter T_h in the hth stratum, obtained from the two sample first-stage units; and similarly for u_{h1} and u_{h2}.
3. For other notes, see section 9.3, notes (2)–(4) and section 14.4, notes (1)–(5).

20 Stratified Multi-stage Simple Random Sampling

20.1 Introduction

In this chapter we will consider the estimating methods for totals, means, and ratios of the values of study variables and their variances in a stratified multi-stage sample with simple random sampling at each stage: the methods for two- and three-stage designs are illustrated in some detail, and the procedures for four and higher stages indicated. The methods of estimation of proportion of units in a category and the use of ratio estimators will also be considered.

20.2 Stratified two-stage srs: estimation of totals, means, ratios, and their variances

20.2.1 Universe totals and means

The universe is divided on some criteria into L mutually exclusive and exhaustive strata. Let the number of first-stage units (fsu's) in the hth stratum ($h = 1, 2, \ldots, L$) be N_h, and the total number of fsu's in all the strata taken together $N = \sum^{L} N_h$; the number of second-stage units (ssu's) in the ith fsu ($i = 1, 2, \ldots, N_h$) in the hth stratum is denoted by M_{hi}, the total number of ssu's in the hth stratum and in all strata taken together being respectively $\sum_{i=1}^{N_h} M_{hi}$ and $\sum_{h=1}^{L} \sum_{i=1}^{M_h} M_{hi}$. Denoting by Y_{hij} the value of the study variable of the jth ssu ($j = 1, 2, \ldots, M_{hi}$) in the ith fsu in the hth stratum, we define the following universe totals and means:

For the hth stratum

Total of the values of the study variable in the ith fsu:

$$Y_{hi} = \sum_{j=1}^{M_{hi}} Y_{hij} \tag{20.1}$$

Mean for the ith fsu:

$$\overline{Y}_{hi} = Y_{hi}/M_{hi} \tag{20.2}$$

Total for the hth stratum:

$$\begin{aligned} Y_h &= \sum^{N_h} \sum^{M_{hi}} Y_{hij} \\ &= \sum^{N_h} Y_{hi} = \sum^{N_h} M_{hi} \overline{Y}_{hi} \end{aligned} \tag{20.3}$$

Mean per fsu:

$$\overline{Y}_h = Y_h/N_h = \sum^{N_h} M_{hi} \overline{Y}_{hi}/N_h \tag{20.4}$$

Mean per ssu:

$$\bar{\bar{Y}}_h = Y_h \Big/ \left(\sum^{N_h} M_{hi}\right)$$
$$= \sum^{N_h} M_{hi} \bar{Y}_{hi} \Big/ \left(\sum^{N_h} M_{hi}\right) \tag{20.5}$$

For all strata combined

Total of the values of the study variable:

$$Y = \sum^{L} Y_h = \sum^{L} \sum^{N_h} Y_{hi}$$
$$= \sum^{L} \sum^{N_h} \sum^{M_{hi}} Y_{hij} \tag{20.6}$$

Mean per fsu:

$$\bar{Y} = Y/N \tag{20.7}$$

Mean per ssu:

$$\bar{\bar{Y}} = Y \Big/ \left(\sum^{L} \sum^{N_h} M_{hi}\right) \tag{20.8}$$

20.2.2 Structure of a stratified two-stage srs

In the hth stratum, out of the N_h fsu's, n_h are selected; and out of the M_{hi} ssu's in the ith selected fsu ($i = 1, 2, \ldots, n_h$), m_{hi} are selected: sampling at both stages is srs with replacement. The total number of sample ssu's in the hth stratum is $\sum_{i=1}^{n_h} m_{hi}$, and in all the strata taken together is $\sum_{h=1}^{L} \sum_{i=1}^{n_h} m_{hi}$: the latter is the total sample size. The sampling plan is shown in summary form in Table 20.1.

Table 20.1 Sampling plan for a stratified two-stage simple random sample with replacement. In the hth stratum ($h = 1, 2, \ldots, L$)

Stage (t)	*Unit*	*No. in universe*	*No. in sample*	*Selection method*	*Selection probability*	f_t
1	First-stage	N_h	n_h	srswr	Equal = $1/N_h$	n_h/N_h
2	Second-stage	M_{hi}	m_{hi}	srswr	Equal = $1/M_{hi}$	m_{hi}/M_{hi}

20.2.3 Estimation of the totals of a study variable

Unbiased estimators of the stratum totals Y_h and the overall total Y of a study variable and their variance estimators could be obtained from the general estimating equations of section 19.3(5), or by extending the methods of sections 15.2.3(a) and 10.2 for srs (with replacement).

Let y_{hij} ($h = 1, 2, \ldots, L$; $i = 1, 2, \ldots, n_h$; $j = 1, 2, \ldots, m_{hi}$) denote the value of the study variable in the jth selected ssu in the ith selected fsu in the hth stratum. For the hth stratum, the unbiased estimator of the stratum total Y_h, obtained from the ith fsu, is (from equation (15.12))

$$y^*_{hi} = N_h \frac{M_{hi}}{m_{hi}} \sum_{j=1}^{m_{hi}} y_{hij}$$

$$= N_h M_{hi} \bar{y}_{hi} \tag{20.9}$$

where

$$\bar{y}_{hi} = \sum_{j=1}^{m_{hi}} y_{hij}/m_{hi} \tag{20.10}$$

The combined unbiased estimator of Y_h from all the n_h sample fsu's is (from equation (15.13))

$$y^*_{h0} = \frac{1}{n} \sum^{n_h} y^*_{hi} \tag{20.11}$$

For all the strata combined, the unbiased estimator of the total Y is (from equation (19.4)) the sum of unbiased estimators of the stratum totals Y_h,

$$y = \sum^{L} y^*_{h0} = \sum^{L} \frac{N_h}{n_h} \sum^{n_h} \frac{M_{hi}}{m_{hi}} \sum_{j=1}^{m_{hi}} y_{hij} \tag{20.12}$$

20.2.4 Sampling variances of the estimators y^*_{h0} and y

From section 15.2.4, the sampling variance of y^*_{h0} in srswr at both stages is

$$\sigma^2_{y^*_{h0}} = \frac{N_h^2 \sigma^2_{hb}}{n_h} + \frac{N_h}{n_h} \frac{M^2_{hi} \sigma^2_{hwi}}{m_{hi}} \tag{20.13}$$

where

$$\sigma^2_{hb} = \sum_{i=1}^{N_h} (Y_{hi} - \bar{Y}_h)^2/N \tag{20.14}$$

is the variance between the fsu's in the hth stratum; and

$$\sigma^2_{hwi} = \sum_{j=1}^{M_{hi}} (Y_{hij} - \bar{Y}_{hi})^2/M_{hi} \tag{20.15}$$

is the variance between the ssu's within the ith fsu in the hth stratum.

The sampling variance of $y = \sum^{L} y^*_{h0}$ is

$$\sigma^2_y = \sum^{L} \sigma^2_{y^*_{h0}} \tag{20.16}$$

Note: The sampling variances and their unbiased estimators in srs, without replacement at both stages and with replacement at the first stage and without replacement at the second stage, follow from section 17.2.4 for any stratum; for the sampling variance y of all the strata taken together, the stratum sampling variances are summed up and an unbiased estimator is obtained on summing up the unbiased stratum variance estimators.

20.2.5 Estimation of the variances of sample estimators of totals

An unbiased estimator of $\sigma^2_{y^*_{ho}}$, the sampling variance of the stratum estimator y^*_{ho} is, from equation (19.23) or (15.14),

$$s^2_{y^*_{ho}} = \sum^{n_h} (y^*_{hi} - y^*_{ho})^2 / n_h(n_h - 1)$$

$$= \text{SS}y^*_{hi} / n_h(n_h - 1) \qquad (20.17)$$

An unbiased estimator of $\sigma^2_y = \sum^L \sigma^2_{y^*_{ho}}$, the variance of the overall estimator y is, from equation (19.25), the sum of the stratum unbiased estimators of variances

$$s^2_y = \sum^L s^2_{y^*_{ho}} \qquad (20.18)$$

20.2.6 Estimation of the ratio of the totals of two study variables

For another study variable, the unbiased estimators of the stratum total X_h are similarly defined.

From the hith fsu:

$$x^*_{hi} = N_h \frac{M_{hi}}{m_{hi}} \sum^{m_{hi}} x_{hij} \qquad (20.19)$$

Combined from n_h fsu's:

$$x^*_{ho} = \sum^{n_h} x^*_{hi} / n_h \qquad (20.20)$$

$s^2_{x^*_{ho}}$, an unbiased estimator of variance of x^*_{ho}, is defined as for $s^2_{y^*_{ho}}$, given in equation (20.17).

For the whole universe, the unbiased estimator of X is (from equation (20.12),

$$x = \sum^L x^*_{ho} \qquad (20.21)$$

and s^2_x, the unbiased estimator of the variance of x, is defined as for s^2_y, given in equation (20.18).

A consistent but generally biased estimator of the ratio of two universe totals at the stratum level $R_h = Y_h/X_h$ is the ratio of the sample estimators,

$$r_h = y^*_{ho} / x^*_{ho} \qquad (20.22)$$

with a variance estimator (from equation (19.12))

$$s^2_{r_h} = (s^2_{y^*_{ho}} + r^2_h s^2_{x^*_{ho}} - 2 r_h s_{y^*_{ho} x^*_{ho}}) / x^{*2}_{ho} \qquad (20.23)$$

where

$$s_{y^*_{ho} x^*_{ho}} = \sum_{i=1}^{n_h} (y^*_{hi} - y^*_{ho})(x^*_{hi} - x^*_{ho}) / n_h(n_h - 1) \qquad (20.24)$$

is the unbiased covariance estimator of y^*_{ho} and x^*_{ho} (from equation (19.9)).

A consistent but generally biased estimator of two universe totals for all the strata combined, $R = Y/X$, is the ratio of the sample estimators,

$$r = y/x \tag{20.25}$$

with a variance estimator (from equation (19.14))·

$$s_r^2 = (s_y^2 + r^2 s_x^2 - 2rs_{yx})/x^2 \tag{20.26}$$

where

$$s_{yx} = \sum_{h=1}^{L} s_{y^*_{ho} x^*_{ho}} \tag{20.27}$$

is the unbiased covariance estimator of y and x (from equation (19.10)).

20.2.7 Estimation of the means of a study variable

For the estimation of the means of a study variable in a two-stage srs in any stratum, we use the methods of section 15.2.6 adding the subscript h for the hth stratum. Here we consider the estimation of the overall mean, i.e. the mean for all the strata combined.

(a) *Estimation of the mean per first-stage unit.* The unbiased estimator of the overall mean per fsu is

$$\bar{y} = y/N \tag{20.28}$$

with an unbiased variance estimator

$$s_{\bar{y}}^2 = s_y^2/N^2 \tag{20.29}$$

The unbiased estimator y of Y, is defined by equation (20.12), and s_y^2, the unbiased variance estimator of y, is defined by equation (20.18).

(b) *Estimation of the mean per second-stage unit.* Two situations will arise: (i) M_{hi} (the total number of ssu's) is known for all the fsu's; and (ii) M_{hi} is known for the n_h sample fsu's only.

(i) *Unbiased estimator.* When M_{hi} is known for all the fsu's, the unbiased estimator of mean per ssu is

$$\bar{\bar{y}} = y\Big/\left(\sum^{L}\sum^{N_h} M_{hi}\right) \tag{20.30}$$

with an unbiased variance estimator

$$s_y^2\Big/\left(\sum^{L}\sum^{N_h} M_{hi}\right)^2 \tag{20.31}$$

See the notes to section 15.2.6(b-i).

(ii) *Ratio estimator.* When M_{hi} is known for only the sample fsu's, the unbiased estimator of the total number of ssu's in the universe, $\sum^{L}\sum^{N_h} M_{hi}$, is given by an estimating equation of the type (20.12) by putting $y_{hij} = 1$ for the sample ssu's, i.e. by

$$m = \sum^{L} m^*_{ho} \tag{20.32}$$

with an unbiased variance estimator

$$s_m^2 = \sum^{L} s^2_{m^*_{ho}}$$

$$= \sum^{L}\sum^{n_h} (m^*_{hi} - m^*_{ho})^2/n_h(n_h - 1) \tag{20.33}$$

where

$$m^*_{ho} = \sum^{n_h} m^*_{hi}/n_h \tag{20.34}$$

and

$$m^*_{hi} = N_h m_{hi} \tag{20.35}$$

A consistent but generally biased estimator of the universe mean $\bar{\bar{Y}}$ per ssu is (from equation (20.25))

$$r = y/m \tag{20.36}$$

with variance estimator

$$s_r = (s_y^2 + r^2 s_m^2 - 2r s_{ym})/m^2 \tag{20.37}$$

where

$$s_{ym} = \sum^{L} s_{y^*_{ho} m^*_{ho}}$$

$$= \sum^{L}\sum^{n_h} (y^*_{hi} - y^*_{ho})(m^*_{hi} - m^*_{ho})/n_h(n_h - 1) \tag{20.38}$$

is the unbiased covariance estimator of y and m.

See note to section 15.2.6(b-ii).

(iii) *Unweighted mean of means.* An estimator of the mean per ssu in the universe is the simple (unweighted) mean of means,

$$\sum^{L}\sum^{n_h} \bar{y}_{hi}/n \tag{20.39}$$

where $\bar{y}_{hi}$ is the mean of the y_{hij} values over the m_{hi} sample ssu's in the hith sample fsu (equation (20.10)), and $n = \sum^{L} n_h$. This estimator is neither unbiased nor consistent: it will be unbiased and consistent when the design is self-weighting.

Note: For a comparison of the different types of estimators, see section 15.2.6(b-iv).

20.2.8 Estimation for sub-universes

Here as for stratified uni-stage and unstratified multi-stage sampling, unbiased estimators are obtained for the total value of a study variable in the sub-universe Y'_h in the hth stratum or for all the strata taken together Y' by defining:

$y'_{hij} = y_{hij}$ if the sample ssu belongs to the sub-universe
$y'_{hij} = 0$ otherwise.

Unbiased estimators of the variances of the totals for the sub-universe are similarly given.

20.2.9 Estimation of proportion of units

Here we combine the methods of section 10.3 for a stratified uni-stage srs and section 15.4 for an unstratified two-stage srs. Let M'_{hi} be the total number of second-stage units possessing a certain attribute in the ith first-stage unit ($i = 1, 2, \ldots, N_h$) in the hth stratum ($h = 1, 2, \ldots, L$); the total number of such ssu's in the hth stratum is

$$\sum^{N_h} M'_{hi} \tag{20.40}$$

The universe number of ssu's with the attribute for all the strata combined is

$$\sum^{L}\sum^{N_h} M'_{hi} \tag{20.41}$$

The proportion of such ssu's with the attribute in the ith fsu in the hth stratum is

$$P_{hi} = M'_{hi}/M_{hi} \tag{20.42}$$

and for all the N_h fsu's in the stratum

$$\begin{aligned} P_h &= \sum^{N_h} M'_{hi}\Big/\sum^{N_h} M_{hi} \\ &= \sum^{N_h} M_{hi}P_{hi}\Big/\sum^{N_h} M_{hi} \end{aligned} \tag{20.43}$$

The proportion for all the strata combined is

$$P = \sum^{L}\sum^{N_h} M'_{hi}\Big/\sum^{L}\sum^{N_h} M_{hi} \tag{20.44}$$

As for unstratified two-stage srs (section 15.4), so also here, unbiased estimators of the total number of ssu's in a stratum and for all strata combined are obtained respectively from the estimating equations (20.11) and (20.12) by putting $y_{hij} = 1$ if the sample ssu has the attribute, and 0 otherwise, and so, also, for the unbiased variance estimators.

For the estimators for any stratum, we use the methods of section 15.4 adding the subscript h for the hth stratum. Here we consider the estimation of the overall universe proportion, i.e. for all the strata combined.

The unbiased estimator of $M' = \sum^{L} M'_h$ is the sum of the stratum estimators m'^{*}_{h0}

$$m' = \sum^{L} m'^{*}_{h0} \tag{20.45}$$

where

$$m'^{*}_{h0} = \sum^{n_h} m'^{*}_{hi}/n_h \tag{20.46}$$

from equation (15.54),

$$m'^{*}_{hi} = N_h M_{hi} p_{hi} \tag{20.47}$$

from equation (15.53), and

$$p_{hi} = m'_{hi}/m_{hi} \tag{20.48}$$

from equation (15.52), m'_{hi} of the m_{hi} sample fsu's in an srs having the attribute in the hth stratum.

An unbiased variance estimator of m' is

$$s_{m'}^2 = \sum^{L} s_{m'^*_{ho}}^2$$

$$= \sum^{L} \sum^{n_h} (m'^*_{hi} - m'^*_{ho})^2 / n_h (n_h - 1) \qquad (20.49)$$

If all the M_{hi} values are known for the universe, then the unbiased estimator of the universe proportion P is

$$m' \Big/ \sum^{L} \sum^{N_h} M_{hi} \qquad (20.50)$$

with an unbiased variance estimator

$$s_{m'}^2 \Big/ \left(\sum^{L} \sum^{N_h} M_{hi} \right)^2 \qquad (20.51)$$

If $\sum^{L} \sum^{N_h} M_{hi}$ is not known, it is estimated unbiasedly by m in equation (20.32) with an unbiased variance estimator given by equation (20.33).

A consistent but generally biased estimator of the universe proportion P is then

$$p = m'/m \qquad (20.52)$$

with a variance estimator

$$s_p^2 = (s_{m'}^2 + p^2 s_m^2 - 2p s_{mm'})/m^2 \qquad (20.53)$$

where

$$s_{mm'} = \sum^{L} s_{m^*_{ho} m'^*_{ho}}$$

$$= \sum^{L} \sum^{n_h} (m^*_{hi} - m^*_{ho})(m'^*_{hi} - m'^*_{ho}) / n_h (n_h - 1) \qquad (20.54)$$

is an unbiased covariance estimator of m and m'.

See notes to sections 10.3 and 15.4

Example 20.1

In each of the simple random samples of two villages each from the three zones selected in Example 10.2 from the list of villages in Appendix IV, select five households at random, given the listing of households in the sample villages. On the basis of the collected data for the sample households on the size, number of adults, and possession of transistor radios, estimate the total number of persons, the average household size, the proportion of adults, and the proportion of households with transistor radios for the three zones separately and also combined, along with their standard errors. The sample data are given in Table 20.2 on page 222. The use of the last column will be illustrated in Example 20.2

Table 20.2 Size, number of adults (18 years or over), possession of transistor radios, and knowledge of development plans by the selected adult in the sample households in each of the two sample villages in the three zones of Example 10.2: Stratified two-stage simple random sample

Household serial no.	*Household size* (y_{hij})	*Number of adults* (x_{hij})	*Possession of transistor radios* (y'_{hij} = 1, *for Yes;* = 0, *for No*)	*Knowledge of development plans by selected adult* (q'_{hij} = 1, *for Yes;* = 0, *for No*)
Zone I: Village serial no. 2; total number of households = 18				
7	6	4	1	0
11	4	2	0	0
12	4	2	0	0
15	3	1	0	1
18	4	2	1	1
Zone I: Village serial no. 10; total number of households = 22				
1	4	1	0	0
4	5	2	0	0
6	4	2	1	1
9	3	2	1	0
17	4	3	0	0
Zone II: Village serial no. 5; total number of households = 20				
5	6	3	1	0
7	7	4	1	0
8	5	3	0	1
12	6	4	0	0
16	6	2	1	0
Zone II: Village serial no. 11: total number of households = 18				
4	4	2	0	1
7	5	3	0	0
9	5	3	1	0
12	7	3	1	0
16	5	3	0	0
Zone III: Village serial no. 2; total number of households = 21				
7	6	3	0	0
10	5	3	0	0
17	5	2	1	0
19	5	3	0	0
20	7	4	1	1
Zone III: Village serial no 3: total number of households = 18				
6	6	4	0	0
10	6	3	0	1
11	6	4	0	0
16	5	3	0	0
18	4	2	1	0

As there are two sample first-stage units, namely villages, in each stratum, we follow the simplified procedure mentioned in note 2 of section 19.3, illustrated also in Example 10.2 for the stratified single-stage srs. We denote the household size by y_{hij}, the number of adults by x_{hij}, possession of transistor radios by y'_{hij} (= 1, for yes; and = 0 for no) in the sample households (h = 1, 2, 3 for the zones; i = 1, 2 for the sample villages; and j = 1, 2, 3, 4, 5 for the sample households). The respective means per sample household in the sample villages are denoted by $\bar{y}_{hi}$, $\bar{x}_{hi}$, and $\bar{m}'_{hi}$, defined by equations of the type (20.10). Unbiased estimates of the total number of households in the three zones separately and combined have already been obtained in Example 10.2. These, along with the other required computations, are shown in Tables 20.3–20.5.

The final estimates are given in Table 20.6. The coefficients of variation of the estimated proportion of households with transistor radios are rather high. To reduce standard errors, information on items such as this, that do not require detailed probing, could be collected for all the households in the sample villages when a list of households is prepared before a sample of households could be drawn for collecting other information. The present computations are merely illustrative.

Table 20.3 Computation of the estimated total numbers of persons, adults, and households with transistor radios: data of Table 20.2†

Zone	*Number of villages*		*Sample village*	*Number of households*			*Number of persons*		
	Total	*Sample*		*Total*	*Sample*	*Stratum estimate**	*Sample number*	*Sample mean*	*Stratum estimate*
h	N_h	n_h	i	M_{hi}	m_{hi}	$m^*_{hi} = N_h M_{hi}$	$\sum_{j=1}^{m_{hi}} y_{hij}$	$\bar{y}_{hi}$	$y^*_{hi} = N_h M_{hi} \bar{y}_{hi}$
(1)	(2)	(3)	(4)	(5)	(6)	(7)	(8)	(9)	(10)
I	10	2	1	18	5	180	21	4.2	756
			2	22	5	220	20	4.2	880
					Total	400			1636
					Mean (= stratum estimate of total)	800 (m^*_{h0})			818 (y^*_{h0})
					Difference	–40 (d_{mh})			–156 (d_{yh})
					$\frac{1}{2}$|Difference| (= estimated standard error)	20 ($s_{m^*_{h0}}$)			78 ($s_{y^*_{h0}}$)
II	11	2	1	20	5	220	30	6.0	1320
			2	18	5	198	26	5.2	1029.6
					Total	418			2349.6
					Mean (= stratum estimate of total)	209 (m^*_{h0})			1174.8 (y^*_{h0})
					Difference	22 (d_{mh})			290.4 (d_{yh})
					$\frac{1}{2}$|Difference| (= estimated standard error)	11 ($s_{m^*_{h0}}$)			145.2 ($s_{y^*_{h0}}$)
III	9	2	1	21	5	189	28	5.6	1058.4
			2	18	5	162	27	5.4	874.8
					Total	351			1933.2
					Mean (= stratum estimate of total)	175.5 (m^*_{h0})			966.6 (y^*_{h0})
					Difference	27 (d_{mh})			183.6 (d_{yh})
					$\frac{1}{2}$|Difference| (= estimated standard error)	13.5 ($s_{m^*_{h0}}$)			91.8 ($s_{y^*_{h0}}$)

* The estimated numbers of households were obtained in Table 10.4, denoted by x^*_{hi}. In this example, x^*_{hi} denotes the estimated numbers of adults, and m^*_{hi} those of households.

† This table is continued on page 224.

Table 20.3 continued.

Zone	*Sample village*	*Number of adults*			*Number of households with radio*		
		Sample number	*Sample mean*	*Stratum estimate*	*Sample number*	*Sample mean*	*Stratum estimate*
h	i	$\sum_{j=1}^{m_{hi}} x_{hij}$	$\bar{x}_{hi}$	$x_{hi}^* = N_h M_h \bar{x}_{hi}$	$\sum_{j=1}^{m_{hi}} y'_{hij}$	$\bar{m}'_{hi}$	$m'^*_{hi} = N_h M_{hi} \bar{m}'_{hi}$
(1)	(4)	(11)	(12)	(13)	(14)	(15)	(16)
I	1	11	2.2	396	1	0.2	36
	2	10	2.0	440	2	0.4	88
Total				836			124
Mean (= stratum estimate of total)				418 (x_{h0}^*)			62 (m'^*_{h0})
Difference				−44 (d_{xh})			−52 ($d_{m'h}$)
$\frac{1}{2}$|Difference| (= estimated standard error)				22 ($s_{x_{h0}^*}$)			26 ($s_{m'^*_{h0}}$)
II	1	16	3.2	704	3	0.6	132
	2	14	2.8	554.4	2	0.4	79.2
Total				1258.4			211.2
Mean (= stratum estimate of total)				629.2 (x_{h0}^*)			105.6 (m'^*_{h0})
Difference				149.6 (d_{xh})			52.8 ($d_{m'h}$)
$\frac{1}{2}$|Difference| (= estimated standard error)				74.8 ($s_{x_{h0}^*}$)			26.4 ($s_{m'^*_{h0}}$)
III	1	15	3.0	567	2	0.4	75.6
	2	16	3.2	518.4	1	0.2	32.4
Total				1085.4			108.0
Mean (= stratum estimate of total)				542.7 (x_{h0}^*)			54.0 (m'^*_{h0})
Difference				48.6 (d_{xh})			43.2 ($d_{m'h}$)
$\frac{1}{2}$|Difference| (= estimated standard error)				24.3 ($s_{x_{h0}^*}$)			21.6 ($s_{m'^*_{h0}}$)

Table 20.4 Computation of estimated total numbers of households, persons, adults, and households with transistor radios: Tables 20.2 and 20.3

Zone	*Number of households*★			*Number of persons*		
	Estimate	*Standard error*	*Variance*	*Estimate*	*Standard error*	*Variance*
h	m^*_{h0}	$s_{m^*_{h0}}$	$s^2_{m^*_{h0}}$	y^*_{h0}	$s_{y^*_{h0}}$	$s^2_{y^*_{h0}}$
I	200	20	400	818	78	6084
II	209	11	121	1174.8	145.2	21083.04
III	175.5	13.5	182.25	966.6	91.8	8427.24
All zones combined	584.5	26.52	703.25	2959.4	188.66	35594.28

Zone	*Number of adults*			*Number of households with radio*		
	Estimate	*Standard error*	*Variance*	*Estimate*	*Standard error*	*Variance*
h	x^*_{h0}	$s_{x^*_{h0}}$	$s^2_{x^*_{h0}}$	m'^*_{h0}	$s_{m'^*_{h0}}$	$s^2_{m'^*_{h0}}$
I	418	22	484	62	26	676
II	629.2	74.8	5595.04	105.6	26.4	696.96
III	542.7	24.3	590.49	54	21.6	466.56
All zones combined	1589.9	81.67	6669.53	221.6	42.89	1839.52

★Also from Example 10.2.

Table 20.5 Computation of estimated ratios and variances: data of Table 20.3

Zone	*For the average household size*			
	$r_h = y^*_{h0}/m^*_{h0}$	$s_{y^*_{h0} m^*_{h0}} = \frac{1}{4} d_{yh} d_{mh}$	$s^2_{y^*_{h0}} + r^2_h s^2_{m^*_{h0}} - 2r_h s_{y^*_{h0} m^*_{h0}}$	$s^2_{r_h}$ = col. (4)/m^{*2}_{h0}
(1)	(2)	(3)	(4)	(5)
I	4.09	1560.0	14.4400	0.000361
II	5.62	1597.2	6952.2244	0.149159
III	5.51	1239.3	303.2822	0.009847
All zones combined	5.06 (r)	4396.5 (s_{ym})	9107.4317 ($s^2_y + r^2 s^2_m - 2rs_{ym}$)	0.026658 (s^2_r)

Zone	*For the proportion of adults*			
	$p_h = x^*_{h0}/y^*_{h0}$	$s_{y^*_{h0} x^*_{h0}} = \frac{1}{4} d_{yh} d_{xh}$	$s^2_{x^*_{h0}} + p^2_h s^2_{y^*_{h0}} - 2p_h s_{y^*_{h0} x^*_{h0}}$	$s^2_{p_h}$ = col. (8)/y^{*2}_{h0}
(1)	(6)	(7)	(8)	(9)
I	0.5110	1716.00	318.9082	0.00047661
II	0.5356	10 860.96	8.8081	0.00000638
III	0.5615	2230.74	742.3261	0.00079451
All zones combined	0.5372 (p)	14 807.70 (s_{yx})	1032.0768 ($s^2_x + p^2 s^2_y - 2ps_{yx}$)	0.00011784 (s^2_p)

Zone	*For the proportion of households with transistor radios*			
	$p'_h = m'^*_{h0}/m^*_{h0}$	$s_{m^*_{h0} m'^*_{h0}} = \frac{1}{4} d_{mh} d_{m'h}$	$s^2_{m'^*_{h0}} + p'^2_h s^2_{m^*_{h0}} - 2p'_h s_{m^*_{h0} m'^*_{h0}}$	$s^2_{p'_h}$ = col. (12)/m^{*2}_{h0}
(1)	(10)	(11)	(12)	(13)
I	0.3100	520.0	392.0400	0.00980100
II	0.5053	290.4	434.3764	0.00994429
III	0.3077	291.6	304.3646	0.00988189
All zones combined	0.3791 (p')	1102.0 ($s_{mm'}$)	1105.0526 ($s^2_{m'} + p'^2 s^2_m - 2p's_{mm'}$)	0.00323455 ($s^2_{p'}$)

20.2.10 Ratio method of estimation

(a) *Estimation procedures for totals.* As for unstratified single- or multi-stage and stratified single-stage, so also for stratified multi-stage sampling, the ratio method of estimation may increase the efficiency of estimators under favourable conditions if information on ancillary variables is available for the second-stage units in the universe.

For the hth stratum

The combined unbiased estimator of the stratum total of the ancillary variable $Z_h = \sum^{N_h} \sum^{M_{hi}} z_{hij}$ is first obtained from equation (20.11)

$$z^*_{h0} = \sum^{n_h} z^*_{hi}/n_h \qquad (20.55)$$

with an unbiased variance estimator

$$s^2_{z^*_{ho}} = \sum^{n_h} (z^*_{hi} - z^*_{h0})^2/n_h(n_h - 1) \qquad (20.56)$$

where

$$z^*_{hi} = N_h \frac{M_{hi}}{m_{hi}} \sum^{m_{hi}} z_{hij} \qquad (20.57)$$

The ratio estimator of the total Y_h, using the ancillary information, is, from equation (15.66),

$$y^*_{h\mathrm{R}} = Z_h y^*_{h0}/z^*_{h0} = Z_h r_{1h} \qquad (20.58)$$

where

$$r_{1h} = y^*_{h0}/z^*_{h0} \qquad (20.59)$$

A consistent but generally biased estimator of the variance of $y^*_{h\mathrm{R}}$ is (from equation (15.68))

$$s^2_{y^*_{h\mathrm{R}}} = Z^2_h s^2_{r_{1h}} = (s^2_{y^*_{ho}} + r^2_{1h} s^2_{z^*_{ho}} - 2r_{1h} s_{y^*_{ho} z^*_{ho}}) \qquad (20.60)$$

where

$$s_{y^*_{ho} z^*_{ho}} = \sum^{n_h} (y^*_{hi} - y^*_{h0})(z^*_{hi} - z^*_{h0})/n_h(n_h - 1) \qquad (20.61)$$

is the unbiased covariance estimator of y^*_{h0} and z^*_{h0}.

For all strata combined (see also section 10.4)

For the estimation of universe total $Y = \sum^{L} Y_h$, two types of ratio estimators can be used:

(i) *Separate ratio estimator.* The stratum ratio estimators $y^*_{h\mathrm{R}}$ are summed up:

$$y_{\mathrm{RS}} = \sum^{L} y^*_{h\mathrm{R}} \qquad (20.62)$$

the stratum ratio estimators $y^*_{h\mathrm{R}}$ being given by equation (20.58), the additional subscript S in y_{RS} standing for *separate* ratio estimator.

The variance estimator of y_{RS} is the sum of the stratum variance estimators $s^2_{y^*_{h\mathrm{R}}}$, given by equation (20.60):

$$s^2_{y_{\mathrm{RS}}} = \sum^{L} s^2_{y^*_{h\mathrm{R}}} \qquad (20.63)$$

(ii) *Combined ratio estimator.* The ratio method is applied to the estimators of the overall totals, thus

$$y_{RC} = Zy/z = Zr \qquad (20.64)$$

the additional subscript C in y_{RC} standing for *combined* ratio estimator; y is the unbiased estimator of the universe total Y and is given by equation (20.12), z is the unbiased estimator of $Z = \sum^{L} Z_h$ given by

$$z = \sum^{L} z^*_{h0} \qquad (20.65)$$

and

$$r = y/z \qquad (20.66)$$

A variance estimator of y_{RC} is

$$s^2_{y_{RC}} = Z^2 s^2_r = (s^2_y + r^2 s^2_z - 2rs_{yz}) \qquad (20.67)$$

where

$$s_{yz} = \sum^{L} s_{y^*_{h0} z^*_{h0}} \qquad (20.68)$$

is the unbiased covariance estimator of y and z.

See notes to sections 10.4 and 15.5.

• Example 20.2

For the data of Example 20.1, given the additional information on the previous census population of all the villages as in Example 10.5, obtain ratio estimates of the total population for the three zones separately and also combined, along with their standard errors.

Here the ratio method of estimation is applied at the level of the first-stage units. The unbiased estimates of the previous census population in the three zones along with their standard errors have already been obtained in Example 10.5. The rest of the computations is shown in Table 20.7 and the final results given in Table 20.8.

Notes

1. The ratio estimates of the total numbers of households have already been obtained in Example 10.2.
2. The ratio estimates of population are more efficient than the unbiased estimates, as shown by Table 20.6. No marked difference appears between the separate and combined ratio estimates for the whole universe.

Table 20.6 Estimates and standard errors computed from the data of stratified two-stage srs in Table 20.2

Item	*Zone I*	*Zone II*	*Zone III*	*All zones combined*
1. *Total number of persons*				
(a) Estimate	818	1 174.8	966.6	2 959.4
(b) Standard error	78	145.2	91.8	188.66
(c) Coefficient of variation (%)	9.54	12.36	9.50	6.37
2. *Average household size*				
(a) Estimate	4.09	5.62	5.51	5.06
(b) Standard error	0.019	0.399	0.099	0.163
(c) Coefficient of variation (%)	0.46	7.10	1.80	3.22
3. *Proportion of adults*				
(a) Estimate	0.5110	0.5356	0.5615	0.5372
(b) Standard error	0.02183	0.00253	0.02819	0.01086
(c) Coefficient of variation (%)	4.27	0.47	5.02	2.02
4. *Proportion of households with transistor radios*				
(a) Estimate	0.3100	0.5053	0.3077	0.3791
(b) Standard error	0.0990	0.0997	0.0994	0.0569
(c) Coefficient of variation (%)	31.94	19.73	32.30	15.01

Table 20.7 Computation of ratio estimates of total population for a stratified two-stage srs: data of Tables 20.2, and 10.12

Zone	Z_h	$r_{1h} = y^*_{h0}/z^*_{h0}$	$y^*_{hR} = Z_h r_{1h}$	$s_{y^*_{h0}z^*_{h0}} = \frac{1}{4}d_{yh}d_{zh}$	$s^2_{y^*_{hR}} = s^2_{y^*_{h0}} + r^2_{1h}s^2_{z^*_{h0}} - 2r_{1h}s_{y^*_{h0}z^*_{h0}}$	$s_{y^*_{hR}}$
(1)	(2)	(3)	(4)	(5)	(6)	(7)
I	863	1.0161	876.89	390.0	6902.369475	83.08
II	1010	1.1609	1172.51	9583.2	4788.032436	69.20
III	942	1.0278	968.19	6196.5	502.844801	22.42
Total	2815		3017.67(y_{RS})	15389.7	12 193.246712 ($s^2_{y_{RS}}$)	110.42 ($s_{y_{RS}}$)
Combined	2815 (Z)	1.0732 ($r_1 = y/z$)	3021.06 ($y_{RC} = Zr_1$)	15389.7 (s_{yz})	12 855.394980 ($s^2_{y_{RC}} = s^2_y + r^2_1 s^2_z - 2r_1 s_{yz}$)	113.38 ($s_{y_{RC}}$)

Table 20.8 Ratio estimates of population and their standard errors computed from the data of the stratified two-stage srs in Table 20.2, using the previous census population of the sample villages (data of Tables 20.7)

Zone	*Ratio estimate*	*Standard error*	*CV (%)*
I	876.9	83.1	9.48
II	1172.5	69.2	5.90
III	968.2	22.4	2.31
All zones:			
Separate ratio estimate	3017.7	110.4	3.66
Combined ratio estimate	3021.1	113.4	3.75

•(b) *Ratio of ratio estimators of totals of two study variables.* Ratio estimators of totals of two study variables may be used in estimating their ratios. However, as noted in section 10.5 for stratified single-stage sampling, for any stratum (and also for all strata if the combined ratio estimators are used), the ratio of the ratio estimators of totals of two study variables becomes the same as the ratio of the corresponding unbiased estimators. For the separate ratio estimators of totals of two study variables, the ratio of the estimators and its variance estimators have been defined in section 10.5 for stratified single-stage srs. The same methods will apply for stratified multi-stage designs, noting that the appropriate estimating formulae have to be used. For an example, see exercise 3 of this chapter. •

• 20.3 Stratified three-stage srs

20.3.1 Universe totals and means

Following on the formulation in section 20.2.1, let the number of third-stage units (tsu's) in the jth second-stage unit (ssu) in the ith first-stage unit (fsu) of the hth stratum be denoted by Q_{hij} ($h = 1, 2, \ldots, L$; $i = 1, 2, \ldots, N_h$); $j = 1, 2, \ldots, M_{hi}$). The total number of tsu's in the ith fsu of the hth stratum is $\sum_{j=1}^{M_{hi}} Q_{hij}$. The total number of tsu's in the hth stratum for all the N_h fsu's is $\sum_{i=1}^{N_h} \sum_{j=1}^{M_{hi}} Q_{hij}$, and in the whole universe $\sum_{h=1}^{L} \sum_{i=1}^{N_h} \sum_{j=1}^{M_{hi}} Q_{hij}$. Denoting by Y_{hijk} the value of the study variable of the kth tsu in the jth ssu in the ith fsu in the hth stratum, we define the following universe totals and means:

For the hth stratum

Total of the values of the study variable in the jth ssu in the ith fsu in the hth stratum:

$$Y_{hij} = \sum_{k=1}^{Q_{hij}} Y_{hijk} \tag{20.69}$$

Mean for the jth ssu in the ith fsu:

$$\bar{Y}_{hij} = Y_{hij}/Q_{hij} \tag{20.70}$$

Total for the ith fsu:

$$Y_{hi} = \sum_{j=1}^{M_{hi}} Y_{hij} \tag{20.71}$$

Mean for the ith fsu per ssu:

$$\overline{\overline{Y}}_{hi} = Y_{hi}/M_{hi} \tag{20.72}$$

Total for the hth stratum:

$$\begin{aligned} Y_h &= \sum^{N_h} \sum^{M_{hi}} \sum^{Q_{hij}} Y_{hijk} \\ &= \sum^{N_h} \sum^{M_{hi}} Y_{hij} \\ &= \sum^{N_h} Y_{hi} \\ &= \sum^{N_h} \sum^{M_{hi}} Q_{hij} \overline{Y}_{hij} \\ &= \sum^{N_h} M_{hi} / \overline{Y}_{hi} \end{aligned} \tag{20.73}$$

Mean per fsu in the hth stratum:

$$\overline{Y}_h = Y_h/N_h = \sum^{N_h} M_{hi} \overline{Y}_{hi}/N_h \tag{20.74}$$

Mean per ssu:

$$\begin{aligned} \overline{\overline{Y}}_h &= Y \Big/ \left(\sum^{N_h} M_{hi} \right) \\ &= \sum^{N_h} M_{hi} \overline{Y}_{hi} \Big/ \left(\sum^{N_h} M_{hi} \right) \end{aligned} \tag{20.75}$$

Mean per tsu:

$$\begin{aligned} \overline{\overline{\overline{Y}}}_h &= Y \Big/ \left(\sum^{N_h} \sum^{M_{hi}} Q_{hij} \right) \\ &= \sum^{N_h} \sum^{M_{hi}} Q_{hij} \overline{Y}_{hij} \Big/ \left(\sum^{N_h} \sum^{M_{hi}} Q_{hij} \right) \end{aligned} \tag{20.76}$$

For all strata combined

Total of the values of the study variable:

$$\begin{aligned} Y &= \sum^{L} Y_h \\ &= \sum^{L} \sum^{N_h} Y_{hi} \\ &= \sum^{L} \sum^{N_h} \sum^{M_{hi}} Y_{hij} \\ &= \sum^{L} \sum^{N_h} \sum^{M_{hi}} \sum^{Q_{hij}} Y_{hijk} \end{aligned} \tag{20.77}$$

Mean per fsu:

$$\bar{Y} = Y/N \tag{20.78}$$

Mean per ssu:

$$\bar{\bar{Y}} = Y\Big/\left(\sum^{L}\sum^{N_h} M_{hi}\right) \tag{20.79}$$

Mean per tsu:

$$\bar{\bar{\bar{Y}}} = Y\Big/\left(\sum^{L}\sum^{N_h}\sum^{M_{hi}} Q_{hij}\right) \tag{20.80}$$

20.3.2 Structure of a three-stage srs

In the hth stratum, out of the N_h fsu's, n_h are selected; out of the M_{hi} ssu's in the ith selected fsu ($i = 1, 2, \ldots, n_h$), m_{hi} are selected; and finally, out of the Q_{hij} tsu's in the jth selected ssu ($j = 1, 2, \ldots, m_{hi}$), q_{hij} are selected; sampling at the three stages is srs with replacement. The total number of sample tsu's in the hth stratum is $\sum_{i=1}^{n_h} \sum_{j=1}^{m_{hi}} q_{hij}$, and in all the strata taken together is $\sum^{L} \sum^{n_h} \sum^{m_{hi}} q_{hij}$: the latter is the total sample size. The sampling plan is shown in summary form in Table 20.9.

Table 20.9 Sampling plan for a stratified three-stage simple random sample with replacement. In the hth stratum ($h = 1, 2, \ldots, L$)

Stage *(t)*	*Unit*	*No. in universe*	*No. in sample*	*Selection method*	*Selection probability*	f_t
1	First-stage	N_h	n_h	srswr	Equal = $1/N_h$	n_h/N_h
2	Second-stage	M_{hi}	m_{hi}	srswr	Equal = $1/M_{hi}$	m_{hi}/M_{hi}
3	Third-stage	Q_{hij}	q_{hij}	srswr	Equal = $1/Q_{hij}$	q_{hij}/Q_{hij}

20.3.3 Estimation of the totals of a study variable

Unbiased estimators of the stratum totals Y_h and the overall total Y of a study variable and their variance estimators could be obtained from the general estimating equations of section 19.3(5) or by extending the methods of sections 15.6.3 and 10.2 for srs with replacement.

Let y_{hijk} ($h = 1, 2, \ldots, L$; $i = 1, 2, \ldots, n_h$; $j = 1, 2, \ldots, m_{hi}$; $k = 1, 2, \ldots, q_{hij}$) denote the value of the study variable in the kth selected tsu in the jth selected ssu in the ith selected fsu in the hth stratum.

For the hth stratum, the unbiased estimator of Y_h from the ith selected fsu is (from section 15.6.3)

$$y^*_{hi} = N_h \frac{M_{hi}}{m_{hi}} \sum_{j=1}^{m_{hi}} \frac{Q_{hij}}{q_{hij}} \sum_{k=1}^{q_{hij}} y_{hijk}$$

$$= N_h \frac{M_{hi}}{m_{hi}} \sum_{j=1}^{m_{hi}} Q_{hij}\bar{y}_{hij} \tag{20.81}$$

where

$$\bar{y}_{hij} = \sum_{k=1}^{q_{hij}} y_{hijk}/q_{hij} \tag{20.82}$$

is the simple arithmetic mean of the y_{hijk} values in the *hij*th ssu.

The combined unbiased estimator of Y_h is the arithmetic mean

$$y^*_{ho} = \frac{1}{n_h}\sum_{i=1}^{n_h} y^*_{hi}$$

$$= \frac{N_h}{n_h}\sum_{i=1}^{n_h}\frac{M_{hi}}{m_{hi}}\sum_{j=1}^{m_{hi}}\frac{Q_{hij}}{q_{hij}}\sum_{k=1}^{q_{hij}} y_{hijk} \tag{20.83}$$

For all the strata combined, the unbiased estimator of the total Y is the sum of the estimators of the stratum totals

$$y = \sum_{h=1}^{L} y^*_{ho} = \sum_{h=1}^{L}\frac{N_h}{n_h}\sum_{i=1}^{n_h}\frac{M_{hi}}{m_{hi}}\sum_{j=1}^{m_{hi}}\frac{Q_{hij}}{q_{hij}}\sum_{k=1}^{q_{hij}} y_{hijk} \tag{20.84}$$

20.3.4 Estimation of the variances of the sample estimators of totals

An unbiased variance estimator of the stratum total estimator y^*_{ho} is

$$s^2_{y^*_{ho}} = \sum^{n_h} (y^*_{hi} - y^*_{ho})^2/n_h(n_h - 1) \tag{20.85}$$

and that of the universe total estimator y

$$s^2_y = \sum^{L} s^2_{y^*_{ho}} \tag{20.86}$$

20.3.5 Estimation of the ratio of the totals of two study variables

For another study variable, the stratum and universe totals may be defined and estimated similarly. The unbiased estimator x^*_{ho} of the stratum total X_h is obtained as for y^*_{ho} (equation (20.83)), as also its unbiased variance estimator $s^2_{x^*_{ho}}$ (by equation (20.85)). For the whole universe, the unbiased estimator x is obtained in the same manner as for y (equation (20.84)), as also its unbiased variance estimator (equation (20.86)).

A consistent but generally biased estimator of the ratio of the stratum totals $R_h = Y_h/X_h$ is the ratio of the sample estimators

$$r_h = y^*_{ho}/x^*_{ho} \tag{20.87}$$

with a variance estimator

$$s^2_{r_h} = (s^2_{y^*_{ho}} + r^2_h s^2_{x^*_{ho}} - 2r_h s_{y^*_{ho}x^*_{ho}})/x^{*2}_{ho} \tag{20.88}$$

where

$$s_{y^*_{ho}x^*_{ho}} = \sum^{n_h} (y^*_{hi} - y^*_{ho})(x^*_{hi} - x^*_{ho})/n_h(n_h - 1) \tag{20.89}$$

is the unbiased covariance estimator of y^*_{ho} and x^*_{ho}.

For the whole universe, a consistent but generally biased estimator of the ratio of totals $R = Y/X$ is the ratio of the sample estimators

$$r = y/x \tag{20.90}$$

with a variance estimator

$$s_r^2 = (s_y^2 + r^2 s_x^2 - 2r s_{yx})/x^2 \tag{20.91}$$

where

$$s_{yx} = \sum^{L} s_{y^*_{ho} x^*_{ho}} \tag{20.92}$$

is the unbiased covariance estimator of y and x.

20.3.6 Estimation of the means of a study variable

For the estimation of the means of a study variable in a three-stage srs in any stratum, see section 15.6.5, and for the estimation of the means per fsu and ssu in all the strata combined, see sections 20.2.7(a) and 20.2.7(b). We consider here the estimating method for the mean per tsu for the whole universe.

Two situations will arise: (i) the total number of the tsu's Q_{hij} is known for all the ssu's in the universe; and (ii) Q_{hij} is known only for the sample ssu's.

(i) *Unbiased estimator.* When Q_{hij} is known for all the ssu's in the universe, the unbiased estimator of the universe mean per tsu $\bar{\bar{\bar{Y}}}$ is

$$\bar{\bar{\bar{y}}} = y \Big/ \left(\sum^{L} \sum^{N_h} \sum^{M_{hi}} Q_{hij}\right) \tag{20.93}$$

with an unbiased variance estimator

$$s_y^2 \Big/ \left(\sum^{L} \sum^{N_h} \sum^{M_{hi}} Q_{hij}\right)^2 \tag{20.94}$$

The unbiased estimator y of Y is defined by equation (20.84), and its unbiased variance estimator s_y^2 by equation (20.86).

(ii) *Ratio estimator.* When Q_{hij} is known only for the sample ssu's, the unbiased estimator of the total number of tsu's in the universe, namely, $\sum^{L} \sum^{N_h} \sum^{M_{hi}} Q_{hij}$, is given by an estimation equation of the type (20.84) by putting $y_{hijk} = 1$ for the sample tsu's, i.e. by

$$q = \sum^{L} q^*_{ho} \tag{20.95}$$

with an unbiased variance estimator

$$s_q^2 = \sum^{L} s^2_{q^*_{ho}} \tag{20.96}$$

where

$$q^*_{ho} = \sum^{n_h} q^*_{hi}/n_h \tag{20.97}$$

$$q^*_{hi} = N_h \frac{M_{hi}}{m_{hi}} \sum^{m_{hi}} Q_{hij} \tag{20.98}$$

and

$$s^2_{q^*_{ho}} = \sum^{n_h} (q^*_{hi} - q^*_{ho})^2/n_h(n_h - 1) \tag{20.99}$$

A consistent but generally biased estimator of $\bar{\bar{\bar{Y}}}$ is the ratio

$$r = y/q \tag{20.100}$$

with a variance estimator

$$s_r^2 = (s_y^2 + r^2 s_q^2 - 2rs_{yq})/q^2 \tag{20.101}$$

where

$$s_{yq} = \sum^{L} s_{y_{ho}^* q_{ho}^*}$$
$$= \sum^{L}\sum^{n_h}(y_{hi}^* - y_{ho}^*)(q_{hi}^* - q_{ho}^*)/n_h(n_h - 1) \tag{20.102}$$

(iii) *Unweighted mean of means.* An estimator of $\bar{\bar{\bar{Y}}}$ is the simple (unweighted) mean of means,

$$\sum^{L}\sum^{n_h}\sum^{m_{hi}} \bar{y}_{hij} \Big/ \left(\sum^{L}\sum^{n_h} m_{hi}\right) \tag{20.103}$$

where $\bar{y}_{hij}$ is the sample mean in the *hij*th ssu (equation (20.82)).

This estimator will be both biased and inconsistent: it will be unbiased and consistent when the design is self-weighting.

For a comparison of the estimators and other observations, see notes to section 15.6.5.

20.3.7 Estimation for sub-universe

Here also, unbiased estimators are obtained for the total values of a study variable in the sub-universe Y_h' in the *h*th stratum or for all the strata taken together Y' by defining

$y'_{hijk} = y_{hijk}$ if the *hijk*th third-stage sample unit belongs to the sub-universe
$y'_{hijk} = 0$ otherwise.

Unbiased variance estimators of the totals are similarly obtained.

20.3.8 Estimation of proportion of units

Here we combine the methods of section 10.3 for a stratified single-stage srs and section 15.8 for an unstratified three-stage srs. Let Q'_{hij} be the total number of third-stage units possessing a certain attribute in the *j*th ssu in the *i*th fsu in the *h*th stratum ($h = 1, 2, \ldots, L$; $i = 1, 2, \ldots, N_h$; $j = 1, 2, \ldots, M_{hi}$). The total number of such units in the *hi*th fsu is

$$\sum^{M_{hi}} Q'_{hij} \tag{20.104}$$

in the *h*th stratum

$$\sum^{N_h}\sum^{M_{hi}} Q'_{hij} \tag{20.105}$$

and in the whole universe

$$\sum^{L}\sum^{N_h}\sum^{M_{hi}} Q'_{hij} \tag{20.106}$$

The proportion of such tsu's with the attribute in the hith fsu is

$$P_{hi} = \sum^{M_{hi}} Q'_{hij} \Big/ \sum^{M_{hi}} Q_{hij} \qquad (20.107)$$

in the hth stratum

$$P_h = \sum^{N_h} \sum^{M_{hi}} Q'_{hij} \Big/ \sum^{N_h} \sum^{M_{hi}} Q_{hij} \qquad (20.108)$$

and in the whole universe

$$P = \sum^{L} \sum^{N_h} \sum^{M_{hi}} Q'_{hij} \Big/ \sum^{L} \sum^{N_h} \sum^{M_{hi}} Q_{hij} \qquad (20.109)$$

Here again, as in section 15.6 for an unstratified three-stage srs, the unbiased estimators of the total number of tsu's in the hth stratum and in the whole universe can be obtained respectively from equations (20.83) and (20.84) by putting $y_{hijk} = 1$ if the third-stage sample unit has the attribute, and 0 otherwise. And so also for their unbiased variance estimators.

For estimating the proportion of the third-stage units possessing the attribute in any stratum, we use the method of section 15.8 for an unstratified three-stage srs, adding the subscript h for the hth stratum. For all the strata combined, the unbiased estimator of the total number of such units is

$$q' = \sum^{L} q'^{*}_{h0} \qquad (20.110)$$

with an unbiased variance estimator

$$s^2_{q'} = \sum^{L} s^2_{q'^{*}_{h0}} = \sum^{L} \sum^{n_h} (q'^{*}_{hi} - q'^{*}_{h0})^2 / n_h(n_h - 1) \qquad (20.111)$$

where

$$q'^{*}_{h0} = \sum^{n_h} q'^{*}_{hi} / n_h \qquad (20.112)$$

and

$$q'^{*}_{hi} = N_h \frac{M_{hi}}{m_{hi}} \sum^{m_{hi}} \frac{Q_{hij}}{q_{hij}} q'_{hij} \qquad (20.113)$$

q'_{hij} being the number of tsu's possessing the attribute out of the sample number q_{hij} of tsu's in the jth ssu in the ith fsu in the hth stratum.

If all the Q_{hij} values are known for the universe, then the unbiased estimator of the universe proportion P (equation (20.109)) is

$$q' \Big/ \left(\sum^{L} \sum^{N_h} \sum^{M_{hi}} Q_{hij} \right) \qquad (20.114)$$

with an unbiased variance estimator

$$s^2_{q'} \Big/ \left(\sum^{L} \sum^{N_h} \sum^{M_{hi}} Q_{hij} \right)^2 \qquad (20.115)$$

If $\sum^{L} \sum^{N_h} \sum^{M_{hi}} Q_{hij}$ is not known, it is estimated unbiasedly by q (equation (20.95)), with an unbiased variance estimator given by equation (20.96).

A consistent but generally biased estimator of the universe proportion P is the ratio

$$p = q'/q \tag{20.116}$$

with a variance estimator

$$s_p^2 = (s_{q'}^2 + p^2 s_q^2 - 2ps_{q'q})/q^2 \tag{20.117}$$

where

$$s_{qq'} = \sum^{L} s_{q^*_{h0} q'^*_{h0}}$$

$$= \sum^{L} \sum^{n_h} (q^*_{hi} - q^*_{h0})(q'^*_{hi} - q'^*_{h0})/n_h(n_h - 1) \tag{20.118}$$

is the unbiased covariance estimator of q and q'.

See notes to sections 10.3 and 15.8.

Example 20.3

In Table 20.2 on page 222 are given the number of adults in the sample households in each of the two sample villages, selected from the three zones. In each selected household, one adult member (18 years or over) was further selected at random from the total number of adults in the household, and asked about his knowledge of the development plans in the country. The information is given also in Table 20.2. Estimate the total number of adults with knowledge

Table 20.10 Computation of the estimated total number of adults with knowledge of development plans from a stratified three-stage srs with two sample first-stage units (villages) in each stratum: data of Table 20.2

Zone	*Number of villages*		*Sample village*	*For number of adults with knowledge of development plans*		
h	*Total* N_h	*Sample* n_h	i	$\sum_{j=1}^{m_{hi}} Q_{hij} q'_{hij}$	$\frac{1}{5}\sum_{j=1}^{m_{hi}} Q_{hij} q'_{hij}$	$q'^*_{hi} = N_h M_{hi}$. Col. (6)
(1)	(2)	(3)	(4)	(5)	(6)	(7)
I	10	2	1	3	0.6	108
			2	2	0.4	88
					Total	196
					Mean (= stratum estimate of total)	98 (q'^*_{h0})
					Difference	20 ($d_{q'h}$)
					$\frac{1}{2}$|Difference|(= estimated standard error)	10 ($s_{q'^*_{h0}}$)
II	11	2	1	3	0.6	132
			2	2	0.4	79.2
					Total	211.2
					Mean (= stratum estimate of total)	105.6 (q'^*_{h0})
					Difference	52.8 ($d_{q'h}$)
					$\frac{1}{2}$|Difference|(= estimated standard error)	26.4 ($s_{q'^*_{h0}}$)
III	9	2	1	4	0.8	151.2
			2	3	0.6	97.2
					Total	284.4
					Mean (= stratum estimate of total)	124.2 (q'^*_{h0})
					Difference	54.0 ($d_{q'h}$)
					$\frac{1}{2}$|Difference| (= estimated standard error)	27.0 ($s_{q'^*_{h0}}$)

of development plans and also the proportion they constitute of the total number of adults in each of three zones and also in all the zones combined, along with their standard errors.

Extending the notation of Example 20.1, let Q_{hij} (= x_{hij} in Example 20.1) denote the total number of adults in the ijth sample household in the hth stratum. The number of adults selected for the interview (i.e. the number of third-stage sample units) is $q_{hij} = 1$. We can therefore dispense with the subscript k in y_{hijk} by which we had denoted the value of the study variable for the ijkth third-stage sample unit in the hth stratum.

Following the method of section 20.3.8, we put $q'_{hij} = 1$ if the selected adult in the ijth sample household in the hth stratum knows of development plans, and $q'_{hij} = 0$ otherwise.

From estimating equation (20.113), we have the

Table 20.11 Computation of the estimated total numbers of adults with knowledge of development plans and their variances: data of Tables 20.2 and 20.10,

Zone	*Number of adults with knowledge of development plans*		
	Estimate	*Standard error*	*Variance*
h	q'^*_{h0}	$s_{q'^*_{h0}}$	$s^2_{q'^*_{h0}}$
I	98.0	10.0	100.00
II	105.6	26.4	696.96
III	124.2	27.0	729.00
All zones	327.8 (q')	39.05 ($s_{q'}$)	1525.96 ($s^2_{q'}$)

Table 20.12 Computation of the estimated proportions of adults with knowledge of development plans and their variances

Zone h	$p'_h = q'^*_{h0}/x^*_{h0}$	$s_{x^*_{h0}q'^*_{h0}} = \frac{1}{4} d_{xh} d_{q'h}$	$s^2_{q'^*_{h0}} + p^2_h s^2_{x^*_{h0}} - 2p_h \cdot s_{x^*_{h0}q'^*_{h0}}$	$s^2_{p_h}$ = col. (4)/x^{*2}_{h0}
(1)	(2)	(3)	(4)	(5)
I	0.2344	−220.00	229.728896	0.00131481
II	0.1678	1974.72	191.783509	0.00048433
III	0.2289	656.10	459.576144	0.00156047
Total	0.2062 (p)	2410.82 ($s_{xq'}$)	815.312908 ($s^2_{q'} + p^2 s^2_x - 2p s_{xq'}$)	0.00032254 s^2_p = col. (4)/x^2

Table 20.13 Estimated numbers and proportions of adults with knowledge of development plans, computed from the data of a stratified three-stage srs in Table 20.2

Adults with knowledge of development plans	*Zone I*	*Zone II*	*Zone III*	*All zones combined*
(a) *Number*				
Estimate	98.0	105.6	124.2	327.8
Standard error	10.0	26.4	27.0	39.05
Coefficient of variation (%)	11.36	25.00	21.74	11.91
(b) *Proportion to total adults*				
Estimate	0.2344	0.1678	0.2289	0.2062
Standard error	0.03626	0.02201	0.03950	0.01796
Coefficient of variation (%)	15.47	13.12	9.62	8.71

unbiased estimator of the stratum total of the number of adults with knowledge of development plans

$$q'^*_{hi} = N_h \frac{M_{hi}}{m_{hi}} \sum_{j=1}^{m_{hi}} Q_{hij} q'_{hij}$$

from the ith first-stage unit (here a village) in the hth stratum. The required computations follow the methods of section 20.3.8, and are shown in Tables 20.10–20.12, and the final results in Table 20.13.

20.3.9 Ratio method of estimation

As for a stratified two-stage srs (section 20.2.10) so also for a stratified three-stage srs, the ratio method of estimation may, under favourable conditions, improve the efficiency of estimators.

The procedures are the same as for stratified two-stage srs, given in section 20.2.11, except that the unbiased estimator of the stratum total Z_h of the ancillary variable will be computed on the basis of the appropriate estimating equation, namely, an equation of the type (20.83), thus:
From the ith fsu:

$$z^*_{hi} = N_h \frac{M_{hi}}{m_{hi}} \sum^{m_{hi}} \frac{Q_{hij}}{q_{hij}} \sum^{q_{hij}} z_{hijk} \qquad (20.119)$$

From all the n_h fsu's:

$$z^*_{h0} = \sum^{n_h} z^*_{hi}/n_h \qquad (20.120)$$

where z_{hijk} is the value of the ancillary variable in the $hijk$th selected third-stage unit.

An unbiased variance estimator of z^*_{h0} will be given by an estimating equation of the type (20.85).

Example 20.4

For the data of Example 20.3, given the additional information on the previous census population of all the villages as in Example 20.2, obtain ratio estimates of the number of adults with knowledge of development plans in the three zones separately and also combined, along with their standard errors.

Here the ratio method is applied, as in Example 20.2, at the level of the first-stage units.

Table 20.14 Computation of ratio estimates of the total number of adults with knowledge of development plans for a stratified three-stage srs: data of Tables 20.2, 10.8 and 10.9

Zone	Z_h	$r_{3h} = q'^*_{h0}/z^*_{h0}$	$q'^*_{hR} = Z_h r_{3h}$	$s_{q'^*_{h0} z^*_{h0}} = \frac{1}{4} d_{q'h} d_{zh}$	$s^2_{q'^*_{hR}} = s^2_{q'^*_{h0}} + r^2_{3h} s^2_{z^*_{h0}} - 2r_{3h} s_{q'^*_{h0} z^*_{h0}}$	$s_{q'^*_{hR}}$
(1)	(2)	(3)	(4)	(5)	(6)	(7)
I	863	0.1217	105.03	50.0	88.200275	9.39
II	1010	0.1043	105.34	1676.4	394.221928	19.86
III	942	0.1321	124.44	1822.5	327.002062	18.08
Total	2815		334.81 (q'^*_{RS})	3548.9	809.424265 ($s^2_{q'_{RS}}$)	28.45 ($s_{q'_{RS}}$)
Combined	2815 (Z)	0.1164 ($r_3 = q'/z$)	327.67 ($q'_{RC} = Zr_3$)	3548.9 ($s_{q'z}$)	920.866880 ($s^2_{q'_{RC}} = s^2_{q'} + r^2_3 s^2_z - 2r_3 s_{q'z}$)	30.35 ($s_{q'_{RC}}$)

Table 20.15 Ratio estimates of number of adults with knowledge of development plans from the data of the stratified three-stage srs in Table 20.2, using the previous census population of the sample villages (data of Table 10.8)

Zone	*Ratio estimate*	*Standard error*	*CV (%)*
I	105.0	9.39	8.94
II	105.3	19.86	18.86
III	124.4	18.08	14.53
All zones:			
Separate ratio estimate	334.8	28.45	8.50
Combined ratio estimate	327.7	30.35	9.26

The unbiased estimates of the previous census population in the three zones have already been obtained in Example 10.2, along with their standard errors.

We define

$$r_{3h} = q'^*_{ho}/z^*_{ho}$$

as the ratio of the combined unbiased estimator of the stratum number of adults with knowledge with development plans $\sum^{N_h} \sum^{M_{hi}} Q'_{hij}$, to that of the stratum estimator of the previous census population Z_h. The computations follow the methods of sections 20.3.10 and 20.2.11 and are shown in Table 20.14. The final results are given in Table 20.15.

Note: The ratio estimates obtained are more efficient than the unbiased estimates of the number of adults with knowledge of development plans in Table 20.13. There is no marked difference between the separate and combined ratio estimates for the whole universe. ●

20.4 Miscellaneous Notes

1. *Stratified four- and higher-stage srs.* Estimating procedures can be derived for stratified four- and higher-stage srs either directly from the general estimating procedures given in section 19.3 or by extending the procedures for a stratified three-stage design given in section 20.3. All that is required are the unbiased estimators of the stratum totals of the study variables, from which estimators of ratios, means, variances etc. follow.

2. *Gain due to stratification.* As for stratified srs (section 10.6), so also for stratified multi-stage srs it is possible to estimate the gain, if any, due to stratification, as compared with an unstratified srs with the same total number of ultimate-stage units. For this, see Sukhatme and Sukhatme, section 7.17.

3. *Stratification after sampling.* The same principles as for a single-stage srs, given in section 10.8, hold for multi-stage srs. If the sample number first-stage units is large, say over twenty in each stratum, and the errors in weights negligible, the method of post-stratification can give results almost as precise as proportional stratified sampling.

4. *Stratification at lower stages of samples.* As observed in note 2 of section 9.2, stratification may be introduced at different stages of sampling. In household budget and consumption surveys, for example, households are often stratified on the basis of size or a rough measure of the total income (which information can be readily obtained while listing all the households in the selected penultimate stages), and the required number of sample households selected from the different strata with a different sampling fraction in each. The same principles for estimation outlined earlier will apply in such cases.

Further reading

Hansen *et al.*, Vols. I and II, Chapters 7–10; Kish, sections 5.6, 6.4, and 6.5; Sukhatme and Sukhatme, sections 7.15 and 7.17.

Exercises

1. For a survey on the yield of corn in a district, the villages were grouped into 10 strata, and from each stratum two villages were selected at random. From each selected village, two fields were selected at random from all fields on which corn was grown. In each selected field, a rectangular plot of area 1/160 acre was located at random, and the yield of corn in the plot was measured in ounces. The sample data are given in Table 20.16, which also shows for each stratum the area under corn. Estimate for the district the total yield of corn and its standard error (Chakravarti *et al.*, Exercise 3.6, adapted).

2. For the same data of Example 20.1, estimate for all the zones combined the average number of adults per household and its standard error.

3. For the data of Example 20.2, estimate the average household size from the separate ratio estimates of numbers of persons and households and its standard error.

Table 20.16 Yield of corn in sample plots

Stratum no.	*Area under corn (acres)*	*Yield of corn (oz. from plot of $\frac{1}{160}$ acre)*			
		Sample Village 1		*Sample Village 2*	
		Plot 1	*Plot 2*	*Plot 1*	*Plot 2*
1	2767	90	102	346	166
2	1577	87	69	85	99
3	1778	82	195	40	236
4	6669	206	176	63	110
5	368	60	72	58	67
6	2875	28	241	120	34
7	5305	367	378	339	328
8	7219	48	236	208	180
9	3782	198	241	195	149
10	1457	32	148	160	112

21 Stratified Multi-stage Varying Probability Sampling

21.1 Introduction

In this chapter, we will consider the estimating procedures for a stratified multi-stage sample with varying probabilities of selection: the methods combine those for a stratified single-stage varying probability sample (Chapter 11) and for an unstratified multi-stage varying probability sample (Chapter 16). The estimating methods follow from the fundamental theorems of section 19.3 for a stratified multi-stage varying probability sample.

The general estimating procedures for stratified two- and three-stage designs will be considered first, followed by stratified two- and three-stage designs with pps at the first stage or the first two stages and srs at other stages. The special case of crop surveys will also be mentioned.

The method of ratio estimation and the estimating procedures for proportion of units and sub-universes etc. can be derived on similar lines as those for stratified multi-stage srs.

21.2 Stratified two- and three-stage pps designs

21.2.1 General

The general estimating equations for totals of study variables and the ratio of two totals, as well as those for their respective variances, follow from the general estimating equations in section 19.3. The cases of stratified two- and three-stage pps designs are mentioned briefly.

21.2.2 Two-stage pps design

The structure of a stratified two-stage design is given in Table 21.1. The universe is sub-divided into L strata, and in the hth

Table 21.1 Sampling plan for a stratified two-stage pps sample design with replacement. In the hth stratum ($h = 1, 2, \ldots, L$)

Stage (t)	*Unit*	*No. in universe*	*No. in sample*	*Selection method*	*Selection probability*	f_t
1	First-stage	N_h	n_h	ppswr	$\pi_{hi} = z_{hi}/Z_h$	$n_h \pi_{hi}$
2	Second-stage	M_{hi}	m_{hi}	ppswr	$\pi_{hij} = w_{hij}/W_{hi}$	$m_{hi}\pi_{hij}$

stratum ($h = 1, 2, \ldots, L$), a sample of n_h first-stage units is selected out of the total N_h units with pps (and replacement), the (initial) probability of selection of the ith first-stage unit ($i = 1, 2, \ldots, N_h$) at any draw being

$$\pi_{hi} = z_{hi}/Z_h \tag{21.1}$$

where z_{hi} is the value ('size') of an ancillary variable, $Z_h = \sum_{i=1}^{N_h} z_{hi}$ being known.

In each of the n_h first-stage units thus selected, a sample m_{hi} ($i = 1, 2, \ldots, n_h$) of second-stage units is selected with pps (and replacement) from the total M_{hi} units, the initial probability of selection of the jth second-stage unit ($j = 1, 2, \ldots, N_h$) being

$$\pi_{hij} = w_{hij}/W_{hi} \tag{21.2}$$

where w_{hij} is the value of another ancillary variable, and $W_{hi} = \sum_{j=1}^{M_{hi}} w_{hij}$.

The universe totals and means are defined as for a stratified two-stage srs in section 20.2.1.

If y_{hij} is the value of the study variable in the jth selected ssu in the ith selected fsu in the hth stratum ($i = 1, 2, \ldots, n_h$; $j = 1, 2, \ldots, m_{hi}$), then from equation (16.3) or (19.21), an unbiased estimator of the stratum total Y_h (as defined by equation 20.3) from the ith selected fsu is

$$y_{hi}^* = \frac{1}{\pi_{hi} m_{hi}} \sum_{j=1}^{m_{hi}} \frac{y_{hij}}{\pi_{hij}}$$

$$= \frac{Z_h W_{hi}}{z_h m_{hi}} \sum_{j=1}^{m_{hi}} \frac{y_{hij}}{w_{hij}} \tag{21.3}$$

and the combined unbiased estimator of Y_h from all the n_h fsu's is the arithmetic mean

$$y_{ho}^* = \frac{1}{n_h} \sum^{n_h} y_{hi}^* \tag{21.4}$$

with an unbiased variance estimator

$$s_{y_{ho}^*}^2 = \sum^{n_h} (y_{hi}^* - y_{ho}^*)^2 / n_h(n_h - 1) \tag{21.5}$$

For the whole universe, an unbiased estimator of the total $Y = \sum^{L} Y_h$ is

$$y = \sum^{L} y_{ho}^* \tag{21.6}$$

with an unbiased variance estimator

$$s_y^2 = \sum^{L} s_{y_{ho}^*}^2 \tag{21.7}$$

Estimators of the ratios of totals of two study variables and their variances follow directly from the fundamental theorems of section 19.3.

Notes

1. *Sampling variance of the estimator y.* From equation (16.6) the sampling variance of y_{ho}^* in ppswr at both stages is

$$\sigma_{y_{ho}^*}^2 = \frac{1}{n_h} \left(\sum_{i=1}^{N_h} \frac{Y_{hi}^2}{\pi_{hi}} - Y_h^2 \right) + \frac{1}{n_h} \sum_{i=1}^{N_h} \frac{1}{\pi_{hi} m_{hi}} \left(\sum_{j=1}^{M_{hi}} \frac{Y_{hij}^2}{\pi_{hij}} - Y_{hi}^2 \right) \tag{21.8}$$

which is estimated unbiasedly by $s_{y_{ho}^*}^2$, defined by equation (21.5).

The sampling variance of y is

$$\sigma_y^2 = \sum^{L} \sigma_{y_{h0}^*}^2 \tag{21.9}$$

which is estimated unbiasedly by s_y^2, defined by equation (21.7).

2. As noted in section 16.2.1, great simplifications result if the same size variable is used for pps sampling in both the stages.

21.2.3 Three-stage pps design

The sampling plan for a stratified three-stage pps design is given in summary form in Table 21.2. If y_{hijk} is the value of the study variable in the kth selected third stage unit of the jth selected second-stage unit of the ith selected first-stage unit

Table 21.2 Sampling plan for a stratified two-stage pps sample design with replacement. In the hth stratum ($h = 1, 2, \ldots, L$)

Stage (t)	*Unit*	*No. in universe*	*No. in sample*	*Selection method*	*Selection probability*	f_t
1	First-stage	N_h	n_h	ppswr	$\pi_{hi} = z_{hi}/Z_h$	$n_h \pi_{hi}$
2	Second-stage	M_{hi}	m_{hi}	ppswr	$\pi_{hij} = w_{hij}/W_{hi}$	$m_{hi}\pi_{hij}$
3	Third-stage	Q_{hij}	q_{hij}	ppswr	$\pi_{hijk} = v_{hijk}/V_{hij}$	$q_{hij}\pi_{hijk}$

in the hth stratum ($h = 1, 2, \ldots, L$; $i = 1, 2, \ldots, n_h$; $j = 1, 2, \ldots, m_{hi}$; $k = 1, 2, \ldots, q_{hij}$), then from the general estimating equation (19.21) or from equation (16.12), an unbiased estimator of the stratum total Y_h, obtained from the ith selected fsu is

$$y_{hi}^* = \frac{1}{\pi_{hi} m_{hi}} \sum_{j=1}^{m_{hi}} \frac{1}{\pi_{hij} q_{hij}} \sum_{k=1}^{q_{hij}} \frac{y_{hijk}}{v_{hijk}}$$

$$= \frac{Z_h W_{hi}}{z_{hi} m_{hi}} \sum_{j=1}^{m_{hi}} \frac{V_{hij}}{w_{hij} q_{hij}} \sum_{k=1}^{q_{hij}} \frac{y_{hijk}}{v_{hijk}} \tag{21.10}$$

and the combined unbiased estimator is

$$y_{h0}^* = \sum^{n_h} y_{hi}^*/n_h \tag{21.11}$$

with an unbiased variance estimator

$$s_{y_{h0}^*}^2 = \sum_{h=1}^{n_h} (y_{hi}^* - y_{h0}^*)^2/n_h(n_h - 1) \tag{21.12}$$

where $Z_h = \sum_{i=1}^{N_h} z_{hi}$; $W_{hi} = \sum_{j=1}^{M_{hi}} w_{hij}$; and $V_{hij} = \sum_{k=1}^{Q_{hij}} v_{hijk}$.

For the whole universe, an unbiased estimator of the total Y is

$$y = \sum^{L} y_{h0}^* \tag{21.13}$$

with an unbiased variance estimator $s_y^2 = \sum^{L} s_{y_{h0}^*}^2$ (21.14)

Estimators of the ratios of totals of two study variables and their variances follow directly from the fundamental theorems of section 19.3.

Note: As noted in section 16.2.2, great simplifications result if the same size variable is used for pps sampling in all the three stages.

21.3 Stratified two-stage design with pps and srs

21.3.1 General case

We illustrate this with a sampling plan for a rural sample survey, where in each stratum into which the universe is divided, villages are the first-stage units selected with ppswr (the 'size' being population in a previous census for a demographic inquiry and area for a crop survey) and households or fields the second-stage units selected as an srs or systematically (Table 21.3). This is a very common type of nation-wide sample inquiries.

Table 21.3 Sampling plan for a stratified two-stage design with pps sampling at the first-stage and srs at the second-stage. In the hth stratum ($h = 1, 2, \ldots, L$)

Stage t	*Unit*	*No. in universe*	*No. in sample*	*Selection method*	*Selection probability*	f_t
1	First-stage	N_h	n_h	ppswr	$\pi_{hi} = z_{hi}/Z_h$	$n_h \pi_{hi}$
2	Second-stage	M_{hi}	m_{hi}	srs	Equal = $1/M_{hi}$	m_{hi}/M_{hi}

An unbiased estimator of the hth stratum total Y_h ($h = 1, 2, \ldots, L$), obtained from the ith fsu ($i = 1, 2, \ldots, n_h$) is (from equation (21.3))

$$y_{hi}^* = \frac{M_{hi}}{\pi_{hi} m_{hi}} \sum_{j=1}^{m_{hi}} y_{hij} = \frac{M_{hi}}{\pi_{hi}} \bar{y}_{hi} \tag{21.15}$$

where y_{hij} is the value of the study variable in the jth selected ssu (household or field) of the ith selected fsu (village) in the hth stratum, and

$$\bar{y}_{hi} = \sum_{j=1}^{m_{hi}} y_{hij}/m_{hi} \tag{21.16}$$

is the mean of the y_{hij} values in the hith sample fsu.

The combined unbiased estimator of Y_h is

$$y_{ho}^* = \sum^{n_h} y_{hi}^*/n_h \tag{21.17}$$

and an unbiased estimator of the overall total Y is

$$y = \sum^{L} y_{ho}^* \tag{21.18}$$

Unbiased variance estimators of the stratum and the universe estimators of totals are given by equations (19.23) and (19.25) respectively: estimators of ratios of two totals and their variances follow from the methods of section 19.3. These will be illustrated by Examples 21.1 and 21.2.

21.3.2 First-stage units selected with probability proportional to the number of second-stage units

If M_{hi}, the number of ssu's, is known beforehand for all the N_h fsu's, and the n_h fsu's in the hth stratum are selected with probability proportional to M_{hi}, then, as shown in section 16.3.2 for one stratum, the estimating procedure becomes simpler.

Example 21.1

In each of the two sample villages selected in each of the three-zones in Example 11.1 from the list of villages in Appendix IV with probability proportional to previous census population and with replacement, select five households at random, given the listing of the households in the sample villages. On the basis of the collected data for the sample households on the size and number of adults (persons aged eighteen years or over), estimate the total number of persons, the number and the proportion of adults for three zones separately and also combined, along with their standard errors. The sample data are given in Table 21.4. The use of the last column will be illustrated in Example 21.3.

As there are two sample first-stage units, i.e.

Table 21.4 Size, number of adults (18 years or over), and knowledge of development plans by the selected adult in the srs of households in each of the two pps sample villages in the three zones of Example 11.1: Stratified two-stage pps and srs sample

Household serial no.	*Household size* (y_{hij})	*Number of adults* (x_{hij})	*Knowledge of development plans by the selected adult* (q'_{hij} = 1 *for Yes*; = 0 *for No*)
Zone I: Village serial no. 5; total number of households = 24			
1	5	2	0
6	5	2	0
16	4	2	1
18	6	3	0
22	3	2	0
Zone I: Village serial no. 7; total number of households = 20			
4	5	3	1
8	4	2	0
13	5	3	0
16	4	2	0
18	4	2	0
Zone II: Village serial no. 3; total number of households = 17			
3	5	3	0
5	6	3	0
7	3	2	0
14	5	2	1
15	3	2	0
Zone II: Village serial no. 7; total number of households = 19			
4	3	1	0
12	6	4	0
14	7	4	0
16	5	3	1
18	5	2	0
Zone III: Village serial no.2; total number of households = 21			
3	5	3	0
5	6	3	0
11	6	4	1
14	5	2	0
15	8	5	0
Zone III: Village serial no. 4; total number of households = 21			
7	5	2	0
8	6	3	1
12	5	3	0
13	7	3	0
15	8	5	0

Table 21.5 Computation of the estimated total number of persons and adults: data of Tables 11.2 and 21.9

Zone	Sample village	Reciprocal of probability $1/\pi_{hi}$	Total number of households M_{hi}	$m^*_{hi} = M_{hi}/\pi_{hi}$ *	For number of persons			For number of adults		
					$\sum_{j=1}^{5} y_{hij}$	$\bar{y}_{hi} = \frac{1}{5}\sum_{j=1}^{5} y_{hij}$	$y^*_{hi} = (M_{hi}/\pi_{hi})\bar{y}_{hi}$	$\sum_{j=1}^{5} x_{hij}$	$\bar{x}_{hi} = \frac{1}{5}\sum_{j=1}^{5} x_{hij}$	$x^*_{hi} = (M_{hi}/\pi_{hi})\bar{x}_{hi}$
(1)	(2)	(3)	(4)	(5)	(6)	(7)	(8)	(9)	(10)	(11)
I	1	9.3804	24	225.1296	23	4.6	1035.60	11	2.2	495.29
	2	11.9861	20	239.7220	22	4.4	1054.78	12	2.4	575.33
Total				464.8516			2090.38			1070.62
Mean (= Stratum estimate of total)				232.45 (m^*_{ho})			1045.19 (y^*_{ho})			535.31 (x^*_{ho})
Difference				−14.59 (d_{mh})			−19.18 (d_{yh})			−80.04 (d_{xh})
$\frac{1}{2}$ |Difference| (= Estimated standard error)				7.295 ($s_{m^*_{ho}}$)			9.59 ($s_{y^*_{ho}}$)			40.02 ($s_{x^*_{ho}}$)
II	1	13.8356	17	235.2052	22	4.4	1034.90	12	2.4	564.49
	2	11.8824	19	225.7656	26	5.2	1173.98	14	2.8	620.26
Total				460.9708			2208.88			1184.75
Mean (= Stratum estimate of total)				230.49 (m^*_{ho})			1104.44 (y^*_{ho})			592.38 (x^*_{ho})
Difference				9.44 (d_{mh})			−139.08 (d_{yh})			−55.47 (d_{xh})
$\frac{1}{2}$ |Difference| (= Estimated standard error)				4.72 ($s_{m^*_{ho}}$)			69.54 ($s_{y^*_{ho}}$)			27.885 ($s_{x^*_{ho}}$)
III	1	8.4107	21	176.6247	30	6.0	1059.75	17	3.4	600.52
	2	8.0513	21	169.0773	31	6.2	1048.28	16	3.2	541.05
Total				345.7020			2108.03			1141.57
Mean (= Stratum estimate of total)				172.85 (m^*_{ho})			1054.02 (y^*_{ho})			570.78 (x^*_{ho})
Difference				7.55 (d_{mh})			11.73 (d_{yh})			59.47 (d_{xh})
$\frac{1}{2}$ |Difference| (= Estimated standard error)				3.775 ($s_{m^*_{ho}}$)			5.865 ($s_{y^*_{ho}}$)			29.735 ($s_{x^*_{ho}}$)

* These are the estimated numbers of households in Tables 11.2, denoted by x^*_{hi}. In this example, x^*_{hi} denotes the estimated numbers of adults, and m^*_{hi} those of households.

Table 21.6 Computation of estimated numbers of households, persons and adults: data of Tables 21.4 and 21.5

Zone	*Number of households*★			*Number of persons*			*Number of adults*		
h	*Estimate* m^*_{ho}	*S.E.* $s_{m^*_{ho}}$	*Variance* $s^2_{m^*_{ho}}$	*Estimate* y^*_{ho}	*S.E.* $s_{y^*_{ho}}$	*Variance* $s^2_{y^*_{ho}}$	*Estimate* x^*_{ho}	*S.E.* $s_{x^*_{ho}}$	*Variance* $s^2_{x^*_{ho}}$
I	232.43	7.295	53.2170	1045.19	9.59	91.9681	535.31	40.020	1601.6004
II	230.49	4.720	22.2784	1104.44	69.54	4835.8116	592.38	27.885	777.5732
III	172.85	3.775	14.2506	1054.02	5.865	34.3982	570.78	29.735	884.1702
All zones combined	635.77 (m)	9.473 (s_m)	89.7460 (s^2_m)	3203.65 (y)	70.443 (s_y)	4962.1779 (s^2_y)	1698.47 (x)	57.126 (s_x)	3263.3438 (s^2_x)

★ Also from Example 11.1.

Table 21.7 Computation of estimated ratios and variances: data of Table 21.5

Zone	*For the household size*			
	$r_h = y^*_{ho}/m^*_{ho}$	$s_{y^*_{ho}m^*_{ho}} = \frac{1}{4}d_{yh}d_{mh}$	$s^2_{y^*_{ho}} + r^2_h s^2_{m^*_{ho}} - 2r_h s_{y^*_{ho}m^*_{ho}}$	$s^2_{r_h}$ = col. (4)$/m^{*2}_{ho}$
(1)	(2)	(3)	(4)	(5)
I	4.497	70.0070	538.5325	0.009971
II	4.791	−328.2288	8493.3551	0.159859
III	6.100	22.1116	541.1492	0.018123
All zones combined	5.039 (r)	−255.0108 (s_{ym})	9811.2035 $(s^2_y + r^2 s^2_m - 2rs_{ym})$	0.024278 (s^2_r)

Zone	*For the proportion of adults*			
	$p_h = x^*_{ho}/y^*_{ho}$	$s_{y^*_{ho}x^*_{ho}} = \frac{1}{4}d_{yh}d_{xh}$	$s^2_{x^*_{ho}} + p^2_h s^2_{y^*_{ho}} - 2p_h s_{y^*_{ho}x^*_{ho}}$	$s^2_{p_h}$ = col. (8)$/y^{*2}_{ho}$
(1)	(6)	(7)	(8)	(9)
I	0.5122	383.8398	1232.9229	0.00112859
II	0.5363	139.9123	2018.3693	0.00165480
III	0.5416	174.3958	705.3547	0.00063493
All zones combined	0.5302 (p)	698.1479 (s_{yx})	3918.3563 $(s^2_x + p^2 s^2_y - 2ps_{yx})$	0.00038179 (s^2_p)

villages, in each stratum, we follow, as in Example 20.1, the simplified procedures mentioned in note 2 of section 19.3, illustrated also in Example 11.1 for the stratified single-stage pps sampling. As in Example 20.1, we denote by y_{hij} the household size and by x_{hij} the number of adults in the sample households (h = 1, 2, 3 for the zones; i = 1, 2 for the sample villages; and j = 1, 2, 3, 4, 5 for the sample households): note that in Example 11.1, the total number of households in a sample village was denoted by x_{hi}, which we denote here by M_{hi}, and the sample number by m_{hi}. Unbiased estimates of the total number of households in the three zones separately and combined have already been obtained in Example 11.1, along with their standard errors. These, along with the other required computations, are shown in Tables 21.5–21.7, and the final estimates in Table 21.8.

Example 21.2

For estimating the total yield of paddy in a district, a stratified two-stage sample design was adopted, where four villages were selected from each stratum, with ppswr, the 'size' being the geographical area, and four plots were selected from each sample village circular systematically for ascertaining the yield of paddy. Using the data given in Table 21.9, estimate unbiasedly the total yield of paddy in the district and its standard error (Murthy, Problem 9.7).

The required computations are given in Tables 21.10 and 21.11 on page 250 denoting by y_{hij} the yield of paddy (in kilogrammes) in the jth sample plot of the ith sample village of the hth stratum (h = 1, 2, 3, for the strata, i = 1, 2, 3, 4 for the sample villages, and j = 1, 2, 3, 4 for the sample plots).

The estimated total yield of paddy in the district is 3976 tonnes (1 tonne = 1000 kg). The estimated standard error of estimate of this total is $\sqrt{(118\,352)}$ = 344 tonnes, i.e. a CV of 8.65 per cent.

Table 21.8 Estimates and standard errors computed from the data of stratified two-stage (pps and srs) design in Table 21.4

Item	*Zone I*	*Zone II*	*Zone III*	*All zones combined*
1. *Number of households**				
(a) Estimate	232.4	230.5	172.8	635.8
(b) Standard error	7.30	4.72	3.78	9.47
(c) Coefficient of variation (%)	3.14	2.05	2.18	1.49
2. *Number of persons*				
(a) Estimate	1045.2	1104.4	1054.0	3203.6
(b) Standard error	9.59	69.54	5.86	70.44
(c) Coefficient of variation (%)	0.92	6.30	0.56	2.20
3. *Number of adults*				
(a) Estimate	535.3	592.4	570.8	1698.5
(b) Standard error	40.02	27.88	29.74	57.13
(c) Coefficient of variation (%)	7.48	4.71	5.21	3.36
4. *Average household size*				
(a) Estimate	4.50	4.79	6.10	5.04
(b) Standard error	0.0999	0.3998	0.1346	0.1558
(c) Coefficient of variation (%)	2.22	8.34	2.21	2.61
5. *Proportion of adults*				
(a) Estimate	0.5122	0.5363	0.5416	0.5302
(b) Standard error	0.0336	0.0407	0.0252	0.0195
(c) Coefficient of variation (%)	6.56	7.59	4.65	3.69

* Also from Example 11.1.

Table 21.9 Yield of paddy in sample plots, selected circular systematically in the sample villages, and selected with probability proportional to geographical area in each of the three strata of a district

Stratum h	*Sample village* i	*Reciprocal of probability* $1/\pi_{hi}$	*Total number of plots* M_{hi}	*Yield of paddy (in kg)* y_{hij} 1	2	3	4
I	1	440.21	28	104	182	148	87
	2	660.43	14	108	64	132	156
	3	31.50	240	100	115	50	172
	4	113.38	76	346	350	157	119
II	1	21.00	256	124	111	135	216
	2	16.80	288	123	177	106	138
	3	24.76	222	264	78	144	55
	4	49.99	69	300	114	68	111
III	1	67.68	189	110	281	120	114
	2	339.14	42	80	61	118	124
	3	100.00	134	121	212	174	106
	4	68.07	161	243	116	314	129

Table 21.10 Computation of the estimated total yield of paddy: data of Table 21.9

Stratum h	*Sample village* i	(M_{hi}/π_{hi})	$\sum_{j=1}^{4} y_{hij}$ *(kg)*	$\bar{y}_{hi} = \frac{1}{4}\sum_{j=1}^{4} y_{hij}$ *(kg)*	$y^*_{hi} = (M_{hi}/\pi_{hi})\,\bar{y}_{hi}$ *(tonnes; 1 tonne = 1000 kg)*
(1)	(2)	(3)	(4)	(5)	(6)
I	1	12 325.88	521	130.25	1605
	2	9246.03	460	115.00	1063
	3	7560.00	437	109.25	826
	4	8616.88	972	243.00	2094
Total					5588
Mean (= stratum estimate of total)					1397 (y^*_{ho})
II	1	5376.00	586	146.50	788
	2	4838.40	544	136.00	658
	3	5496.72	541	135.25	743
	4	3449.31	593	148.25	511
Total					2700
Mean (= stratum estimate of total)					675 (y^*_{ho})
III	1	12 791.52	625	156.25	1999
	2	14 243.88	383	95.75	1364
	3	13 400.00	613	153.25	2054
	4	10 959.27	802	200.50	2197
Total					7614
Mean (= stratum estimate of total)					1903.5 (y^*_{ho})

Table 21.11 Computation of the estimated variance of the estimated total yield of paddy: data of Tables 21.9 and 21.10

Stratum h	y^*_{ho}	$\sum_{i=1}^{4} y^{*2}_{hi}$	$\frac{1}{4}\left(\sum_{i=1}^{4} y^*_{hi}\right)^2$	SSy^*_{hi} = col. (3)– col. (4)	$s^2_{y^*_{ho}} = SSy^*_{hi}/12$
(1)	(2)	(3)	(4)	(5)	(6)
I	1397.0	8773.106	7806.436	966.670	80 555.83
II	675.0	1867.078	1822.500	44.578	3714.83
III	1903.5	14 902.222	14 493.249	408.973	34 081.08
Total	3975.5 (y)				118 351.74 (s^2_y)

21.4 Stratified three-stage design with pps, srs, and srs

The example of section 21.3 will be extended in this section by taking a simple random sample of q_{hij} third-stage units out of the total number Q_{hij} of such units in the selected jth ssu of the selected ith fsu in the hth stratum ($h = 1, 2, \ldots, L$; $i = 1, 2, \ldots, n_h$; $j = 1, 2, \ldots, m_{hi}$). The sampling plan is shown in summary form in Table 21.12.

Table 21.12 Sampling plan for a three-stage design with pps sampling at the first stage, and srs at the second and third stage. In the hth stratum ($h = 1, 2, \ldots, L$)

Stage (t)	*Unit*	*No. in universe*	*No. in sample*	*Selection method*	*Selection probability*	f_t
1	First-stage (village)	N_h	n_h	ppswr	$\pi_{hi} = z_{hi}/Z_h$	$n_h \pi_{hi}$
2	Second-stage (household or field)	M_{hi}	m_{hi}	srswr	Equal = $1/M_{hi}$	m_{hi}/M_{hi}
3	Third-stage (person or plot)	Q_{hij}	q_{hij}	srswr	Equal = $1/Q_{hij}$	q_{hij}/Q_{hij}

An unbiased estimator of the stratum total Y_h (as defined by equation (20.3)), obtained from the ith sample fsu is, from equation (16.25) or (19.21) or (21.10),

$$y_{hi}^* = \frac{M_{hi}}{\pi_{hi} m_{hi}} \sum_{j=1}^{m_{hi}} \frac{Q_{hij}}{q_{hij}} \sum_{k=1}^{q_{hij}} y_{hijk}$$

$$= \frac{M_{hi}}{\pi_{hi} m_{hi}} \sum_{j=1}^{m_{hi}} Q_{hij} \bar{y}_{hij} \qquad (21.19)$$

where y_{hijk} is the value of the study variable in the kth selected tsu of the jth selected ssu of the ith selected fsu in the hth stratum ($h = 1, 2, \ldots, L$; $i = 1, 2, \ldots, n_h$; $j = 1, 2, \ldots, m_{hi}$; $k = 1, 2, \ldots, q_{hij}$) and

$$\bar{y}_{hij} = \sum_{k=1}^{q_{hij}} y_{hijk}/q_{hij} \qquad (21.20)$$

is the arithmetic mean of the y_{hijk} values in the hijth ssu.

The combined unbiased estimator of Y_h is

$$y_{ho}^* = \sum^{n_h} y_{hi}^*/n_h \qquad (21.21)$$

and an unbiased estimator of the universe total Y is

$$y = \sum^{L} y_{ho}^* \qquad (21.22)$$

Unbiased variance estimators of y_{ho}^* and y are given respectively by estimating equations (19.23) and (19.25). An estimator of the ratio of the totals of two study variables and its variance estimator follow from estimating equations (19.11)–(19.14). These will be illustrated with Examples 21.3 and 21.4.

Example 21.3

In Table 21.3 are given the number of adults in the sample households in each of the two sample villages, selected with pps from the three zones. In each selected household, one adult member (eighteen years or over) was further selected at random from the total number of adults in the household and asked about his knowledge of development plans in the country. The information is given also in Table 21.4. Estimate the total number of adults with knowledge of development plans, and also the proportion they constitute of the total number of adults in each of the three zones and in all the zones combined, along with their standard errors.

Extending the notation of Example 21.1, let Q_{hij} (= x_{hij} in Example 21.1) denote the total number of adults in the ijth sample household in the hth stratum. The number of adults selected for interview (i.e. the number of third-stage sample units) is

$q_{hij} = 1$. We can therefore dispense with the subscript k in y_{hijk} by which we had denoted the value of the study variable for ijkth third-stage sample unit in the hth stratum.

Similarly to the procedures adopted in Examples 16.2 and 20.3, we put $q'_{hij} = 1$ if the selected adult in the ijth sample household in the hth stratum knows of development plans, and 0 otherwise.

From estimating equation (21.19), we have an unbiased estimator of the stratum total Q'_{hi} of the number of adults with knowledge of development plans

$$q'^{*}_{hi} = \frac{M_{hi}}{\pi_{hi} m_{hi}} \sum_{j=1}^{m_{hi}} Q_{hij} q'_{hij}$$

from the ith first-stage unit (i.e. village) in the hth stratum. The required computations are shown in Tables 21.13–21.15, and the final results in Table 21.16.

Table 21.13 Computation of the estimated total number of adults with knowledge of development plans from a stratified three-stage design with two pps first-stage units (villages) in each stratum, srs of second-stage units (households), and srs of one adults in the selected households: data of Tables 11.2, 21.4, and 21.5

Zone h	*Sample village* i	$\sum_{j=1}^{5} Q_{hij'} q'_{hij}$	$\frac{1}{5}\sum_{j=1}^{5} Q_{hij'} q'_{hij}$	$q'^{*}_{hi} = (M_{hi'}/\pi_{hi})$.col.(4)
(1)	(2)	(3)	(4)	(5)
I	1	2	0.4	90.05
	2	3	0.6	143.83
			Total	233.88
			Mean (= Stratum estimate of total)	116.94 (q'^{*}_{h0})
			Difference	−53.78
			½ |Difference| (= Estimated standard error)	26.89 ($s_{q'^{*}_{h0}}$)
II	1	2	0.4	94.08
	2	3	0.6	135.46
			Total	229.54
			Mean (= Stratum estimate of total)	114.77 (q'^{*}_{h0})
			Difference	−41.38
			½ |Difference| (= Estimated standard error)	20.69 ($s_{q'^{*}_{h0}}$)
III	1	4	0.8	141.30
	2	3	0.6	101.45
			Total	242.75
			Mean (= Stratum estimate of total)	121.38 (q'^{*}_{h0})
			Difference	39.85
			½ |Difference| (= Estimated standard error)	19.925 ($s_{q'^{*}_{h0}}$)

Table 21.14 Computation of the estimated total number of adults with knowledge of development plans: data of Tables 21.4 and 21.13

Zone h	*Estimate* q'^{*}_{h0}	*Standard error* $s_{q'^{*}_{h0}}$	*Variance* $s^2_{q'^{*}_{h0}}$
I	116.94	26.89	723.0721
II	114.77	20.69	428.0761
III	121.38	19.92	397.0056
All zones combined	353.09 (q')	39.347 ($s_{q'}$)	1548.1538 ($s^2_{q'}$)

Table 21.15 Computation of the estimated proportions of adults with knowledge of developments and their variances

Zone	$p'_h = q'^*_{h0}/x^*_{h0}$	$s_{x^*_{h0} q'^*_{h0}} = \frac{1}{4} d_{xh} d_{q'_h}$	$s^2_{q'_{h0}} + p'^2_h s^2_{x^*_{h0}} - 2p'_h s_{x^*_{h0} q'^*_{h0}}$	$s^2_{p'_h} =$ col.(4)$/x^{*2}_{h0}$
(1)	(2)	(3)	(4)	(5)
I	0.2185	1076.1378	329.2635	0.00114903
II	0.1937	576.9406	233.7438	0.00066610
III	0.2127	592.4699	184.9696	0.00056776
All zones combined	0.2079 (p')	2245.5483 ($s_{xq'}$)	755.4598 ($s^2_{q'} + p'^2 s^2_x - 2p' s_{xq'}$)	0.00026189 ($s^2_{p'}$)

Table 21.16 Estimated numbers and proportions of adults with knowledge of development plans, computed from the data of stratified three-stage (pps, srs, and srs) design in Table 21.9

Adults with knowledge of development plans	*Zone I*	*Zone II*	*Zone III*	*All zones combined*
1. *Number*				
(a) Estimate	116.9	114.8	121.4	353.1
(b) Standard error	26.9	20.7	19.9	39.35
(c) Coefficient of variation (%)	2.99	18.03	16.42	11.14
2. *Proportion of total adults*				
(a) Estimate	0.2185	0.1937	0.2127	0.2079
(b) Standard error	0.0339	0.0258	0.0238	0.0162
(c) Coefficient of variation (%)	15.51	13.31	11.19	7.79

Example 21.4

For the household inquiries in the Indian National Sample Survey (1953–4) in the rural sector, the design was stratified three-stage. The total number of 2522 *tehsils* were grouped into 240 strata on the basis of consumer expenditure (as estimated in earlier surveys) and geographical continuity, such that each stratum contained approximately equal population, as in the census of 1951. In each stratum, two *tehsils* and in each selected *tehsil*, two sample villages, were selected sampling at both stages being with probability proportional to the 1951 Census population or area and with replacement; within each selected village, nine households on an average were selected by the enumerators systematically with a random start from the lists of households in the village, which they had prepared in the villages. The sample comprised 8235 households and 49 177 persons.

With the previous notations, the reader will see that an unbiased estimator of the stratum total Y_h in the hth stratum ($h = 1, 2, \ldots, 240$), obtained from the ith ($i = 1, 2$) first-stage unit is (from equation (21.19))

$$y^*_{hi} = \frac{1}{2} \cdot \frac{1}{\pi_{hi}} \sum_{j=1}^{2} \frac{Q_{hij}}{\pi_{hij} q_{hij}} \sum_{k=1}^{q_{hij}} y_{hijk}$$

where y_{hijk} is the value of the study variable in the kth selected third-stage unit of the jth selected ssu of the ith selected fsu in the hth stratum. The combined unbiased estimator of Y_h is

$$y^*_{h0} = \tfrac{1}{2}(y^*_{h1} + y^*_{h2})$$

and the unbiased estimator of the universe total $Y = \sum_{h=1}^{240} Y_h$ is

$$y = \sum_{h=1}^{240} y^*_{h0}$$

The unbiased variance estimators of the strata and the universe estimators of totals are given by equations (19.23) and (19.25) respectively: estimators of ratios of two totals and their variance estimators follow from section 19.3. As two fsu's are selected with replacement in each stratum, the simplified procedure of note 2 to section 19.3 could be used for the estimation of totals and ratios of totals and their variances.

Some results are given in Table 21.17 on the rates of births, deaths, marriage, and sickness. The coefficients of variations of the estimates are rather large, but the sample was not designed specifically to provide demographic estimates: for other results of a later survey, see Example 10.3. The other point to note is that the reported birth, death, and sickness rates were obvious under-estimates, as could be seen from available external evidence. The method of obtaining adjusted birth and death rates from such defective data is described briefly in section 25.5.6.

Table 21.17 Estimates of vital rates per 1000 persons: Indian National Sample Survey, rural sector, 1953–4

	Estimate	*Standard error*	*CV (%)*
Births (annual)	34.6	1.0	2.9
Deaths (annual	16.6	1.1	6.6
Marriages (annual)	7.1	0.5	7.0
Prevalence of sickness (monthly)	64.8	4.4	6.8

21.5 Special cases of crop surveys

21.5.1 Introduction

The estimating procedures in crop surveys are greatly simplified if the sampling units at each stage in a multi-stage design are selected with probability proportional to area (section 16.5). The extension to stratified multi-stage design is straightforward and will be illustrated with a stratified two-stage design.

21.5.2 Area surveys of crops

In a stratified design with L strata, in each stratum the fsu's (villages) are selected with probability proportional to their (geographical) areas and in the selected fsu's, the ssu's (fields) are selected with probability proportional to their areas. We assume that the same number n_0 of fsu's is selected in each stratum and the same number m_0 of ssu's selected in the selected fsu's.

If a_{hij} is the total area, y_{hij} the area under a crop, and $p_{hij} = y_{hij}/a_{hij}$ the proportion of the area under the crop in the hijth sample field ($h = 1, 2, \ldots, L$; $i = 1, 2, \ldots, n_0$; $j = 1, 2, \ldots, m_0$), then an unbiased estimator of proportion of area under the crop in the hth stratum is (from equation (16.31))

$$\bar{p}_h = \sum_{i=1}^{n_0} \sum_{j=1}^{m_0} p_{hij}/n_0 m_0 \tag{21.23}$$

and an unbiased estimator of the total area under the crop in the hth stratum is (from equation (16.29))

$$y_{h0}^* = A_h \bar{p}_h \tag{21.24}$$

(where A_h is the total area of the hth stratum) and in the whole universe

$$y = \sum^{L} y_{h0}^* \tag{21.25}$$

Unbiased variance estimators of $\bar{p}_h$ and y_{h0}^* are given respectively by equations of the type (16.32) and (16.33), and an unbiased variance estimator of y is given by equation (21.7).

21.5.3 Yield surveys of crops

If x_{hij} is the yield of a crop and $r_{hij} = x_{hij}/y_{hij}$ is the average yield per unit area in the hijth sample field, then an unbiased estimator of the total yield in the hth stratum is (from equation (16.34))

$$x_{h0}^* = A_h \bar{r}_h \tag{21.26}$$

where

$$\bar{r}_h = \sum_{i=1}^{n_0} \sum_{j=1}^{m_0} r_{hij}/n_0 m_0 \qquad (21.27)$$

is an unbiased estimator of the average yield per unit area in the stratum.

An unbiased estimator of the total yield in the whole universe is

$$x = \sum^{L} x^*_{h0} \qquad (21.28)$$

Unbiased variance estimators of x^*_{h0} and $\bar{r}_h$ are given respectively by estimating equations of the type (16.33) and (16.32), and an unbiased variance estimator of x is given by an estimating equation of the type (21.7).

See notes to section 16.5.

Further reading

Hansen *et al.*, Vols. I and II, Chapters 7–10; Kish, Chapter 7; Sukhatme and Sukhatme, sections 8.11–8.13.

Exercises

1. The sampling design for a socio-economic survey was stratified two-stage. In a stratum, two villages were selected with ppswr, and households within a selected village were selected with equal probability and without replacement. Table 21.18 gives the data collected on household size and monthly consumption of cereals in the sample households. Estimate the total and *per capita* monthly expenditures on cereals and the average household size in the region covered by the strata, along with their standard errors (Chakravarti *et al.*, Exercise 3.10, adapted).

2. For the data of Example 21.1, estimate for the three zones combined, the average number of adults per household and its standard error.

Table 21.18 Summary data on household size and per household consumption of cereals in sample villages

Stratum	*Sample village*	*Selection probability*	*Number of households*	*Average per household*	
				Number of persons	*Monthly expenditure on cereals (Ind. Rs.)*
h	i	π_{hi}	M_{hi}	$\bar{y}_{hi}$	$\bar{x}_{hi}$
1	1	0.0537	996	3.23	37.18
	2	0.0423	761	5.21	81.05
2	1	0.0634	1165	6.43	63.00
	2	0.0990	3108	4.26	108.04
3	1	0.0157	802	7.13	29.46
	2	0.0482	2324	5.68	59.54
4	1	0.0646	3085	6.28	42.74
	2	0.1285	7981	4.39	81.34
5	1	0.2092	8592	7.55	34.86
	2	0.0965	5921	3.68	60.37
6	1	0.0785	1119	4.10	70.48
	2	0.0167	1742	5.54	60.37
7	1	0.0268	1356	7.76	47.45
	2	0.0712	4098	5.09	43.76
8	1	0.1159	8976	6.07	45.12
	2	0.0297	1387	7.02	83.89

22 Size of Sample and Allocation to Different Strata and Stages

22.1 Introduction

The procedures for the allocation of the total sample to the different strata and stages in a stratified multi-stage design follow from those for a stratified design (Chapter 12) and a multi-stage design (Chapter 17). The total sample size is fixed in general by the availability of resources, especially the available number of trained enumerators and the number of sample units that can be surveyed by the enumerators during the survey period.

22.2 Optimum allocations

Extending the notations in sections 12.3 and 17.2, for a stratified two-stage design (and sampling with replacement), and a fixed number m_{0h} of sample ssu's in each selected fsu, the variance of the unbiased estimator of a universe total can be expressed as

$$\sum_h \frac{V_{1h}}{n_h} + \sum_h \frac{V_{2h}}{n_h m_{0h}} \tag{22.1}$$

where V_{1h} is the variability between the fsu's, V_{2h} that between the ssu's within the fsu's, and n_h the sample number of fsu's in the hth stratum ($h = 1, 2, \ldots, L$).

If the cost of travel between the fsu's within a stratum is small and is not taken into account, the following cost function may be adopted:

$$\sum_h n_h c_{1h} + \sum_h n_h m_{0h} c_{2h} \tag{22.2}$$

where c_{1h} is the cost per fsu and c_{2h} the cost per ssu in the hth stratum.

The optimum sample number of ssu's in the hth stratum is obtained on minimizing the variance for a fixed cost or *vice versa*, and is given by

$$m_0 = \sqrt{(V_{2h} c_{1h} / V_{1h} c_{2h})} \tag{22.3}$$

which is of the same form as expression (17.11) for an unstratified two-stage design. The optimum sample number of fsu's is similarly given by

$$n_h \propto N_h \sqrt{(V_{1h}/c_{1h})} \tag{22.4}$$

which again is of a similar structure as expression (12.6) for a stratified single-stage design when the cost is fixed.

The above formulae indicate the following rules for allocating the sample:

1. Select more fsu's in a stratum where (a) the number of fsu's is large, and (b) the ssu's within fsu's are heterogeneous.
2. Select more ssu's in a stratum when (a) the cost of selecting fsu's is greater than that of selecting the ssu's, (b) the average number of ssu's per fsu is large, and (c) the variability within ssu's is larger than that between fsu's.

22.3 Gain due to stratification

It can be shown that a stratified two-stage srs will be more efficient than an unstratified two-stage srs, with the same total sample size, when the variation of the stratum means is large.

To combine the efficiencies of both stratification and multi-stage sampling, the strata have to be made heterogeneous with respect to each other, with the fsu's within a stratum internally homogeneous. A small number of sample fsu's in each stratum will then provide an efficient sample. When strata are formed to contain an approximately equal sum of the sizes of the fsu's in each, two additional advantages ensue – first, an equal allocation of sample size per stratum, and second, achievement of a self-weighting design.

Further reading

Cochran, section 10.10; Hansen *et al*, Vols. I and II, Chapters 7 and 9; Kish, Chapter 8; Raj, section 6.9; Sukhatme and Sukhatme, section 7.16.

23 Self-weighting Designs in Stratified Multi-stage Sampling

23.1 Introduction

The problem of how to make a stratified multi-stage design self-weighting will be considered in this chapter. The results follow from a combination of those of Chapter 13 for stratified self-weighting designs and Chapter 18 for multi-stage self-weighting designs.

23.2 General case

Taking the stratified multi-stage design of section 19.3(5), an unbiased estimator of the universe total Y of a study variable is

$$y = \sum_{\text{sample}} \left(\prod_{t=1}^{u} \frac{1}{f_{ht}} \right) y_{h12\ldots u} \tag{23.1}$$

(see also equation (19.18)) where $y_{h12\ldots u}$ is the value of the study variable in the $(12\ldots u)$th ultimate stage unit in the hth stratum, and the factor

$$\prod_{t=1}^{u} \frac{1}{f_{ht}} = w_{h12\ldots u} \tag{23.2}$$

(see also (19.19)) is the multiplier for the $(12\ldots u)$th ultimate stage sample unit in the hth stratum, and $f_{ht} \equiv n_{h12\ldots(t-1)} \pi_{h12\ldots t}$, there being $n_{h12\ldots(t-1)}$ sample units each selected with probability $\pi_{h12\ldots t}$ out of the total $N_{h12\ldots(t-1)}$ units at the tth stage.

The design will be self-weighting with respect to y when the multiplier $w_{h12\ldots u}$ defined above is a pre-determined constant, w_0, for all the ultimate-stage sample units. It will be so when

$$f_{hu} \propto \prod_{t=1}^{u-1} \frac{1}{f_{ht}} \tag{23.3}$$

In particular, a stratified multi-stage srs will be self-weighting with proportional allocation for the first-stage units, and for each of the subsequent stages, a constant sampling fraction for selecting the next-stage units. That is, the first-stage sampling fraction n_{h1}/N_{h1} should be a constant, the second-stage sampling fraction n_{h12}/N_{h12} another constant, and so on.

If, further, a fixed number of ultimate-stage units, n_0, is to be selected in each of the selected penultimate-stage units, then the same procedure as for an unstratified multi-stage sample (section 18.2) will apply in each stratum.

See the notes to section 18.2.

23.3 Stratified two-stage design with pps and srs

The procedure for an unstratified two-stage design with pps selection at the first stage and srs (or systematic selection) at the second stage, given in section 18.3, can be extended readily to a stratified design. The multiplier for the *ij*th sample ssu in the *h*th stratum is

$$w_{hij} = \frac{1}{f_{h1} f_{h2}} = \frac{M_{hi}}{n_h \pi_{hi} m_{hi}} \tag{23.4}$$

and the design will be self-weighting with a constant multiplier w_0, then

$$m_{hi} = M_{hi}/(w_0 n_h \pi_{hi}) \tag{23.5}$$

In a household inquiry, if the number of households to be sampled in each village is to remain a constant m_0, then with a self-weighting design,

$$m_0 = M_{hi}/(n_h \pi_{hi} w_0) \tag{23.6}$$

or

$$\pi_{hi} = M_{hi}/(w_0 n_h m_0) = M_{hi}/M_h \tag{23.7}$$

where $M_h = \sum^{N_h} M_{hi}$ is the total number of households in the *h*th stratum, i.e. in any stratum the villages are to be selected with probability proportional to the total number of households; also

$$n_h = M_h/(w_0 m_0) \tag{23.8}$$

i.e. the number of sample villages is to be allocated to the different strata in proportion to the total number of households. It will also be seen that

$$\begin{aligned} w_0 &= M/(nm_0) \\ &= \text{total number of households/number of sample households} \\ &= 1/\text{sampling fraction} \end{aligned} \tag{23.9}$$

where $n = \sum^{L} n_h$ is the total number of sample villages and $M = \sum^{L} M_h$ is the total number of households in the universe.

As the total number of households existing at the time of actual enumeration will not in general be known unless the inquiry is conducted simultaneously with a census, the following operational steps may be prescribed in order to restrict the variation of the number of sample households in a sample village:

1. Make a first estimate (M'') of the present total number of households in the universe, and fix the overall sampling fraction, which is the ratio of the required number of households to be sampled (nm_0) to the first estimate of the total number of households in the universe (M''). The reciprocal of this will be the constant overall multiplier w_0.
2. Divide the whole universe into L strata by using some suitable criteria.
3. Allocate the number of sample villages to the different strata in proportion to the number of households in them as per the latest census.
4. Select the sample villages in any stratum with probability proportional to the number of census households.
5. Compute the estimated increase in the number of households in the universe from the latest census, $I = M''/M'$; and compute $m_0' = m_0/I$, the expected number of sample households per village, where M' is the total number of households in the universe as per the census.

6. Give to the enumerators the values m_0' and M_{hi}' (the number of households in the villages in the census) with instruction to select at random (or systematically) $m_{hi} = m_0' M_{hi}/M_{hi}'$ households out of the M_{hi} households listed in the ith village in the hth stratum.

   See the notes to section 18.3.

23.4 Opinion and marketing research

Extending the procedures of section 18.4, it can be seen that with the selection of one individual from a household, the sampling design can be made self-weighting if the households are selected with probability proportional to the respective sizes.

Considering Example 21.3, if households are selected with probability proportional to size (number of adults in this case), the multiplier for the adult in the ijth sample household in the hth stratum becomes

$$Q_{hi}/(n_h \pi_{hi} m_{hi}) \qquad (23.10)$$

If this is to be a constant, then

$$m_{hi} = Q_{hi}/(w_0 n_0 \pi_{hi}) \qquad (23.11)$$

In practice, the factors $1/(w_0 n_h \pi_{hi})$ will be given to the enumerators, who will select the required number m_{hi} of households which is the product of this factor and Q_{hi} the total number of adults listed in the ith block in the hth stratum.

If the blocks are selected with equal probability, then the factor $1/(w_0 n_h \pi_{hi}) = N_h/(w_0 n_h)$ will be a constant in the ith stratum. If, in addition, a constant sampling fraction for the first-stage units is taken in all the strata, then the factor $N_h/(w_0 n_h)$ becomes a constant for all strata. In this case, the number of households to be selected in any block will bear a constant ratio to the current number of adults of the block.

23.5 Self-weighting *versus* optimum designs

With simple random sampling, and assuming that $M_{hi} = M_{h0}$, and $m_{hi} = m_{h0}$, it can be shown that the design will be self-weighting when $n_h m_{h0}/N_h M_{h0} = f_0$ is a constant in all strata. On the other hand, the optimum allocation gives

$$f_{0h} = n_h m_{h0}/N_h M_{h0} \propto \sqrt{(V_{2h}/c_{2h})}$$

While the cost per ssu (c_{2h}) may be approximately equal in fsu's of different sizes (i.e. with different total number of ssu's), the variation among the ssu's (as measured by V_{2h}) may be greater in larger fsu's. A self-weighting design will not thus be necessarily optimum. But as the theoretically optimum design is not generally attainable while the optimum values are generally broad, a self-weighting design will often be as efficient as the optimum.

As the above formulation assumes the same field cost function for both the optimum and the self-weighting designs, while the latter would entail considerably less cost for tabulation, there is an added justification in trying to achieve a self-weighting design.

Further reading

Cochran, section 10.10; Hansen *et al.*, Vol. I, sections 7.12 and 9.11; Murthy, section 12.3; Som, 1958–59, 1959; Sukhatme and Sukhatme, section 8.14.

Part V :

Miscellaneous Topics

24 Miscellaneous Sampling Topics

24.1 Introduction

In this chapter will be considered miscellaneous topics such as multi-phase sampling; sampling on successive occasions; estimation of mobile populations; inverse sampling; interpenetrating networks of samples; simplified methods of variance computations; method of collapsed strata; controlled selection; and the use of Bayes's theorem in sampling.

24.2 Multi-phase sampling

24.2.1 Reasons for multi-phase sampling

It is sometimes convenient and economical to collect certain items of information from all units of a sample, and other items of information from a sub-sample of the units which constitute the original sample. This plane is termed *two-phase sampling or double sampling:* when extended to three or more phases, it is termed *multi-phase sampling.* Multi-phase sampling can be used with stratified multi-stage designs, but its use will be illustrated for double sampling with srs (with replacement) at both phases in an unstratified design.

Multi-phase sampling can be used for various reasons:

1. When the numbers of units needed to give the required precision on different items is widely different, either owing to the fact that the variabilities of the variables are different or because the precisions required are different. If no use is made of the relations between the different variables, such multi-phase sampling is equivalent to taking samples of different sizes for the different items. Thus in a household budget inquiry, information on irregular purchases, such as of clothings and furniture, could be collected, because of their larger variability, from a large sample, and information on regular purchases, such as of foodstuffs, from a sub-sample.
2. When it is comparatively difficult or costly to collect information on some items, this can be done at the second-phase sample, while the first-phase sample can be canvassed for simpler items only. In a crop survey, it is usual to select a large sample of fields (or farms or plots), generally in a stratified multi-stage design; the acreage of the crop is determined from the whole sample of fields (or farms or plots) and yield rates are determined from a sub-sample.
3. First-phase information may also be used as ancillary information, in order to improve the efficiency of second-phase information by the ratio and regression methods, where ancillary information on the whole universe is not available. If the first-phase information is collected prior to the second-phase information, it may be used as a basis for the sampling plan, e.g. for stratification of the first-phase sample units (for the selection of the second-phase units) or for pps selection of the second-phase units or for both.

Notes

1. For selecting the second-phase sample, the first-phase sample serves as the universe, but since the first-phase information is based on a sample, it is itself subject to sampling errors, and this must be taken into account in the estimating procedure.

2. Multi-phase sampling is structurally different from multi-stage sampling: in the former the same sampling units are used throughout, but in the latter a hierarchy of sampling units is used. The two may of course be combined.
3. An example of double sampling is found in the Survey of Level of Living among rural Africans in Cameroon, 1961–5. For the demographic sample survey, which was conducted first, the design was stratified single-stage, a systematic sample of villages (fsu's) being taken in each stratum. For the socio-economic inquiry, a sub-sample of the original sample fsu's was taken with probability proportional to the existing number of households, and in each such selected unit a fixed number of households (second-stage units) were selected with equal probability and without replacement. For obtaining estimates of totals and means, the ratio method of estimation was used.

24.2.2 Double sampling

We shall consider simple random sampling with replacement at both the phases, a sub-sample of n' ($< n$) being taken of the original sample of n units from a universe of N units. In the second phase, if no account is taken of the information collected at the first phase, then an unbiased estimator of the universe mean $\bar{Y} = \sum^{N} Y_i/N$ is the sample mean

$$\bar{y}' = \sum^{n'} y_i'/n' \tag{24.1}$$

with sampling variance (from equation (2.17))

$$\sigma^2_{\bar{y}'} = \sigma^2_y/n' \tag{24.2}$$

an unbiased estimator of which is (from equation (2.20))

$$s^2_{\bar{y}'} = \sum^{n'} (y_i' - \bar{y}')^2/n'(n' - 1) \tag{24.3}$$

However, account may be taken of the first-phase information by the regression or the ratio method of estimation. If the regression method is used, then analogous to equation (3.24), which applies in single-phase sampling, we have the regression estimator of the universe mean

$$\bar{y}'_{\mathbf{Reg}} = \bar{y}' + \hat{\beta}'(\bar{x} - \bar{x}') \tag{24.4}$$

where $\bar{x}$ and $\bar{x}'$ are the sample means of the ancillary variable, obtained respectively from the first phase of n units and the second phase of n' units, and $\hat{\beta}'$ is an estimator of β, the regression coefficient of y and x, based on the second phase sample.

The variance of this regression estimator is approximately

$$\begin{aligned}\sigma^2_{\bar{y}'_{\mathbf{Reg}}} &= \frac{\sigma^2_y \rho^2}{n} + \frac{\sigma^2_y (1 - \rho^2)}{n'} \\ &= \frac{\sigma^2_y}{n'}\left[1 - \rho^2\left(1 - \frac{n'}{n}\right)\right]\end{aligned} \tag{24.5}$$

where ρ is the universe correlation coefficient between the study and the ancillary variables.

Clearly, the variance of the regression estimator will be less than that for the unbiased estimator based on an srs of n' units (namely, σ^2_y/n'), unless

$\rho = 0$. When $\rho = 0$, the two become equal, and there is no advantage in double sampling.

Optimum sizes. Considering the cost aspect, it is necessary to determine the optimum sizes of the original sample (n) and that of the sub-sample (n') which will provide estimators with the minimum error at a given cost or *vice versa.*

Taking the simple cost function

$$C = nc_1 + n'c_2 \tag{24.6}$$

then for a fixed cost C, the optimum values of n and n' are respectively

$$n = C\left[c_1 + c_2\sqrt{\left(\frac{1-\rho^2}{\rho^2}\cdot\frac{c_1}{c_2}\right)}\right]^{-1} \tag{24.7}$$

and

$$n' = (C - nc_1)/c_2 = n\sqrt{\left[\frac{1-\rho^2}{\rho^2}\cdot\frac{c_1}{c_2}\right]} \tag{24.8}$$

The minimum variance in this case is

$$\frac{\sigma_y^2\,[\sqrt{\{(1-\rho^2)c_2\}} + \rho\sqrt{c_1}]^2}{C} \tag{24.9}$$

where $\rho > 0$.

In the comparison of single-phase and double sampling, it may be noted that if a single-phase sample has to provide the same information as a double sample, the cost per unit in the former would also be c_2, so that with the given total cost C, the size of such a single-phase sample is

$$n = \frac{C}{c_2} = n\frac{c_1}{c_2} + n' \tag{24.10}$$

and the value of ρ at which the variance of the optimum double sample becomes equal to that of an equivalent single-phase sample is

$$\sqrt{\left[\frac{4c_1c_2}{(c_1+c_2)^2}\right]} \tag{24.11}$$

ρ must be considerably large in order to benefit from a double sample. For example, if $c_2/c_1 = 10$, i.e. the unit cost of collecting data on y is ten times that for observing x, then the two variances are equal when $\rho = 0.58$; when $\rho = 0.8$, the variance in double sampling is about 0.75, and when $\rho = 0.9$, the variance is about 0.5, that in an equivalent single sample.

Estimating equations. In the regression estimator in double sampling in srs (equation (24.4)), β' is estimated by (*cf.* equation (3.23))

$$\hat{\beta}' = \mathrm{SP}y_i'x_i'/\mathrm{SS}x_i' \tag{24.12}$$

and a variance estimator of $\bar{y}'_{\mathbf{Reg}}$ is

$$\frac{s_{y'}^2\,\hat{\rho}'^2}{n} = \frac{s_{y'}^2\,(1-\rho'^2)}{n'} \tag{24.13}$$

where $s_{y'}$, $s_{x'}$, and $\hat{\rho}' = \hat{\beta}' s_{x'}/s_{y'}$ are computed from the sub-sample.

Example 24.1

A simple random sample of 75 308 (= n) farms out of a total 1 200 000 (= N) farms gave the estimated average area per farm as 31.25 acres (= $\bar{x}$). A sub-sample of 2055 (= n') gave the following data (y'_i denoting the number of cattle and x'_i the area of the ith farm): $\sum^{n'} y'_i = 25\,751$; $\sum^{n'} y'^2_i = 597\,737$; $\sum^{n'} x'_i = 62\,989$; $\sum^{n'} x'^2_i = 2\,937\,851$; and $\sum^{n'} y'_i x'_i = 1\,146\,391$.

Estimate the average number of cattle per farm and its standard error using the second-phase sample data (United Nations, *Manual,* Example 30; also *see* exercise 3, Chapter 3 of this book).

Here $\bar{y}' = 25\,751/2055 = 12.5309$; $\bar{x}' = 62\,989/2055 = 30.6516$ acres; $s_{y'} = 11.551$; $s_{x'} = 22.143$ farms; and $\hat{\beta}' = 0.354551$.

The regression estimate of the average number of cattle per farm in double sampling is (from equation (24.4)) $\bar{y}_{\text{Reg}} = 12.7431$.

As $\hat{\rho}' = \hat{\beta}' s_{x'}/s_{y'} = 0.679666$, the variance estimator of $\bar{y}'_{\text{Reg}}$ is (from equation (24.13)) 0.03573, so that the estimated standard error of $\bar{y}'_{\text{Reg}}$ is 0.1891, with estimated CV of 1.48 per cent.

If no account is taken of the information in the original sample, then the unbiased estimate of the average number of cattle per farm is $\bar{y}' = 12.5309$, with an estimated standard error of 0.2548 and CV of 2.03 per cent.

Note: In double sampling, the ratio method of estimation may also be used, i.e.

$$\bar{y}'_{\text{R}} = \bar{x}\bar{y}'/\bar{x}' \qquad (24.14)$$

but this estimator does not have any general superiority over the regression estimator, except for simplicity of computation; it may however be better than the unbiased estimator.

See also the notes to sections 3.2 and 3.3.

24.3 Sampling on successive occasions

24.3.1 Introduction

Sampling inquiries are often carried out at successive intervals of time on a continued basis covering the same universe, such as in the Current Population Survey of the U.S.A., and in the National Sample Survey of India. Various methods are employed for this purpose, e.g. by the selection of a new sample on each occasion, by the partial replacement of the sample, and by sub-sampling the initial sample. In such surveys, interest centres on measuring one or more of the following:

1. the change in a universe parameter, such as the mean value of a study variable;
2. the average of the mean values for all the occasions combined;
3. the mean value of the most recent occasion.

The methods of estimation for the above will be illustrated for sampling on two occasions with srs. Owing to the generally high positive correlation between the values on two occasions, it is better to retain the same sample to estimate the change in the mean values, and to draw a new sample to estimate the average for both the occasions combined. For estimating the mean for the second occasion, the same initial sample or a new sample of the same size would give equally precise estimators, but a more efficient scheme would be to replace a part of the sample on the second occasion and to use the double sampling method with the regression estimator.

24.3.2 Estimating the change in the mean value

If $\bar{y}_1$ and $\bar{y}_2$ are the sample means (which are unbiased estimators of the respective universe means $\bar{Y}_1$ and $\bar{Y}_2$ in srs) on two occasions (the subscripts referring to the occasions), both being based on samples of size n, the change in the universe mean values ($\bar{Y}_2 - \bar{Y}_1$) is estimated unbiasedly by the change in the sample means, namely,

$$\bar{y}_2 - \bar{y}_1 \quad (24.15)$$

If the samples are independent, then the sampling variance of the estimator $(\bar{y}_2 - \bar{y}_1)$ in equation (24.15) is

$$(\sigma_1^2 + \sigma_2^2)/n \quad (24.16)$$

If, however, the same sample is used on both the occasions, the sampling variance of the estimator in equation (24.15) is

$$(\sigma_1^2 + \sigma_2^2 - 2\rho\sigma_1\sigma_2)/n \quad (24.17)$$

As ρ, the correlation coefficient between the values on the same units on successive occasions, is likely to be positive and very high, the variance in expression (24.17) will be less than the variance in expression (24.16), and it is therefore better to retain the same sample to estimate more efficiently the change in the mean values.

24.3.3 Estimating the mean for both the occasions combined

The estimator of the overall mean is $\frac{1}{2}(\bar{y}_1 + \bar{y}_2)$. Its sampling variance is

$$(\sigma_1^2 + \sigma_2^2 + 2\rho\sigma_1\sigma_2)/4n \quad (24.18)$$

When ρ is positive it is better to draw a new sample in order to estimate the overall mean more efficiently.

24.3.4 Estimating the mean for the second occasion

With the reasonable assumption that $\sigma_1^2 = \sigma_2^2 = \sigma^2$, the means obtained from the initial sample and from an independent second sample will have the same sampling variance σ^2/n, i.e. will be equally efficient.

Let us, however, consider that the sample is replaced in part, with m out of the initial n units matching (i.e. retained) and a fresh sample of $n - m = u$ unmatching units (i.e. replaced) taken on the second occasion. Unbiased estimators of the universe mean $\bar{Y}_2$ for the second occasion are:

$$\left.\begin{array}{l}\text{from the unmatched } u \text{ units: } \bar{y}_{2u}\text{, with variance } \text{var}(y_{2u}) = \sigma^2/u \\ \text{from the matched } m \text{ units: } \bar{y}_{2m}\text{, with variance } \text{var}(y_{2m}) = \sigma^2/m\end{array}\right\} \quad (24.19)$$

However, for the matched part, a more efficient estimator than the unbiased estimator $\bar{y}_{2m}$ is obtained by using the double sampling method with the regression estimator, i.e. from equation (24.4),

$$\bar{y}_{2\text{Reg}} = \bar{y}_{2m} + \hat{\beta}(\bar{y}_1 - \bar{y}_{1m}) \quad (24.20)$$

where $\bar{y}_1$ is the sample mean for all the n units and $\bar{y}_{1m}$ that for the m matched units on the first occasion. From equation (24.5), the sampling variance of $\bar{y}_{2\text{Reg}}$ is approximately

$$\text{var}(\bar{y}_{2\text{Reg}}) = \frac{\sigma^2\rho^2}{n} + \frac{\sigma^2(1-\rho^2)}{m} \quad (24.21)$$

A still better estimator of $\bar{Y}_2$ is obtained on weighting the two independent estimators $\bar{y}_{2u}$ and $\bar{y}_{2\text{Reg}}$ inversely as their variances, i.e.

$$\bar{y}_{2\text{comb}} = [\bar{y}_{2u}\,\text{var}(\bar{y}_{2\text{Reg}}) + \bar{y}_{2\text{Reg}}\,\text{var}(\bar{y}_{2u})]/[\text{var}(\bar{y}_{2u} + \text{var}(\bar{y}_{2\text{Reg}}))] \quad (24.22)$$

The variance of this combined estimator is approximately

$$\frac{\sigma^2(n - u\rho^2)}{n^2 - u^2\rho^2} \tag{24.23}$$

The optimum matching fraction (not taking the cost and other factors into consideration) is

$$\frac{m}{n} = \frac{\sqrt{(1 - \rho^2)}}{1 + \sqrt{(1 - \rho^2)}} \tag{24.24}$$

and the minimum variance in this case is

$$V_{\min} = \frac{\sigma^2}{2n}[1 + \sqrt{(1 - \rho^2)}] \tag{24.25}$$

The optimum percentage to match never exceeds 50 per cent and decreases steadily as $|\rho|$ increases. Considering more than one item in a survey, a good matching fraction will be one-third or one-fourth. Unless ρ is very high, the gain from using the above procedure is modest.

Notes

1. In sampling on more than two occasions, the optimum matching fraction tends to $\frac{1}{2}$, irrespective of the value of ρ.
2. Sampling the same units on successive occasions reduces the field cost. But in surveys of human populations, the co-operation of the respondents may diminish after a time and the process of repetitive surveys itself may condition response and behaviour; it is better, therefore, to rotate the sample by partial replacement of the units according to a specific pattern. In the Current Population Survey in the U.S.A., and the Labour Force Surveys in Israel, an individual household remains in the sample during four consecutive months, and after an eight-month interval is brought back for another four months. In the Micro-Census of the Federal Republic of Germany, where the sample design is stratified single-stage (see exercise 5, Chapter 10), one-third of the sample enumeration districts are replaced every year. Partial replacement of sampling units is also made in the labour force surveys in Canada and Egypt.
3. A relevant question in sampling on successive occasions is the choice of the sampling unit. Problems, often intractable, arise from the addition and depletion of units, such as buildings, dwelling units, households, and families. In a household or demographic inquiry, for example, over a period of time households change their composition (due to births, deaths, and migration), re-form or dissolve; taking a specific geographical area, some households move out of it and some new households move in. Similar problems arise in agricultural inquiries where the farmers are the ultimate-stage sampling units. To overcome these problems, it is preferable to use areal units as the ultimate-stage sampling units and enumerate completely the elementary (or recording) units in the sample areal units: areal sampling units have the other advantage of reducing response errors mentioned in section 25.7. On the other hand, to avoid the bias arising from repeated visits of the same recording units (note 2 above), a compromise would be either to have a partial replacement of the areal units as in the Micro-Census of the Federal Republic of Germany, or to retain the same penultimate stage areal units (such as villages and urban-blocks) with a partial rotation of the elementary units within them.

24.4 Estimation of mobile populations

The problem of estimation of the number and characteristics of mobile populations such as animals, birds, fish and insects, requires the adoption of special sampling techniques. One is the 'capture-recapture' method.

24.4.1 Capture-recapture method

Let N_A birds be captured, marked (banded), and released. If afterwards n birds are recaptured, of which n_A are found to be marked, from simple consideration of probability an estimator of the total number of birds in the universe is

$$N^* = N_A n/n_A \tag{24.26}$$

This estimator is known as 'Lincoln Index' after F. C. Lincoln (1930) who had used it for estimating the total number of waterfowl on the basis of banding returns in the U.S.A. in 1920–6. It is a maximum likelihood estimator and consistent but biased.

The estimator (24.26) can be derived from the ratio estimator of the total of a study variable from a simple random sample (without replacement) of n out of N units (Hájek and Dupač). Denoting by y and z the sample totals of the study and the ratio variables respectively and Z the total of the ratio variable for the N units, the ratio estimator of the total Y is

$$y_R^* = Zy/z \tag{24.27}$$

If $Y_i = 1$ for all i and $z_i = 1$ for the marked units and $z_i = 0$ for the other units, then $Y = N$, $y = n$, $Z = N_A$, $z = n_A$, and from formula (24.27) we obtain as an estimator of the total number of units in the universe

$$N^* = Zn/z = N_A \,.\, n/n_A \tag{24.28}$$

A variance estimator of N^* is

$$s_{N^*}^2 = \frac{Z}{z}\left(\frac{Z}{z} - 1\right)\frac{n}{n-1}\sum^{n}\left(1 - \frac{n}{z}z_i\right)^2$$
$$= \frac{n^2}{n-1}\frac{N_A}{n_a}\left(\frac{N_A}{n_A} - 1\right)\left(\frac{n}{n_A} - 1\right) \tag{24.29}$$

The assumption in this method is that 'capture and recapture' represent a simple random sample without replacement.

24.4.2 Enumeration of nomads

For estimating mobile human population groups, such as nomads, experiences in some African and West Asian countries show the hierarchical approach through the tribal chiefs and administration to have considerable promise; use of water-holes as sampling units presents certain difficulties and requires progressive modification of the sample design, while aerial photography is much less successful, besides being very costly. The use of the 'capture-recapture' method for this purpose has yet to be explored.

• 24.5 Inverse sampling

In inverse sampling, sampling is continued until certain specified conditions, dependent on the results of the sampling, have been fulfilled. We shall consider inverse sampling for proportions and for continuous data.

24.5.1 Inverse sampling for proportions

When the universe proportion $P(= N'/N)$ of the number of units possessing an attribute (N') is very small, i.e. when the attribute is rare, an estimate of P

obtained from even a very large (simple random) sample by the method of section 2.13, may not be reasonably precise. And a bad guess of P, required to estimate the sample size to provide an estimate with a given precision, may result in considerable under- or over-sampling.

With inverse sampling for an attribute in a simple random sample without replacement, the sample size n is not fixed in advance, but sampling is continued until a predetermined number n' possessing the attribute are returned. An unbiased estimator of P is

$$p = \frac{n' - 1}{n - 1} \qquad (24.30)$$

where n is the total sample size with the n' required units.

An unbiased variance estimator of p is

$$s_p^2 = \frac{p(1-p)}{n-2}\left(1 - \frac{n-1}{N}\right) \qquad (24.31)$$

When N is large, the variance estimator of p is

$$s_p^2 = \frac{p(1-p)}{n-2} \qquad (24.32)$$

Further, with a small p, the sample size n will be fairly large, so that the variance estimator s_p^2 in equation (24.32) could be approximated by $p(1-p)/(n-1)$, which is the form of the unbiased variance estimator of $p = n'/n$ in srs with replacement (equation (2.74)).

24.5.2 Inverse sampling of continuous data

Suppose samples are taken with equal or varying probabilities (and with replacement), until a predetermined number $(r + 1)$ of distinct units are included, and this takes $(n + 1)$ samples; the last unit, which is the $(r + 1)$th distinct unit, is discarded. The sample, now of size n with r distinct units, may then be analysed as if it were a sample deliberately chosen by the method of section 2.9 for srs of section 5.3 of pps, both with replacement.

The (combined) unbiased estimator of the universe total Y is, as before,

$$y_0^* = \sum^{N} y_i^*/n \qquad (24.33)$$

where

$$y_i^* = Ny_i \quad \text{in srs} \qquad (24.34)$$

$$= y_i/\pi_i \quad \text{in pps} \qquad (24.35)$$

where N is the total number of units, y_i the value of the study variable of ith selected unit ($i = 1, 2, \ldots, n$), and π_i the probability of selection of the ith unit ($i = 1, 2, \ldots, N$ for the universe, and $i = 1, 2, \ldots, n$ for the sample) in pps sampling.

An unbiased variance estimator of y_0^* is, as before,

$$s_{y_0^*}^2 = \sum^{n} (y_i^* - y_0^*)^2/n(n-1) \qquad (24.36)$$

But $s_{y_0^*}^2$, as defined in equation (24.36), is not an unbiased estimator of the sampling variance

$$\sigma^2_{y_0^*} = \frac{\sum^{N} (Y_i - \bar{Y})^2}{nN} \qquad \text{in srswr} \qquad (24.37)$$

$$= \frac{1}{n} \sum^{N} \left(\frac{Y_i}{\pi_i} - Y \right)^2 \pi_i \qquad \text{in ppswr} \qquad (24.38)$$

(see also equations (2.17), (5.4)). The sampling variance of y_0^* in inverse sampling is approximately

$$\sigma^2_{y_0^*} \left(1 - \frac{n}{2N}\right) = \sigma^2_{y_0^*} \left(1 - \frac{1}{2} f\right) \qquad (24.39)$$

where $\sigma^2_{y_0^*}$ is given by equation (24.37) for srswr and by equation (24.38) for ppswr, and $f = n/N$ is the sampling fraction.

The sampling variance in inverse sampling is thus less than that of $\sigma^2_{y_0^*}$. •

24.6 Interpenetrating network of sub-samples

When two or more sub-samples are taken from the same universe by the same sampling plan so that each sub-sample covers the universe and provides estimators of the universe parameters on application of the same estimating procedures, the sub-samples are known as interpenetrating networks of samples. The sub-samples may or may not be drawn independently and there may be different levels of interpenetration corresponding to different stages in a multi-stage sample scheme. The sub-samples may also be distinguished by differences in the survey procedures or of processing features. These are sometimes known as 'replicated samples'. The technique was developed by Mahalanobis in 1936; a variant of this is sometimes called the 'Tukey plan' (Deming, 1966, footnote pp. 186–7).

This technique enables one to:

(a) examine the factors causing variation, e.g. enumerators, field schedules, different methods of data collection and processing;

(b) compute the sampling error from the first-stage units if these comprise one level of interpenetration (both by the standard method as also by a non-parametric test);

(c) provide control in data collection and processing;

(d) supply advanced estimates on the basis of one or more sub-samples and to provide estimates based on one or more sub-samples when the total sample cannot be covered due to some emergency;

(e) provide the basis of analytical studies by the method of fractile graphical analysis.

The technique may be incorporated as an integral part of the standard sample designs and has been used in a number of sample surveys in India, including the Indian National Sample Survey, Peru, Southern Rhodesia, the Philippines, and the U.S.A.

The first three uses of the technique are illustrated in section 25.5.4.

24.6.1 Advanced estimates from sub-samples

Provisional, advanced estimates can be obtained as required on earlier processing of one or more of the sub-samples, as was done in crop surveys in Bengal (India) and in the 1961 Census of Agriculture in Peru. In emergencies also, one or more sub-samples can be jettisoned and the other(s) used.

24.6.2 Fractile graphical analysis

A simple graphical representation can supply a visual (or geometric) means of assessing and controlling errors and also a measure of the margin of error (or of uncertainty), especially in the case of an interpenetrating network of independent sub-samples. The observed sample units in each sub-sample are ranked in

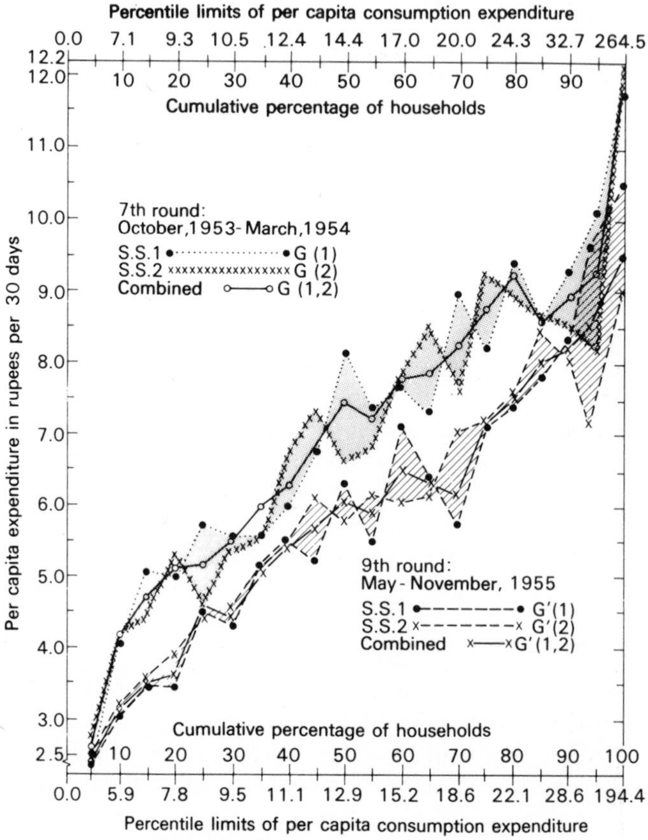

Figure 24.1 Fractile graphical analysis of *per capita* expenditure on food grains and *per capita* household expenditure per 30 days; Indian rural sector 1953–4 and 1955 (Source: Mahanobis, 1960, Figure 1).

ascending order of any suitable variable (or even in order of time of observation) and the whole sub-sample is divided into a suitable number of fractile groups of equal size (i.e. consisting of equal estimated number of universe units). For each sub-sample the average value, or median, or some other estimate of any variable for each fractile group, is plotted on paper and the successive plotted points are connected by straight lines; these polygonal lines are called the fractile graphs of the sub-samples. A fractile graph for the whole sample can be drawn in the same way, retaining the same number of fractile groups, as in the case of the sub-samples. The area between the fractile graphs for the sub-samples provides a visual and geometric estimate of the error associated with the fractile graph of the whole sample. This method can be used not only to assess the significance of observed deviation from a statistical hypothesis relating to the results of the survey for any particular part or the whole range of observed data, but also for comparisons between results based on two or more surveys carried out in different regions or at different periods of time in the same region using the fractile graphical error for each survey. This method is being extensively used in the Indian National Sample Survey, and Figure 24.1 shows the fractile graphs for *per capita* expenditure on food grains and *per capita* household expenditure in 1953–4 and in 1955 (Mahalanobis, 1960). The shaded area for each period provides a measure of the error associated with the graph for the estimates from the two sub-samples combined, and the 'separation' between the two periods is given by the area lying between the two graphs for the sub-samples combined. From the figure, it can be said that (a) the general trend of the increase in expenditure is significant in comparison with the associated error for both the periods; and (b) the decrease in the expenditure on food grains in 1955 as compared with that in 1953–4 for up to 80 per cent of the households (counting from the bottom) may be considered statistically significant, for the 'separation' for these households is seen to be roughly greater than the sum of the two measures of error provided by the shaded areas.

24.7 Simplified methods of variance computations

As has been described before, the standard method of estimating the sampling variance of the estimator of a universe total is to base it on the estimators of the total obtained from each of the sample first-stage units selected with replacement. If a fixed number of sample fsu's are selected in each stratum, the computation of variance is simplified, the more so if two sample fsu's are selected in each stratum (section 9.5 and note 2 of section 19.3). It has also been noted earlier that the computation of estimates and their variances become very simple if the design is made self-weighting, either at the field or at the tabulating stage. The use of interpenetrating networks of samples in providing variance estimators will be discussed in section 25.5.4.

24.7.1 Method of random groups

As noted in section 2.14, if the sample units in an srs without replacement are divided into a number of groups with an equal number of sub-sample units in each, estimates of sampling variances could be computed from the group means.

A variation of this method, followed in the most recent Agricultural Survey in the U.S.A., is, in each of the eight regions covered, to assign a random group number, 1 to 10, to a schedule before it is punched, and in tabulating to sort the cards first according to the tabulation requirements and then by the random group number, obtaining successive cumulative estimates relating to the random groups. Estimates of variance are obtained by 'decumulating' the data from the auxiliary tapes and performing the appropriate computations.

24.7.2 Selection of a random pair of units from each stratum

Another simplified procedure would be to select a random pair of sample units from each of a large number of strata with approximately equal sizes and to estimate the variance from the random pairs.

24.7.3 Use of 'error graphs'

A sample survey usually contains many items of information, and the process of computation of variance estimators for each of the items and their cross-classifications becomes laborious and expensive. If a relation could be found for a survey between the estimator of Y and its standard error that holds substantially for a large number of items, then a curve (the 'error graph'), fitted on a suitable scale, may be used for estimating the standard errors of less important item totals. Such error graphs have been used by Yates (section 7.20), and in surveys conducted by the U.S. Bureau of the Census (Hansen *et al.*, Vol. II, section 12.B), and the U.S. National Center for Health Statistics.

Note: Whenever a sub-sample is drawn from the original sample for computing the variance estimate of an estimate obtained from the larger, original sample, caution must be exercised in using the appropriate estimating equations. Suppose, for example, that from an srs with replacement of n out of N units, the unbiased estimator y_0^* of $\bar{Y}$ is computed by the method of section 2.9, i.e. $y_0^* = N \sum^n y_i/n$. If a sub-sample of n' units is selected at random from the n units then an unbiased estimator of σ^2, computed from the sub-sample of n' units, will be

$$s'^2 = \sum^{n'} (y_i' - \bar{y}')^2/(n' - 1) \tag{24.40}$$

where $\bar{y}_i' = \sum^n y_i'/n$, and y_i' $(i = 1, 2, \ldots, n')$ are the values of the study variable in the sub-sample. To estimate the sampling variance of y_0^*, defined above, this value of s'^2 is substituted for s^2 in estimating equation (2.44), which then becomes

$$\begin{aligned} s_{y_0^*}^2 &= N^2 s^2/n \\ &= N^2 s'^2/n \\ &= \frac{N^2}{n} \cdot \frac{SSy_i'}{(n' - 1)} \end{aligned} \tag{24.41}$$

24.8 Method of collapsed strata

When the universe is stratified taking into account the characteristics of more than one study variable, stratification may have to be carried to a degree at which there is only one first-stage sample unit in each stratum. Estimators of variances of totals, means etc. cannot then be computed by the standard method. In such a situation, the strata may be combined into a smaller number of groups such that each group contains at least two first-stage sample units. Estimators of variances can then be computed from the sample fsu's in the newly formed groups. This is known as the *method of collapsed strata.*

The variance estimator, computed from such collapsed strata, over-estimates the sampling variance. A very satisfactory approximation may, however, be obtained when the number of strata is large and the strata are combined into groups such that for each group the strata are about equal in size (total Y). On the other hand, the variance estimator will be a serious under-estimate if the groups are constructed by seeing the sample results and making them differ as little as possible.

• 24.9 Controlled selection

Controlled selection is any process of selection in which, while maintaining the assigned probability for each unit, the probabilities of selection for some or all preferred combinations of n out of N units are larger than in stratified srs (and correspondingly the probabilities of selection for at least some non-preferred combinations are smaller than in stratified srs).

To illustrate from an actual survey situation described by Hess, Reidel, and Fitzpatrick (1961), suppose the universe of nine hospitals is divided into two strata, one containing four large hospitals A, B, C and D, and the other five small hospitals a, b, c, d and e. If it is decided to select one hospital each from each stratum with equal probability, the probability of selection of a hospital from stratum I is 0.25 and that from stratum II is 0.20, and all the $4 \times 5 = 20$ combinations of one hospital each from a stratum has the same probability, 0.05. If, however, hospitals A, B, a, and b are state owned and the others are privately owned, then of the twenty combinations ten will comprise one state and one private hospital. Although each of the twenty combinations will provide an unbiased estimator of a universe total, it is preferable to have in the sample one state hospital and one private hospital; but the chance of selection of such preferred samples is 10/20, i.e. 0.50.

One method of ensuring higher probabilities of selecting such preferred samples is due to Goodman and Kish (1950), which in the above example would increase the probability of selection of a preferred sample from 0.50 to 0.90 in stratified srs.

Controlled sampling is mainly practicable and useful when the sample is made up of a few large first-stage units. It then enables additional control to be introduced beyond what is possible by stratification alone, this being limited by the fact that the number of strata cannot exceed the number of units in the sample. However, in the words one of the originators of the method, it 'has several drawbacks . . . it may be safer for non-specialists to avoid any controlled selection in favour of simpler and more standard probability methods' (Kish, section 12.8). The drawbacks relate mainly to the difficulties of application that require skill and care, arbitrariness involved in selection, and the estimation of the sampling variance. •

• 24.10 Use of Bayes' theorem in sampling

In section 2.8 and elsewhere, probabilistic statements concerning an unknown universe parameter have been made from considerations of confidence intervals. Another approach involves the use of Bayes' theorem, which plays an important rôle in the treatment of the foundations of probability on which the theory of survey sampling is based. The application of Bayes' theorem in sampling is somewhat controversial.

Consider k events $H_1, \ldots, H_k$, which are mutually exclusive and exhaustive (i.e. they exclude each other two-by-two and one necessarily occurs). An event A is observed and is known to have occurred in conjunction with or as a consequence of the event H_j. The joint probability of occurrence of the events A and H_j is

$$P(AH_j) = P(A|H_j)\,P(H_j) = P(H_j|A)\,P(A) \qquad (24.41)$$

where $P(A|H_j)$ is the conditional probability of A given H_j, and $P(H_j|A)$ that of H_j given A.

As the event A can occur only in conjunction with some H_i $(i = 1, 2, \ldots, k)$, and the AH_i are mutually exclusive, the probability of occurrence of A is

$$P(A) = \sum_1^k P(AH_i)$$
$$= \sum_1^k P(A|H_i)\,P(H_i) \qquad (24.42)$$

from equation (24.41). Also, from equation (24.41), the probability of occurrence of H_j, given A, is

$$P(H_j|A) = \frac{P(AH_j)}{P(A)}$$
$$= \frac{P(A|H_j)\,P(H_j)}{\sum_1^k P(A|H_i)\,P(H_i)} \qquad (24.43)$$

That is, the probability of H_j, given A, is proportional to the probability of A, given H_j, multiplied by the probability of H. This is *Bayes' theorem or rule for the probability of causes.* $P(H_j|A)$ is called the *posterior probability,* $P(A|H_j)$ the *likelihood,* and $P(H_j)$ the *prior probability;* $H_1, H_2, \ldots, H_k$ are called the *hypotheses* or *causes* of A.

The theorem has many statistical applications but requires a knowledge of the prior probabilities. Suppose a human population is divided into k strata on the criteria of race, and let p_i $(i = 1, 2, \ldots, k)$ be the probability that an individual chosen at random belongs to stratum H_i. Let the event A denote the possession of blood group A, and let p_j' denote the conditional probability $P(A|H_j)$ that a person belonging to race H_j also has blood group A. The probability that an individual chosen at random has blood group A is, from equation (24.42), $\sum^k p_i p_i'$. Given that a person has blood group A, what is the probability of his belonging to race H_j?

From Bayes' Rule (equation 24.43), the answer is

$$P(H_j|A) = \frac{P(AH_j)}{P(A)} = \frac{P(H_j)\,P(A|H_j)}{\sum^k P(H_i)\,P(A|H_i)} = \frac{p_j p_j'}{\sum^k p_i p_i'}$$

A Bayesian approach has also been applied to determine the optimal design in stratified sampling (Ericson, 1969); and further work has been done on the optimal cluster size etc.

Unlike the usual Bayesian posterior inference problem, where the primary focus is on the posterior distribution of the 'hypothesis' H_i, in sample surveys with the finite universe model it is the predictive distribution of the unobserved components of H_i (or a future sample of $N_h - n_h$ units) which is primarily of interest. ●

Further reading

Section 24.2
Cochran, sections 12.1–12.8; Hansen *et al.*, Vols. I and II, Chapter 11; Kish, sections 12.1 and 12.2; Murthy, sections 9.12 and 11.5; Raj, sections 7.1–7.8; Sukhatme and Sukhatme, section 5.11; Yates, section 3.12.

Section 24.3
Cochran, sections 12.9–12.12; Hansen *et al.*, Vols. I and II, Chapter 11; Kish, sections 12.4–12.7; Murthy, sections 11.6 and 11.7; Narain; Raj, sections 7.9–7.12; Sukhatme and Sukhatme, section 5.13; Yates, sections 6.21, 6.22 and 7.19.

Section 24.4
Bailey (1951), Hájek and Dupač (section VIII.6); Hammersley (1953); Leslie (1952);

Section 24.5
Sampford, sections 7.10 and 12.5; Sukhatme and Sukhatme, section 1.20.

Section 24.6
Cochran, section 13.14; Kish, section 4.4; Lahiri; Mahalanchis (1946; 1960); Murthy, sections 2.12 and 13.10g; Sukhatme and Sukhatme, section 10.10; United Nations (1960, section 11; 1964, section V18); Yates, section 7.24; Zarkovich (1963, section 10.8).

Section 24.7
Cochran, section 5A.12; Hansen *et al.*, Vols. I and II, Chapter 10; Vol. I, section 12.13; Kish, section 8.6; Yates, sections 7.20, 7.21 and 7.25.

Section 24.8
Cochran, section 5A.11; Hansen *et al.*, Vols. I and II, Chapter 10; Kish, section 8.6B; Raj, section 4.8.1; Sukhatme and Sukhatme, section 3.14.

Section 24.9
Kish, section 12.8; Murthy, section 7.15; Raj, sections 4.8.3 and 9.8.

Section 24.10
Ericson (1969) and other relevant articles in Johnson and Smith (1969); Feller, sections V.1 and V.2; Lindley; Schmitt.

Exercise

1. In an area of 4500 m^2, 156 running beetles were captured, marked by punching their wing sheaths by special forceps, and then released; after 24 hours, 191 unmarked and 32 marked beetles were found in a number of earth traps. Estimate the total number of beetles (data of Skuhravý, quoted by Hájek and Dupač.

25 Errors and Biases in Data and Estimates

25.1 Introduction

In the previous chapters, we had assumed implicitly that the basic data collected either through a complete enumeration or a sample are free from any error or bias and that the (unbiased) estimators obtained from a sample are subject only to sampling errors. This is not so in actual survey conditions. From the stage of collection to that of preparation of final tables, the data are generally subject to different types of errors and biases. In this chapter will be reviewed the various types of errors and biases in data and in estimates derived from them, and their measurement and control.

The importance of giving proper attention to errors and biases in data and estimates can best be illustrated by the telling words of Deming (1966): 'For what profiteth a statistician to design a beautiful sample when the questionnaire will not elicit the information desired, or if the universe has not been defined, or the field-force is so badly organized that the results will not be worth tabulating'. This observation applies equally to the designing of censuses.

25.2 Classification of errors in data and estimates

The different types of errors in data and estimates can be classified first according to source: (a) errors having their origin in sampling, and (b) errors which are common to both censuses and samples.

25.2.1 Errors having their origin in sampling

These are: (i) sampling errors, and (ii) sampling biases. Errors other than sampling errors and biases are called non-sampling errors and biases.

(a) *Sampling errors.* A measure of the degree to which the sample estimate differs from the expected survey value (which is obtained on repeated applications of the same survey procedure) is given by the standard error, the square of which is called the sampling variance.

(b) *Sampling biases.* In a sample survey, in addition to sampling errors, sampling biases may arise from inadequate or faulty conduct of the specified probability sample or from faulty methods of estimation of the universe values. The former includes defects in frames (in both the available frames and those prepared by the enumerators in a multi-stage design), wrong selection procedures, and partial or incomplete enumeration of the selected units.

Biases of the estimating procedure may either be deliberate, due to uses of a biased estimating procedure itself, or may be due to the inadvertent use of wrong formulae for estimation. An example of the unavoidable bias in a particular estimate, even though the components may be unbiased, is the ratio of the unbiased estimates of two totals (sections 2.7, 3.2 etc.).

Three examples will be cited of bias due to methods of estimation not based on the sampling plan. First, biased estimates will result if the sample is treated for simplicity of computation as simple random or self-weighting when it is neither of these. In an inquiry when it is not feasible to construct a host of estimates by applying proper multipliers, unweighted estimates of ratios, which may be subject to much smaller biases than the estimators of the numerators and denominators, may be obtained for the less important items. A comparison should in any case be made of the estimates based on proper weights and simplified weights for some important items in order to obtain an idea of the magnitude and direction of the bias of the simpler estimates. Results from the Population Survey conducted in the former State of Mysore in India in 1951–2 by the United Nations and Government of India, and the Indian National Sample Survey, May–August 1953, show that such procedures could be used with proper caution.

Second, biased estimates will result when estimates of variances and covariances are computed from results at higher levels of aggregation. Results are not generally published for the ultimate strata but only for the higher levels of aggregation such as states and provinces. Estimates of variances, covariances, correlation coefficients etc. are sometimes computed from such published results: this is a wrong method and the desired estimates should be computed by a design-based procedure by building up estimates of variances, covariances etc. from the ultimate strata in the sample.

As a last example, biases will arise when the sample is treated as if it had come from a stratified design, when the design is either unstratified or stratified with respect to another variable (section 10.7).

25.2.2 Errors common to both censuses and samples

Errors common to both censuses and samples will be classified under seven broad headings: (i) errors due to inadequate preparation; (ii) errors of non-response; (iii) response or ascertainment or observational errors; (iv) processing errors; (v) errors in constructing substitute estimates; (vi) errors in interpretation; and (vii) errors in publication. These categories are not strictly mutually exclusive, and there may be some amount of overlap, especially for the first three.

(i) *Errors due to inadequate preparation.* These might be due to the failure to state carefully the objectives of the inquiry and to decide on the required statistical information; drawbacks of the questionnaire, and lack of clarity of concepts, definitions and instructions; failure to define the universe with enough precision and to provide precise instructions and definitions; and careless and disorganized field procedure, including faulty methods of selection of enumerators and supervisors and faulty or insufficient training.

One common source is the faulty demarcation of the boundaries of the geographical units leading to wrong coverage, mainly under-coverage; available maps do not often contain the details required and such maps are not generally updated to take account of the new premises that are built between the time the map is prepared and the inquiry is conducted.

In a demographic inquiry, if the question of current births is asked only of the mothers who are alive on the survey date, those occurring to mothers who were dead by the survey date would be omitted; similarly, information on deaths to all the members of a household, or to persons after whose deaths the household is dissolved or re-forms in more than one part, may not be obtained unless specific procedures are laid down to collect it.

In an inquiry designed to estimate the size and characteristics of a specific population group, selected after a preliminary screening (where the mesh cannot perforce be fine) of the general population, bias may be introduced at the stage of screening by the failure to take precautions against the occurrences of (a) 'false negatives', i.e. persons actually belonging to the specific population group but not classified as such, and (b) 'false positives', i.e. persons not actually belonging to the population group but classified as such. For example, inquiries on the prevalence of pulmonary tuberculosis were generally conducted first by screening, by means of miniature X-rays, the general population (excluding children under age five years) in the selected areas (or households) and then by bacteriologically examining sputum and laryngeal swab of only those showing evidence of any pathology in the X-ray films: but a study made in the U.S.A. in 1947 showed that in a single reading of X-rays, 54 per cent of cases showing a moderate degree of tuberculosis and 0.8 per cent of cases with no evidence of tuberculosis were diagnosed as positive (Yerushalmy and Neyman). Although in such a procedure the number of 'false positives' could be estimated, that of 'false negatives' (and, therefore, the true total number of persons with pulmonary tuberculosis) could not be estimated, unless a sample of persons from those not showing evidence of any pathology in the X-ray films are also covered for bacteriological examinations of sputum and laryngeal swab: the latter procedure is now reportedly being followed, and is recommended by the World Health Organization. Estimates from both the groups (those which are classified as 'positive' at the screening stage and those who are not) should be combined to provide the overall estimate for the specific population group under study.

(ii) *Errors of non-response or incomplete samples.* Non-response could be due to the non-coverage of areal units (coverage errors) because of incomplete listing or inaccessibility (due to difficulties of the terrain), the absence from home of respondents when the dwelling units are visited ('not at homes'), the inability to answer some questions, and the refusals to answer a question. Non-response increases both the sampling error (by decreasing the effective sample size) and the non-sampling error.

An example of response errors is the deficiency in the reported number of two-member households in a morbidity inquiry which was attributed to the failure of the enumerators to revisit missed households, in which childless married women working away from home were likely to predominate (Kiser, 1934; see also Brookes and Dick, section 4.7, and Sukhatme and Sukhatme, section 10.2).

In other situations, where information is required on physical characteristics of some items by adopting objective techniques of measurement, the necessary information may be lacking due to the failure to reach the survey units at the right time to collect the data: these difficulties arise mostly in agricultural surveys, because of failure of the enumerators to visit the sample fields selected for crop-cutting experiments before harvesting.

(iii) *Response or observational errors.* These refer to the differences between the individual survey values and the corresponding true values (see section 25.3).

The magnitude and direction of response errors depend on the survey conditions and procedures – the skill of the enumerator, the time given for interview, the inclusion of fully detailed probes in the schedule, the acceptance of proxy interviews, and the understanding and co-operation of the respondents, the general climate of attitudes and opinion etc. The response error may be unintentional or it may be deliberate on the part of the respondents: a person may not know his exact age, or he may report his age wrongly even when he knows it.

The fear of the 'evil eye' is known in some cultures in reporting births, especially of males (Das Gupta, 1958), farm produce (Sukhatme and Sukhatme, section 10.18), etc.

In the 1951 Census of England and Wales, for an appreciable number of women married more than once, whose earlier marriages were not reported and who were thus included among the 'married once only', the total number of children in all the marriages were associated with the current marriages; at marriage ages 35–39 years, where this error was concentrated, the total fertility in the first five years of marriage might have been overstated by as much as 10 per cent (Benjamin, 1955).

The enumerator may also have conscious or unconscious biases. For example, he may like to record numerical answers in round figures, or he may over- or under-record some figures for a particular group of units. In crop surveys, eye estimates are always subject to bias, which would vary from enumerator to enumerator, and the net effect can often be substantial. In inquiries on crop yields, the use of too small a 'cut' would give rise to biases when the demarcation of the boundaries becomes of importance as the size of the unit is decreased and also because of the possibility of influencing the results by small changes in location, e.g. so as to include particularly good plants, is greater the smaller the unit areas.

One major source of response error is that of recall. Many of the items in an inquiry relate to events that occurred in the near or distant past, and the problem is both to remember these events and to place them in the correct time periods. Such a recall error is called the *recall lapse,* when it is systematic and can be expressed as a function of the recall period (i.e. the period that elapsed between the occurrence of the event and the inquiry). The recall lapse may relate to the reporting of births and deaths, reporting of the births of a preferential sex, under-reporting of infant deaths, reporting of sickness, of household expenditures in a specific period, etc., all of which can be seen to be a function of the recall period (Neter and Waksberg, 1965; Som, 1973).

(iv) *Processing errors.* Such errors may be introduced at different stages of data processing, from editing to final presentation in the form of summary tables. However, these errors are generally easier to control on the basis of checking and sample reverification, including the insertion of dummy entries (Mahalanobis, 1946). For the population census data of Yugoslavia, errors made during the individual phases of manual processing were found to be quite considerable when compared with the total of corrected figures (Macura and Balaban, 1961).

(v) *Errors in constructing substitute estimates.* When the inquiry, either a census or a sample, fails to provide accurate estimates, substitute measures are often devised. These are sometimes based on assumptions which cannot be validated or do not apply in a particular situation. Thus a method of estimating the current fertility level, based on the assumption that fertility has not changed in the recent past, cannot be applied to provide estimates aimed at measuring fertility changes.

(vi) *Errors in interpretation.* These relate to misleading interpretations and conclusions that do not follow from the data. Take, for example, the crude death rates for single and married males in Portugal in 1950 (10.5 and 12.6

per 1000 single and married males respectively) and in Chile in 1952 (10.5 and 12.8 per 1000 respectively). From these, it would be wrong to conclude that in these countries a male could increase his chance of survival by remaining single: for in each age-group, the death rate for a single male is higher than that for a married one. This paradox stems from the different age distributions for the single and the married.

(vii) *Errors in publication.* These are mainly mechanical in nature and relate to errors in proof reading etc.; they are not generally as serious as the other types of errors.

This inventory and description of errors and biases should not lead one to suppose that all inquiries are worthless because all have errors. The errors are of varying types and degrees, and can occasionally be measured and subtracted out; this could be the main aim in research on errors and biases (Deming, 1966). On the other hand, a survey in which the magnitude and direction of different types of errors are not evaluated, or at least indicated, would give a false sense of accuracy of the collected data and estimates.

25.3 Simple statistical model

25.3.1 Individual value, true value, and expected survey value

The true value of the study variable for the *i*th unit will be designated by z_i. The survey may be a census with the total number of units N, in which case i runs from 1 to N; or it may be a sample of n units with i running from 1 to n. Thus, the study variable may refer to the number of births in the reference period in the particular household, which may be zero (no births), 1, 2, . . . ; or it may refer to the age in years last birthday of a household member, which may have any of the values 0, 1, 2, . . .; or it may refer to the size of the household, which may be 1, 2, . . .; and so on. The recorded survey value, obtained either from a census or a sample, will be designated by y_i. The error or bias in the individual *i*th unit is defined as

$$b_i = y_i - z_i$$

Thus, if in a particular household no birth is reported to have occurred during the reference period ($y_i = 0$), when in fact a birth had occurred ($z_i = 1$), the error in the survey value is $b_i = y_i - z_i = 0 - 1 = -1$; and so on.

When the survey value coincides (which may be accidentally) with the true value, i.e. when $y_i = z_i$, i.e. $b_i = 0$, we say that the survey value is *accurate*; if $y_i \neq z_i$, i.e. $b_i \neq 0$, the survey value is said to be *inaccurate*.

The value which is obtained on repeated applications of the same survey procedure is called the *expected survey value*, and designated by $E(y)$. Note that the expected survey value would differ from survey to survey, from a sample to a census, and so on. Under the term 'same survey procedure' are included the skill and training of the enumerators, the quality of the questionnaires etc.

In most cases the true individual value can be defined, but it may not be known, e.g. the age of a person. In some cases, however, the true value is difficult to define, e.g. opinions, attitudes etc., but is nevertheless useful conceptually.

Note also that the true values of individual items and derived items are generally unknown. What are obtained in a survey (census or sample) are the survey values.

25.3.2 Census

First we deal with a census or complete enumeration. As before, let the individual true value of an item under study be designated by z_i and the recorded survey value by y_i $(i = 1, 2, \ldots, N)$. The errors in the individual survey values are $b_i = y_i - z_i$ from which

$$y_i = z_i + b_i \tag{25.1}$$

The relation between the survey value and the true value of the universe total is given by summing both sides of equation (25.1),

$$\sum^{N} y_i = \sum^{N} z_i + \sum^{N} b_i$$

or

$$Y = Z + B \tag{25.2}$$

and that between the means by dividing both sides of equation (25.2) by N

$$\overline{Y} = \overline{Z} + \overline{B} \tag{25.3}$$

where B is the error of the survey total (Y), and $\overline{B}$ that of the survey mean ($\overline{Y}$).

When $Y > Z$, i.e. B, the error in the survey total, has a positive sign, Y gives an overestimate of Z; and when $Y < Z$, i.e. B has a negative sign, Y gives an underestimate of Z. And similarly for the means, $\overline{Y}$ and $\overline{Z}$.

From equation (25.1) the variance of an individual survey value is given by

$$\sigma_y^2 = \sigma_z^2 + \sigma_b^2 + 2\sigma_{zb} \tag{25.4}$$

where σ_y^2, σ_z^2, and σ_b^2 are the variances of y, z and b respectively and σ_{zb} is the covariance of z and b, defined respectively by

$$\begin{aligned} \sigma_y^2 &= \sum^{N} (y_i - \overline{Y})^2/N \\ \sigma_z^2 &= \sum^{N} (z_i - \overline{Z})^2/N \\ \sigma_b^2 &= \sum^{N} (b_i - \overline{B})^2/N \\ \sigma_{zb} &= \sum^{N} (z_i - \overline{Z})(b_i - \overline{B})/N \end{aligned} \tag{25.5}$$

The covariance can also be expressed as $\rho_{zb}\sigma_z\sigma_b$, where ρ_{zb} is the correlation coefficient between z and b given by

$$\rho_{zb} = \sum^{N} (z_i - \overline{Z})(b_i - \overline{B})/N\sigma_z\sigma_b$$

Apart from the general case where individual errors are present ($b_i \neq 0$), and do not cancel out ($B \neq 0$), so that the survey total, survey mean, and the variance of a survey value are given respectively by equations (25.2), (25.3) and (25.4), we may have the following special cases:

(i) *Individual errors absent, data accurate.* Here $b_i = 0$, $B = 0$ so that $Y = Z$, $\overline{Y} = \overline{Z}$ and $\sigma_y^2 = \sigma_z^2$ (Here $\sigma_b^2 = 0 = \sigma_{zb}$).

(ii) *Individual errors present, but total error absent.* Here $b_i \neq 0$, but $B = 0$, so that $Y = Z$, $\overline{Y} = \overline{Z}$, but $\sigma_b^2 \neq 0$ and

$$\sigma_y^2 = \sigma_z^2 + \sigma_b^2 + 2\sigma_{zb}$$

Thus in this case, the survey total or mean would be free from error, but the variance will be affected; even if $\sigma_{zb} = 0$, i.e. there is no correlation between z and b, but $\sigma_y^2 > \sigma_z^2$

(iii) *Constant individual error.* Here $b_i = b_0$, a constant, so that $Y = Z + Nb_0$, $\overline{Y} = Z + b_0$, but $\sigma_y^2 = \sigma_z^2$ (since $\sigma_b^2 = 0 = \sigma_{zb}$). In this case, the survey total, mean etc., will be affected, but not their variances.

The above results show that even in a census the survey result may be misleading unless some idea is obtained about the errors in the data.

25.3.3 Sample

Here we assume that a simple random sample of n units is drawn with replacement from the total of N units. The individual survey values are

$$y_i = z_i + b_i \tag{25.6}$$

The relation between the totals is

$$\sum^{n} y_i = \sum^{n} z_i + \sum^{n} b_i \tag{25.7}$$

and that between the means, on dividing both sides of equation (25.7) by n,

$$\overline{y} = \overline{z} + \overline{b} \tag{25.8}$$

where $\overline{y}$ is the survey mean, $\overline{z}$ the true mean and $\overline{b}$ the error in the survey mean, for the n sample units.

When $\overline{y} > \overline{z}$, i.e. $\overline{b}$ has a positive sign, $\overline{y}$ will give an overestimate of $\overline{z}$; and when $\overline{y} < \overline{z}$, i.e. $\overline{b}$ has a negative sign, $\overline{y}$ will give an underestimate of $\overline{z}$.

The expected value of the survey mean $\overline{y}$ is obtained on taking the means of all the possible values of $\overline{y}$s from different samples under the same survey procedure, and is given by

$$E(\overline{y}) = E(\overline{z}) + E(\overline{b})$$

or

$$E(\overline{y}) = \overline{Y} = \overline{Z} + \overline{B} \tag{25.9}$$

and the sampling variance of $\overline{y}$ is

$$\begin{aligned}\sigma_{\overline{y}}^2 &= E[\overline{y} - E(\overline{y})]^2 \\ &= E(\overline{y} - Y)^2 \\ &= (\sigma_z^2 + \sigma_b^2 + 2\sigma_{zb})/n \end{aligned} \tag{25.10}$$

In Figure 25.1 the distribution (I) of the true values z_i around the true mean $\overline{Z}$ has been shown by Curve I; the variance of this distribution is σ_z^2. The distribution (II) of the survey values y_i around the survey mean $\overline{Y}$ is shown by Curve II; the variance of the distribution is σ_y^2. If a sample of n units is taken from the distribution II, the estimate of the mean $\overline{y}$ will have the distribution shown by Curve III around the expected survey mean $\overline{Y}$: a measure of this variation is given by $\sigma_{\overline{y}}^2$, as defined by equation (25.10). This does not, however, give any indication of the expected behaviour of the distribution of the sample estimate $\bar{y}$ around the true mean $\bar{Z}$, which is the basic aim in our survey. That is measured by the mean square error defined below, which, in addition to the sampling error, takes account of the expected value of the bias, measured by $\bar{B} = \bar{Y} - \bar{Z}$.

Consider also the four special cases given in Figure 25.2(a)–(d). It is clear that in the two cases (a) and (b), where the bias is heavy, the values of both the estimate and its variance would be misleading in setting any probability limits to

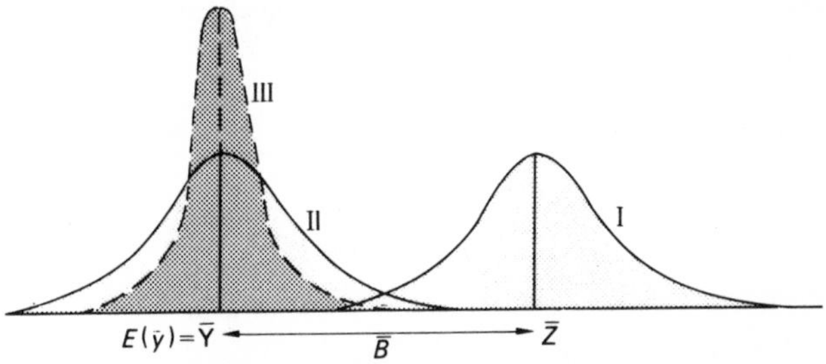

Figure 25.1 Distribution of true and survey individual values and of survey mean (adapted from Zarkovich (1963), Figure 1).

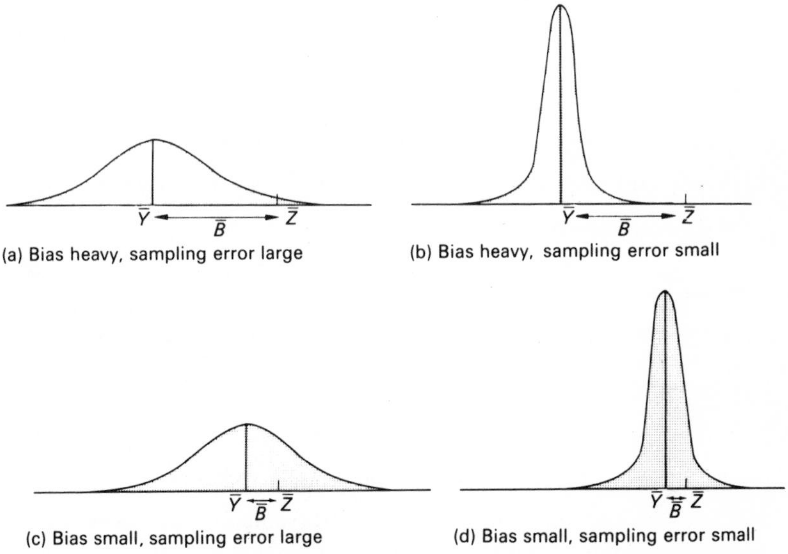

Figure 25.2 Distribution of the sample estimate ($\bar{y}$) in relation to the true value ($\bar{Z}$).

the universe value: the ideal situation is, of course, given by case (d) with both the bias and the sampling error small (see also Deming, 1966, Figure 1).

Mean square error. The variability of the survey mean around the true value is measured by the mean square error which is defined as the expectation of the square of the difference of the survey mean and the true value,

$$\begin{aligned} \text{MSE}_{\bar{y}} &= E(\bar{y} - \bar{Z})^2 \\ &= E[(\bar{y} - \bar{Y}) + (\bar{Y} - \bar{Z})]^2 \\ &= E(\bar{y} - \bar{Y})^2 + 2(\bar{Y} - \bar{Z})\, E(\bar{y} - \bar{Y}) + (\bar{Y} - \bar{Z})^2 \\ &= \sigma_{\bar{y}}^2 + \bar{B}^2 \end{aligned} \qquad (25.11)$$

as $(\overline{Y} - \overline{Z}) = \overline{B}$ is a constant (for the same survey procedure) and $E(\bar{y} - \overline{Y}) = 0$.
Some special cases of the above equations are given in Table 25.1.

Table 25.1 Effects of individual and total errors on the expected value, the sampling variance, and the mean square error of the survey mean.
(Source: Som, 1973, section 1.4.3).

Individual error b_i	*Mean error* $\bar{b}$	*Expected value of survey mean* $E(\bar{y}) = \overline{Y}$	*Sampling variance of survey mean* $\sigma_{\bar{y}}^2$	*Mean square error of survey mean* $MSE_{\bar{y}}$
General case: Individual and total errors present				
$\neq 0$	$\neq 0$	$\bar{Z} + B$	$\frac{1}{n}(\sigma_z^2 + \sigma_b^2 + 2\sigma_{zb})$	$\frac{1}{n}(\sigma_z^2 + \sigma_b^2 + 2\sigma_{zb}) + \bar{B}^2$
Estimate, sampling variance, and mean square error affected.				
Special cases:				
(i) *Individual errors absent, data accurate*				
$\neq 0$	$= 0$	$\bar{Z}$	σ_z^2/n	σ_z^2/n
Estimate, sampling variance, mean square error unaffected. (Note the classical formula for the sampling variance of the mean for simple random samples).				
(ii) *Individual errors present, but mean error absent*				
$\neq 0$	$= 0$	$\bar{Z}$	$\frac{1}{n}(\sigma_z^2 + \sigma_b^2 + 2\sigma_{zb})$	$\frac{1}{n}(\sigma_z^2 + \sigma_b^2 + 2\sigma_{zb}) + \bar{B}^2$
Estimate unaffected, sampling variance and mean square error affected.				
(iii) *Constant individual error*				
$= b_0$ (constant)	b_0	$\bar{Z} + b_0$	σ_z^2/n	$\frac{1}{n}\sigma_z^2 + \bar{B}^2$
Estimate affected, sampling variance unaffected, mean square error affected.				

25.3.4 Effect of bias on errors of estimation

Evidently some error will be involved if the sampling error is used to set probability limits of the true value. Assuming that the estimator $\bar{y}$ is distributed normally about the expected value $E(\bar{y}) = \overline{Y}$, which is at a distance $\overline{B}$ from the true value $\overline{Z}$ (expression (25.9)), it is possible to estimate the effects of the bias on errors of estimate (Cochran, section 1.7).

If the bias is one-tenth of the standard error or less, its effect on the total probability of an absolute error i.e. $|\bar{y} - \overline{Z}|$ of more than $1.96\,\sigma_{\bar{y}}$ is negligible. But when the bias is of the same order as the standard error, the total probability becomes 0.1700, instead of the presumed 0.05 at $\overline{B} = 0$; for a bias which is four times the standard error, the total for a bias which is four times the standard error, the total probability is 0.9793, more than nineteen times the presumed value. Thus, in the above technique of setting limits with the standard error, we presume that we shall be wrong in the direction of underestimate only in 2.5 per cent of cases, when we shall in fact be wrong in 98 per cent of the cases, if the bias is negative and four times the standard error (Som, 1973, section 1.4.6).

• 25.4 More complex models

More complex models of errors in data and estimates have been developed by Cochran (1963), Felligi, Hansen and his colleagues (1951; 1953, Vol. II, Chapter 12; 1961), and Sukhatme and Seth (1952), among others. Following in general the formulation by Cochran, and elaborating the model in section 25.3, we express y_{ij}, the value obtained on the ith unit ($i = 1, 2, \ldots, n$) in the jth independent repetition, as

$$\begin{aligned} y_{ij} &= z_i + b_0 + (b_i - b_0) + d_{ij} \\ &= z_i' + d_{ij} \end{aligned} \tag{25.12}$$

where z_i is the true value; b_0 the constant bias over all the units; $(b_i - b_0)$ the variable component of bias which follows some frequency distribution with mean zero as i varies and may be correlated with the correct value z_i; $d_{ij} = y_{ij} - (z_i + b_i)$ is the fluctuating component of error which follows some frequency distribution with mean zero and variance σ_i^2 as j varies for fixed i; and $z_i' = z_i + b_i$.

The mean $\bar{y}$ of an srs would be unbiased with variance σ_z^2/n (ignoring the finite multiplier) if all the measurements were accurate. However, because of the errors of measurement, the mean may be subject to a bias $\bar{B} = b_0$ and its mean square error is

$$\text{MSE}(\bar{y}_j) = \frac{1}{n}\left[\sigma_{z'}^2 + \sigma_d^2\,(1 + (n-1)\,\rho_w)\right] + b_0^2 \tag{25.13}$$

where $\sigma_d^2 \doteq \sum^{N} \sigma_i^2/N$, and ρ_w is the intrasample correlation of measurements within the same sample defined by analogy with cluster sampling by the equation

$$E(d_{ij}\,d_{i'j}) = \rho_w\,\sigma_d^2\,.$$

It can be shown that when errors of measurements are uncorrelated within the sample and cancel out over the whole universe, the formulae given in the preceding chapters for estimating sampling variances remain valid for stratified and multi-stage sampling, although such errors decrease the efficiency of the estimators. On the other hand, when ρ_w is present (in measurement and in processing, particularly if subjective judgement is involved), it is likely to be positive, and the standard formula for estimating the variance of $\bar{y}_j$ is usually an underestimate. Note also that a constant bias is not detected by survey data.•

25.5 Measurement of response errors

Some types of response errors may be immediately evident in the data. The mis-reporting of ages is a well-known case in point, reflected in the heaping of ages at some individual years and the corresponding deficiency at some others; very high or very low birth and death rates reported in some demographic surveys and civil registration constitute another case.

Some methods of measurement of response errors will be considered: external record checks; resurvey; interpenetrating networks of sub-samples; internal consistency checks; and analysis by recall periods.

25.5.1 External record checks

Such checks can be made only if accurate external data are available for each unit of the universe. An example is the 'Reverse Record Checks' made for coverage errors in the 1960 Census of Population and Housing in the U.S.A., where probability samples of persons were drawn from different sources of records, namely, persons enumerated in the previous census, registered aliens, children born in the intervening period etc., or of special groups such as the aged social security beneficiaries and students enrolled in colleges and universities; these were checked against the census returns. In the U.S.A. and the U.K., data on hospitalization as reported in household interview surveys have been compared with hospital records. In a large number of countries of the world, this method either cannot obviously be used or would be of very limited value.

Where feasible, a related procedure is to monitor responses which are objective in nature. This is illustrated by the checking of the validity of the telephone coincidental method for determining the in-home radio station rating. In this method, telephone calls are made at random times during each quarter hour period of the day to a random sample of individuals in listed telephone households; the selected person is asked if the radio is on, and if it is, to identify the station that was tuned in. In a recent survey conducted in the New York Metropolitan area, this method was tested as follows. Each interviewer who did the calling had at her disposal an electronic device whereby she could transmit, over the telephone she was using, the broadcast currently coming from any of the leading twenty A.M. stations in the area by pressing any of the twenty buttons. In the telephone interview, when the radio was reportedly on and the programme identified by the respondent, the interviewer pressed the test button corresponding to the station reported. If then the respondent reported that what was coming over the telephone was the same as what was on his radio, the orginal response was considered to be correct. Of the total 854 responses validated in the test, 91 per cent of the responses were found correct by this procedure (Frankel, 1969).

25.5.2 Re-survey

A standard method of measuring and adjusting for response errors in censuses and sample surveys is the re-survey of a sub-sample of units of the original survey, preferably using a more detailed schedule, and in a personal interview, with better staff (either the supervisors or the best enumerators). For population and housing censuses, the ultimate sampling unit in re-surveys should be compact areal units. In the form of a post-enumeration check, such re-surveys are almost universal with censuses. A re-survey can form an integrated part of the original sample survey, as in the Demographic Sample Survey of Guinea, in 1954–5, where, in addition, whenever infant deaths were considered to be grossly under-reported in a village, a medical team was sent to re-interview the females. Unless conducted simultaneously with the original survey or immediately after it, the re-survey introduces some operational and technical difficulties.

Unitary checks with one-to-one matching (e.g. of persons in a population census and the post-enumeration survey) is considered as the essential, integral part of post-enumerative checks of censuses to check both gross and net errors; however, the difficulty of such matching with the available resources may be so enormous in some countries that checks at the aggregate levels (e.g. of the total number of persons in the selected areal units) only may be considered practical to check only the net errors.

The statistical model in a re-survey is

$$E(\bar{y}') = \bar{Z}$$

on the assumption that the bias element $\bar{B}' = 0$, the letters with primes denoting the values obtained from the re-survey. The difference $(\bar{y} - \bar{y}')$ therefore gives an estimator of the bias element $\bar{B}$ in the original survey. Adjustments may also be made by regression and ratio estimators.

Some estimates of the under-enumeration of the total population in a number of censuses are as follows: 0.5 per cent in the U.S.S.R. in 1959; 1.1 per cent in Canada in 1956; 1.1 per cent (and 3.5 per cent by analytical methods) in India in 1951; 1.4 per cent (and 3.6 per cent by analytical methods) in the U.S.A. in 1950; 3.5 per cent in Sierra Leone in 1963; and 8.0 per cent in Swaziland in 1956. For the U.K., an estimate at census date in 1951 exceeded the final count by 0.3 per cent; and in 1961 a careful retrawl of sample areas and interviews with a small sub-sample of households, carried out immediately after the census, gave no evidence of significant under- or over-enumeration (Benjamin, 1968).

In agricultural inquiries too, the under-enumeration can be sizable. In an experiment conducted in Greece, the farmers were found to have under-reported by 36 per cent the number of parcels operated by them, and in the 1959 Census of Agriculture in the U.S.A., the total number of farms were undercounted by 8.4 per cent and the area of farms in lands by 6 per cent, whereas the coefficients of variation of the estimated totals were of the order of 1 per cent only (Sukhatme and Sukhatme, section 10.18).

In crop surveys, whenever eye estimates are used, the results should be calibrated by comparison with the physical measurements of a sub-sample: this assumes a high positive correlation between the two sets of figures.

25.5.3 Internal consistency checks

A simple check of consistency of data is to examine items with implausibly high or low figures, such as of consumption of salt in a household. A number of checks may also be introduced in the schedules to ascertain whether consistent replies are obtained from two sets of questions. The total number of children borne by a woman can be asked directly and compared with the number at present alive and living with the same household, the number living away, and the number dead. Information on the sexual life within marriage has been asked in the U.S.A. and that on the use of contraceptives in India from both husband and wife, sometimes with startingly different results. Some of these checks are negative in the sense that they would not reveal any systematic bias.

Another check of internal consistency is the comparison of differential reporting in proxy and self-interviews. In the U.S. National Health Survey, for example, about one-fifth of the minor non-chronic conditions were missed due to the acceptance of proxy interviews (Nisselson and Woolsey, 1959).

In crop-yield surveys, internal consistency checks are provided by measurements of yields from cuts of concentric circles: a comparison could also be made of the data from cuts of the same area but with different shapes such as a circle, triangle, rectangle etc., where a circle with the minimum perimeter would be least subject to the border bias. In any case, the final check will be with the result of harvesting the whole sample field, working under the same survey conditions.

In demographic inquiries, with repeated visits of the same sample of households after an interval of time such as six months or one year, information of pregnancies collected in a visit can be checked with their outcomes in the subsequent visits: this method has been used with considerable success in Pakistan and Turkey. Such repeated visits can also provide generally accurate estimates of births, deaths, and migration in the intervening period as they did in India in 1958–9; these are, however, more expensive than single-time inquiries.

25.5.4 Interpenetrating network of sub-samples

This technique, introduced by Mahalanobis, consists of sub-dividing the total number of units in the survey (census or sample) into a number of parallel, random groups, and permits the testing by the analysis of variance of the differential effects of enumerators and of other factors causing variation such as differences in field schedules, methods of collection etc. (section 24.6).

A simple tabulation may often reveal differences in the estimates obtained by different enumerators covering the same universe. In the above model, the bias element ($\bar{B}_e$) where e refers to the enumerator, may be taken to consist of two parts, a constant bias ($\bar{B}'$) that affects all the enumerators alike, and another component ($\bar{B}_e'$) that affects the eth enumerator. The F-ratio, 'between enumerator' variance divided by the relevant 'error' variance, tests the null hypothesis $\bar{B}_e' =$ constant, i.e. that the biases of the enumerators do not differ.

The net bias common to all enumerators ($\bar{B}'$) passes undetected by this technique. It has to be controlled by improvement in survey methodology, comparison with independently obtained estimates, and adoption of special analytical techniques.

Another formulation would be that due to Cochran (1963, section 13.14). A random sample of n units is divided at random into k sub-samples, each containing $m = n/k$ units, and each of the k enumerators is assigned a different sub-sample. We assume that the correlation we have to deal with is due solely to the biases of the enumerators and that there is no correlation between errors of measurement for different enumerators.

The previous statistical model (25.12) is rewritten as

$$y_{ei} = z_{ei} + d_{ei}$$

where e denotes the sub-sample (enumerator) and i the member (unit) within the sub-sample. The finite multiplier is ignored.

The variance of the mean of the eth random sub-sample is, by equation (25.13),

$$V(\bar{y}_e) = \frac{1}{m} [\sigma_{z'}^2 + \sigma_d^2 (1 + (m-1)\rho_w)]$$

where ρ_w is the correlation between the d_{ei} obtained by the same enumerator. As the errors are assumed to be independent in the different sub-samples,

$$V(\bar{y}) = \frac{1}{k} V(\bar{y}_e)$$

$$= \frac{1}{n} [\sigma_{z'}^2 + \sigma_d^2 (1 + (m-1)\rho_w)] \tag{25.14}$$

From the sample data, analysis of variance could be computed for the variations due to 'between enumerators (sub-samples)' and 'within enumerators', with respective mean squares

$$s_b^2 = m \sum^{k} (\bar{y}_e - \bar{y})^2/(k-1)$$

$$s_w^2 = \sum^{k} \sum^{m} (y_{ei} - \bar{y}_e)^2/k(m-1)$$

As s_b^2/n is an unbiased estimator of $V(\bar{y})$, the technique provides an estimate of error that takes account of the enumerator bias; and the F-ratio s_b^2/s_w^2 tests the null hypothesis $\rho_w = 0$.

The technique is readily extended to stratified multi-stage sampling, by simply ensuring that the sample consists of a number of sub-samples of the same structure in which errors of measurements are independent in different sub-samples. Strictly speaking, the process of interpenetration should be deeper than the field, with not only different enumerator-teams, but also supervisors, data processors etc. used in different sub-samples: in epidemiological surveys, which may require laboratory testing and X-ray reading by independent observers, interpenetration can be extended further and deeper than the field.

An unbiased estimator of $V(\bar{y})$ is provided by $\sum^{k} (\bar{y}_e - \bar{y})^2/k(k-1)$ with $(k-1)$ degrees of freedom. If $k = 2$ in each stratum, each stratum provides one degree of freedom and the computation of estimates is simplified considerably as noted in section 9.5.

This method was followed in computing the variance estimates in the Indian National Sample Survey quoted in this book. Conformance to the fundamental formulae in stratified multi-stage sampling was established by assigning in a stratum each of two or more enumerators, each sub-sample consisting of the same number of first-stage sample units.

Similar estimates of errors (but excluding the enumerator bias) can in many cases be obtained from a sample which was not planned as an interpenetrating network by sub-dividing the sample into sub-samples in some appropriate random manner.

There is a simple non-parametric way of expressing the margin of uncertainty of the estimates from the sub-samples. Each of the independent sub-samples provides an independent and equally valid estimate of the parameters, and therefore gives directly an estimate of the margin of uncertainty involved in the estimator. This is an advantage of particular relevance to surveys in countries where suitable personnel and equipment for estimating the sampling variances of a large number of estimates are scarce. The probability that the median of the distribution of the sample estimators (which is assumed to be symmetrical) will be contained by the range of k estimators obtained from the k independent, interpenetrating sub-sample is $1 - (\frac{1}{2})^{k-1}$: with two sub-sample estimates, for example, the range between the two estimates provides 50 per cent margin of uncertainty for any derived functions worked out from these two samples. The derivation of such limits, besides being simple, involves a minimum of assumptions regarding the sampling distribution of the estimators. An example is provided by the two sub-sample estimates of the birth rate in India in 1958–9, 38.5 and 38.0, with the combined estimate of 38.3 per 1000 persons.

In regard to testing enumerator differences, one example is provided by the Mysore Population Survey in 1951–2, conducted jointly by the Government of India and the United Nations in the then State of Mysore; only one F-ratio came out statistically significant in one 'round' (there were four rounds of work) relating to the number of vacant dwelling units in Bangalore city (United Nations, 1961, Chapter 21).

A limitation of the technique is the increase in travel costs of enumerators. This could be reduced by stratifying the sample into compact areas and assigning, say, two enumerators in each stratum. In the Indian National Sample Survey with this procedure, the increase in travel cost due to interpenetration constituted about $3\frac{1}{2}$ per cent of the total cost (Som, 1965).

Small enumerator differences cannot in general be detected by the technique, unless linked samples are used. It should not of course be considered as a substitute for supervision and other controls of field and processing work.

25.5.5 Analysis by recall periods

In retrospective inquiries, where questions are asked about events which occurred in the near and distant past, a tabulation by recall periods of varying lengths often reveals the existence of recall lapse, which may otherwise lie hidden in the overall estimates.

The statistical model of recall analysis in current vital data is briefly as follows. Estimates (Y_k) are built up from different recall periods k (= 1 month, 2 months, . . ., 12 months, for example), the mathematical expectations of which are given by $E(Y_k) = Z + B_k$, where Z is the 'true' total number of events, and B_k the recall bias, which is a function of k, the recall period.

If there were no recall bias, i.e. were $B_k = 0$, then $E(Y_k) = Z$. In practice, the Y_ks, as obtained from the data, are smoothed by a curve $f(k)$, from which $\hat{Y}_0 = f(0)$, corresponding to the recall period 'zero', can be estimated and taken to be equal to Z on the assumption that $B_0 = 0$. The estimate of the vital rate is then obtained on dividing $\hat{Y}_0$ by the corresponding population estimate.

By applying this analysis, the reported crude birth and death rates in the Indian National Sample Survey (1953–4), 34.3 and 16.6 per 1000, were adjusted to 40.9 and 24.0 respectively; the reported crude birth rates in Upper Volta (1960–1) and Chad (1964) were 49.6 and 45 per 1000, and these were adjusted to 52.6 and 49 respectively (Som, 1973; Nadot, 1966).

25.5.6 Other external analytical checks

The external unitary check most relevant in estimating vital rates is the one-to-one matching of the current vital events, reported in a survey, with those registered. The method for adjustment of the data from such a check in demographic inquiries is based on the assumption that the chance of an event being missed in either list (survey or registration) is independent of its chance of being missed in the other (Chandra Sekar and Deming, 1949): this is the same as the Lincoln Index, used in estimating the size of mobile populations such as animals by the capture-recapture method (section 24.4). It has found applications in nation-wide demographic surveys, using area samples, in Pakistan, Thailand, Turkey, Malawi, and Liberia.

Assuming that the under-reporting in the survey and under-registration operate independently, the adjustment (multiplying) factor for the total number of events reported by either or both the agencies is $1/(1 - p_1 - p_2)$

where p_1 is the probability of an event not being reported in the survey,

$$p_1 = y_{\bar{s}r}/(y_{sr} + y_{\bar{s}r}),$$

and p_2 is the probability of an event not being registered,

$$p_2 = y_{s\bar{r}}/(y_{sr} + y_{s\bar{r}})$$

the subscripts s and $\bar{s}$ denoting whether the event was reported in the survey or not; and similarly for r and $\bar{r}$. For the total number of events, the estimator (the *Chandra Sekar-Deming formula*) is

$$N^* = (y_{sr} + y_{s\bar{r}} + y_{\bar{s}r})/(1 - p_1 - p_2)$$
$$= (y_{sr} + y_{s\bar{r}})\,(y_{sr} + y_{\bar{s}r})/y_{sr}$$

It is a consistent, maximum likelihood estimator but is generally biased. A variance estimator of N^* is

$$N^*(1 - p_1)\,(1 - p_2)/p_1 p_2$$

In the Survey on Population Change in Thailand, this method raised the survey-estimated birth and death rates from 42.2 and 10.9 respectively to 46.0 and 12.9 per 1000 persons (Lauriat and Chintakananda, 1966), and in the Pakistan Growth Estimation Study in 1961–3, the method provided for an additional 5 per cent of births and 7 per cent of deaths missed by the special registrars and by the survey-enumerators (Krotki, 1966).

25.6 Control of non-sampling errors

Response errors can be controlled by the proper selection, training, and supervision of enumerators and the control of enumeration. From expression (25.14) it might appear that, since an increase in the number of enumerators would decrease the intra-enumerator correlation, the contribution to the total variance due to the variability between enumerators would also decrease. But when a very large number of enumerators have to be employed, one has to accept a lower level of staff, training and supervision, resulting in a change in survey conditions that would increase the response errors.

Response errors can also be controlled by introducing internal consistency checks in the questionnaires and by asking screening questions and probes. In the U.S. National Health Survey, for example, only half the chronic conditions were reported in response to the initial questions and the check-list questions were necessary to pick up the other half; for sicknesses without medical care and restricted activity, the proportion of cases missed by the initial probes was as high as two-thirds (Nisselson and Woolsey, 1959). In the morbidity studies conducted in the Indian National Sample Survey, 1960–61, the broad question 'During the last thirty days did you or any member of your households have any sickness or injury?' elicited three-quarters of the total sicknesses, and a reference to specific symptoms and sites included all but less than 4 per cent; additional probes on medical treatment and medicines and on women's complaints reported the remaining sicknesses (Das, 1969).

In the processing of data, quality control techniques can also be usefully introduced. For example, the work of each key-punch operator can be verified one hundred per cent initially, and when his error rate falls below a set level, only a sample verification made for him. This method has been used in practice at the Bureau of the Census of the U.S.A. Owing to the danger of back-sliding, it might not be advisable to withdraw completely verification of punchers and coders whose earlier work had proved satisfactory.

When data are processed manually, estimates for the important items may be calculated by two independent teams of computing clerks.

Some further consideration is given to the subject in sections 26.11, 26.12, and 26.14–26.17 on methods of data collection, questionnaire preparation pilot inquiries and pre-tests, selection and training of enumerators, supervision and processing of data.

25.7 Evaluation and control of sampling *versus* non-sampling errors

For a given total size of sample, the sampling error can be controlled and evaluated in a suitably designed survey with considerations of optimization of stratification, allocation of the total sample size into different strata and stages, probabilities of selection etc., and by using appropriate formulae for estimation, relating all these to the important variables to be studied. The sampling error can be reduced by increasing the sample size, but it may introduce additional non-sampling errors in the estimates unless the survey conditions remain the same. Some requirements of the reduction of sampling errors may, however, come into conflict with that of the response errors, e.g. while sampling considerations might call for a widespread sampling with little clustering, a concern for the response errors might lead the survey designer to confine the sample to a sample of large clusters in which supervisory inspection and control can be more fully exercised (Hansen, Hurwitz, and Pritzker, 1964). In addition, complete enumeration of the areal units may have other advantages, e.g. cross-check of information of the neighbouring households in a social or demographic inquiry, and the psychological effect that the inquiry relates to all persons in the selected area rather than those in some households. The latter factor is of importance both in developing and developed countries, and is one of the factors determining the choice of enumeration districts as the sampling units in the Micro-Census of the Federal Republic of Germany (see exercise 5, Chapter 10).

The effect of increasing sample size may also be represented by Figure 25.3(a). This may, however, be somewhat misleading, as the magnitude of the non-sampling errors has been assumed to be constant, independent of the sample size. A better way of representing the effect of sample size on the sampling error and the bias would be given by Figure 25.3(b); for a particular sample size, the total error, as measured by the root mean square, may be minimum; beyond that, increasing sample size would no doubt decrease the sampling error but would also increase the non-sampling errors, which might more than offset the decrease in sampling error.

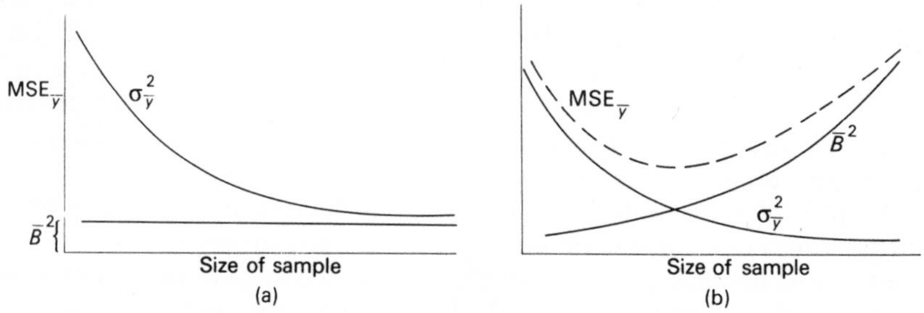

Figure 25.3 Effect of sample size on mean square error.

This is also seen from equation (25.13). In the formula for the mean square error, the two terms σ_z^2/n and $\sigma_d^2(1 - \rho_w)/n$ will decrease with increasing n; but it cannot be assumed that the other two terms, although independent of n, will not be affected by the survey procedures due to a change in the size of the sample, and that the intrasample correlation between errors and the constant bias will remain the same. In large samples, the mean square error will be dominated by these terms and the ordinary sampling variance will be a poor guide to the accuracy of the results, unless the non-sampling errors are controlled substantially.

To try to reduce sampling errors while a bias several times as large is allowed to creep in is not only pointless but also a waste of resources.

Reference has been made in section 1.3 to the study made at the U.S. Bureau of the Census, which showed that for many of the more difficult items in a census, such as occupation, industry, work status, income and education, the enumerator variability is approximately the same as the sampling variance for a 25 per cent sample of households; the census results were further seen to be subject to a bias which varied from one item to another and was 6 per cent on average. The census and the sample will have approximately the same bias if the census enumerators collect the data for the sample of 25 per cent of households as part of the regular census. Under these assumptions, the root mean square errors that can be expected for the complete enumeration, and the sample for various percentages of units possessing a certain attribute, are given in Table 25.2,

Table 25.2 Magnitude of root mean square errors of the estimated percentages of population in complete enumeration and a sample survey of 25 per cent of households for areas of specified size: Experimental Study of the Bureau of the Census, U.S.A.

Percentage of population to be estimated	*Root mean square (in percentage) for an area with a population of*					
	2500		*10 000*		*50 000*	
	Complete enumeration	*Sample*	*Complete enumeration*	*Sample*	*Complete enumeration*	*Sample*
0.5	0.3	0.4	0.1	0.2	0.1	0.1
2.0	0.6	0.8	0.3	0.4	0.2	0.2
5.0	0.9	1.2	0.5	0.7	0.4	0.4
20.0	2.0	2.5	1.4	1.6	1.2	1.3
50.0	3.6	4.1	3.2	3.3	3.0	3.1

for areas with different population sizes. While the root mean square errors for the complete enumeration and the sample are appreciably different in areas with small populations, they converge with increasing population size and become almost identical for areas with 50 000 persons. Thus, when the major census results are published for areas with 50 000 or more persons, it is more advantageous to take a sample for these particular items. This procedure was followed in the 1960 census of population in the U.S.A. (U.S. Bureau of the Census, 1960).

25.8 Adjustment for non-response

25.8.1 Effect of non-response

We assume that the universe of N units can be divided into two strata, the first consisting of N_1 units for which measurements would be obtained, and the second of N_2 units for which no measurement would be obtained, either in a census or a sample: let $\overline{Y}_1$ and $\overline{Y}_2$ denote the two stratum means. When the field work is completed for a simple random sample of n units consisting of n_1 units from the first stratum and n_2 units from the second stratum, we would have measurements only from the n_1 sample units, giving the sample mean $\bar{y}_1$. Assuming that n_1 and n_2 are random samples from the two strata, the amount of bias in the sample mean is

$$\begin{aligned} E(\bar{y}_1) - \overline{Y} &= \overline{Y}_1 - \overline{Y} \\ &= \overline{Y}_1 - (N_1 \overline{Y}_1 + N_2 \overline{Y}_2)/N \\ &= (N_2/N)(\overline{Y}_1 - \overline{Y}_2) \end{aligned}$$

The amount of bias is thus the product of the proportion of non-responding units and the difference between the two stratum means. Any sizable proportion of non-response or difference between the two stratum means would produce a substantial bias. For example, a 5 per cent non-response observed in some African and Indian fertility surveys may not substantially affect fertility rates, but the same percentage will produce a large bias in surveys on income, sales etc. if the non-responding units are those with very high incomes, sales etc.

25.8.2 Adjustment for non-response

Three methods of adjustment for non-response will be considered: (i) selecting random substitutes from the responding units; (ii) 'call-backs' or selecting a random sub-sample of the non-responding units; and (iii) adjustment for bias without call-backs.

(i) *Selecting random substitutes from the responding units.* When the proportion of non-response is low and the difference between the two stratum means is believed not to be substantial, a simple practical method is to replace the non-responding units by a random sub-sample of the responding units, due regard being paid to certain known characteristics of the non-responding units. This is relevant particularly in self-weighting designs in order to keep the multiplier a constant.

(ii) *'Call-backs' or selecting a random sub-sample of the non-responding units (Hansen-Hurwitz method).* Of the total number n_2 of non-responding sample units, suppose a sub-sample n_2' is chosen for call-backs (or remailing of the questionnaire in a mail inquiry), all of whom respond as a result of special efforts. Then an unbiased estimator of the universe mean $\overline{Y}$ is $(n_1\bar{y}_1 + n_2\bar{y}_2)/n$, where $\bar{y}_2$ is the sample mean from the n_2' call backs.

Assume the cost function $C = nc_1 + nPc_2 + n(1 - P)c_3/k$, where c_1 is the cost per unit of making the first attempt, c_2 the cost per unit of processing the data of the first attempt, c_3 the cost per unit of collecting and processing the data of the second attempt, $P(= N_1/N)$ is the proportion of universe units that would have responded at the first attempt, and $k = n_2/n_2'$. Under some broad assumptions, the optimum value of k is given by

$$k = \sqrt{\left(\frac{Pc_3}{c_1 + Pc_2}\right)} \tag{25.15}$$

$$n = n'\,[1 + (k-1)(1-P)] \tag{25.16}$$

Note that although we have assumed that all the n_2' units respond at the second attempt, they may not in fact do so. The process of call-backs could be continued until response is obtained from all but a very insignificant proportion of the units: the Hansen–Hurwitz method has been extended by El-Badry. Successive calls may help to diminish the bias and a graph (Clausen and Ford, 1947) or a regression curve fitted to the estimates and the cumulative rates of response at successive attempts when extrapolated to the hypothetical value of 100 per cent response could give a better approximation to the true value (Hendricks, 1949; 1956, Chapter XI). Some results of the inquiry of fruit growers in North Carolina, U.S.A., are given in Table 25.3; the initial responses

Table 25.3 Results of three repeated mailings to a list of 3241 North Carolina fruit growers

Mailing	*Percentage of questionnaires returned*	*Average trees per farm for reporting farms*	*Cumulative*	
			Percentage of questionnaires returned	*Average trees per farm for reporting farms*
1	9.3	456	9.3	456
2	16.7	382	26.0	406
3	13.4	340	39.4	385
Estimated from fitted regression curve			100	344
True value			100	329

gave a very high average number of fruit trees in North Carolina because a farmer's interest in a fruit survey could be expected to be positively associated with his scale of operation, but a regression curve of the form $y = ax^b$, where y respresents the total number of trees picked by a sample when the (cumulative) percentage return is x (and a and b are estimated from the data), when extrapolated to $x = 100$ per cent, gave an estimate of 344 trees per farm as against the true value of 329, which was known (Hendricks). In the 1946 Family Census of the U.K., the initial response gave a very high birth rate due to the fact that the majority of the initial non-respondents were women with few or no children: of the 230 000 initial non-respondents (who constituted 17 per cent of the total sample), 50 000 responded to the follow-up appeal and the replies of the first 12 000 of them, when combined with the remainder of the sample with a weight of 230/12, gave an adjusted birth rate which corresponded to that already known from other sources (Glass and Grebenik, 1954; cited by Yates, section 5.22, and by Moser and Kalton, section 7.4).

(iii) *Adjustment for bias without call-backs (Hartley-Politz-Simmons method).* For dealing with the not-at-home cases a procedure that saves the cost of call-backs may be adopted which consists of ascertaining from the respondents the chance of their being at home at a particular point of time and weighting the results by the reciprocal of this probability. Suppose the enumerators make calls on households during the evening on six nights of the week. The households are asked whether they were at home at the time of the interview on each of the five preceding evenings. The households may then be classified according to r, the number of evenings they were at home out of five, and the ratio $(r + 1)/6$ taken as an estimate of the probability of the households being at home during the enumeration hours. If n_r is the number of interviews obtained in the group r and $\bar{y}_r$ is the group mean, then the estimator of the universe mean is

$$\bar{y} = \frac{\sum_{r=0}^{5} 6n_r\bar{y}_r/(r + 1)}{\sum_{r=0}^{5} 6n_r/(r + 1)} \tag{25.17}$$

The method is obviously biased as no allowance is made for persons who are away from home during the enumeration hours in all the six evenings: the bias is however negligible if such persons are relatively few. The variance formula for $\bar{y}$ is rather complicated.

25.9 Concluding remarks

Sampling inquiries have assumed a crucial rôle in providing data in areas with defective or non-existent registration systems and also in enabling in-depth studies of socio-economic variables and their interrelations. A sample survey also makes it possible to ensure quality checks and incorporate special techniques to adjust for response biases.

To that end, however, there could be no routine procedure. Methods suitable for a particular set of data have to be applied as critically as possible. Assessment should be made with the existent knowledge on the characteristics of the universe and field procedures that could be obtained. The greater the detail recorded and tabulated in the survey, the more potent the checks for accuracy and internal consistency could be, and hence the greater the likelihood that important discrepancies could be detected and adjusted for in the estimates.

A combination of several methods for evaluating response errors is naturally likely to give better results than the use of a single method. For these methods to be effective, they should be built into the survey design so as to permit the required analysis: one example is the staggering of the total sample over the survey period in order to eliminate seasonality in vital events. These should, of course, be supplemented by *post hoc* techniques wherever possible.

Further reading

For classifications and examples of non-sampling errors: Deming (1960; 1966); Mahalanobis (1946).

For evaluation of non-sampling errors (especially response and non-response errors): Cochran, Chapter 13: Hansen *et al.*, Vol. II, Chapter 12; Hendricks, Chapter XI; Kish, Chapter 13; Murthy, Chapter 13; Raj, Chapter 8; Som (1973, Chapter 1); Sukhatme and Sukhatme, Chapter 10; United Nations (1972, section 11); Yates, Chapter 2 and sections 5.22, 6.15, 7.24, and 10.13; Zarkovich (1963).

For evaluation of non-sampling errors and biases in specific subject fields, see the references given at the end of Chapter 26.

26 Planning, Execution, and Analysis of Surveys

26.1 Introduction

Some aspects of the planning, execution, and analysis of statistical surveys are considered in this chapter. Most of the considerations excepting those relating to the choice of the sample design would apply also to complete enumerations.

The considerations will of necessity be brief here. References to survey methods in some selected subject fields are given at the end of the chapter. A case study dealing with the survey methods and sample design of the Indian National Sample Survey (1964–5) is given in Appendix V. Reference to some selected case studies are given under 'Further reading' in Appendix V.

26.2 Objectives of the survey

The objectives of the survey should be clearly specified to include the following:

1. a clear statement of the desired information in statistical terms, such as estimation of total personal incomes and expenditures on food and other items in a family budget inquiry, or the area and the yield of a certain crop in an agricultural inquiry, or the rates of birth, death and natural increase in a demographic inquiry, or the number of persons with pulmonary tuberculosis in an epidemiological inquiry;
2. a description of the coverage and the definition of the universe such as the geographic region or branch of the economic or social group or other categories constituting parts of the universe covered by the survey. In a survey of human population, for example, it is necessary to specify whether hotel residents, persons in institutions (e.g. boarding houses and sanatoria), persons without fixed abode, military personnel, etc. are to be included and, where possible, to indicate the order of magnitude of the categories omitted; in an agricultural survey, it will be necessary to indicate whether all agricultural holdings have been included;
3. the geographical and classificatory breakdowns by which results are to be tabulated;
4. the degree of precision desired, specifying the sampling error that can be tolerated, as also an indication of the possible inaccuracies in the final estimates;
5. the total cost of the survey.

26.3 Legal basis

Surveys, especially field enumeration, should be conducted within the legal authority obtaining in a country. The confidentiality of individual information should be clearly established and guaranteed by adequate sanctions so as to form a basis for the confident co-operation of the respondents.

26.4 Publicity and co-operation of respondents

The objective of publicity in the form of educational campaigns is twofold; first, to allay any anxiety regarding the purposes of the surveys concerning, for example, taxation, conscription, and repatriation of immigrant labour and forced labour in some countries, and second, to explain the reasons for the various questions to be asked. The publicity media may include films, articles for the press, booklets, talks for radio, television, and general press release.

On the other hand, in some surveys, such as the Social Survey in the U.K., publicity was never used for it was believed better to take the respondent by surprise: if he had time to think out the matter he may decide to refuse or to present some socially acceptable response, but faced with an actual human appeal from an interviewer on the door-step, it may be difficult to take this decision.

26.5 Budget and cost control

Preparation of a preliminary budget estimate is a priority activity that should be planned and executed at an early stage. The budget will depend on the survey design, including the levels of precision desired for various estimates, as well as on the geographical and other classifications for presentation of results, and the operational conditions prevailing in the region. The preliminary budget will have to be re-examined and revised as the survey activities progress. This will inculcate cost consciousness into those responsible for the survey and indicate the existence of any inefficiencies, often acting as an impetus to devise economies and innovations.

As practices vary widely, there does not seem to be a standard way of preparing survey budgets. The budgets for demographic sample surveys in Tunisia and West Cameroon are given as examples in Tables 26.1 and 26.2 respectively on page 301 (United Nations Economic Commission for Africa).

On the basis of these and other available survey budgets for African demographic censuses and samples, the cost estimates for the United Nations sponsored African Census Programme for any country – which consists of a simple head-count of the total population by sex and geographical divisions and sample surveys (with about 100 000 sample persons) for other demographic and associated items – are given in Table 26.3 on page 303: the cost per person enumerated was estimated at U.S. 10 cents for the simple head-count and U.S. $3.50 for the sampling inquiry covering all the items. These are, of course, notional estimates, prepared in order to arrive at an overall figure for the programme covering several countries: for the actual operation in any country, the budget is worked out on the basis of local conditions; the mapping cost in particular would vary widely in different countries.

Thus for an African country with an estimated 10 million persons, the cost of the simple head-count could be estimated at U.S. $1 million, and of the sample inquiry (with 100 000 sample persons) for all the demographic and associated items at U.S. $350 000 per year.

Table 26.1 Budget for Demographic Sample Survey, Tunisia, 1968–9: Sample of 130 000 persons with three rounds at six-monthly intervals (costs for senior personnel and data processing excluded from the budget)

Item	*Cost estimates in Dinars* *
I. *Field personnel*	
(a) *Salaries*	
50 Enumerators for 16 months at 35 D per month	28 000
10 Field supervisors for 16 months at 40 D per month	6400
10 Drivers for 12 months at 40 D per month	4800
Guides for 3200 days at 0.5 D per day	1600
Total salaries (approximately)	41 000
(b) *Allowances*	
40 Enumerators for 8 months at 1 D per day	9600
8 Field supervisors for 8 months at 1.5 D per day	2880
8 Drivers for 8 months at 0.5 D per day	960
Total allowances (approximately)	14 000
Total field personnel	55 000
II. *Equipment and supplies*	
Printing of forms	3000
Supplies	3000
Purchase of vehicles	
2 light estate cars	3600
3 four-wheel drive cars	10 000
1 saloon car	2000
Maintenance	2000
Petrol (50 000 km x 9 vehicles)	5000
Miscellaneous	1400
Total equipment and supplies	30 000
Grand total	85 000

* 1 Tunisian Dinar = 2 US $ or £0.83 approximately.

Table 26.2 Budget for Demographic Sample Survey, West Cameroon, 1965: Sample of 170 000 with single-round team-enumeration†

Item	*Cost estimates in CFA Francs*
I. *Expatriate personnel*	
(a) *Outside France*	
1 Survey director for 14 months at Fr.F. 4800 per month	3 360 000
1 Field organizer for 10 months at Fr.F. 3200 per month	1 600 000
Social security contributions 40%	1 984 000
Subsistence: 2 months at Fr.F. 60 per day	
22 months at Fr.F. 30 per day	1 170 000
Total outside France	8 114 000

† This table is continued on page 302.

Item	*Cost estimates in CFA Francs*
(b) *In France*	
Preparation of mission:	
1 Survey director for 15 days at Fr.F. 2400 per month	60 000
1 Field organizer for 15 days at Fr.F. 1600 per month	40 000
Social security contributions 40%	40 000
Total in France:	140 000
(c) *Miscellaneous charges* 15% of (a) + (b)	1 238 000
Total item I	9 492 000
II. *Local personnel*	
(a) *Training and selection*	
80 persons for 1 month at CFA.F* 12 500 per month	1 000 000
(b) *Field work and data processing*	
1 Assistant field organizer for 13 months at CFA.F. 40 000 per month	520 000
48 Enumerators for 7 months at CFA.F. 25 000 per month	8 400 000
12 Field supervisors for 7 months at CFA.F. 40 000 per month	3 360 000
59 months drivers at CFA.F. 25 000 per month	1 475 000
1 messenger for 10 months at CFA.F. 8000 per month	80 000
1 secretary/Accounts Clerk for 10 months at CFA.F. 30 000 per month	300 000
15 processing clerks and coders for 6 month at CFA.F. 25 000 per month	2 250 000
Miscellaneous allowances	390 000
(a) + (b)	17 775 000
(c) *Miscellaneous charges* 15% of (a + b)	2 666 000
Total item II	20 441 000
III. *Equipment and local transport*	
3 vans (1400 kg)	3 000 000
1 small van (camionnette)	750 000
Running costs, maintenance, repairs, insurance	5 250 000
Printing of questionnaires and data-processing forms	1 500 000
Office equipment and furniture	300 000
Camping equipment for field organizers and supervisors	250 000
Enumerators' equipment	300 000
Bicycle allowances	270 000
Local transport costs	250 000
Rent of offices and accommodation	2 000 000
Total item III	13 870 000
IV. *Air transport*	
3 Round-trip Paris-Yaoundé, tourist	459 000
V. *Miscellaneous*	
Mechanical data processing	2 500 000
Publication of provisional report	800 000
Total item V	3 300 000
VI. *Contingency*	
Approximately 5% of total	2 438 000
Grand total (approximately)	50 000 000

* 1 CFA Franc = French Franc 0.02 or US$ 0.004 or £0.0017 approximately.

Table 26.3 Cost estimates of the African Census Programme

Item	*Cost per person enumerated*
I. *Census* (simple head-count by sex and geographical divisions)	
1. Mapping	US $0.02
2. Enumeration (including training, analysis, publication, etc.)	$0.08
Sub-total: cost for the simple census	$0.10
II. *Sample* (including training, enumeration, analysis, publication, etc.)	
1. Items in the basic minimum list: age, children born alive, children living, live births (by sex) in the past 12 months, deaths (by sex and age) in the past 12 months	$2.00
2. Additional cost for the first priority supplementary items: type of (economic) activity, occupation, educational attainment, marital status	$0.50
3. Additional cost for the rest of the supplementary items: ethnic group (or citzenship), literacy, school attendance, industry, status (employer, employee, etc.), place of usual residence, duration of residence, place of previous residence, religion, number of wives, number of years since first marriage, children below school age, worked any time in the past 12 months	$1.00
Sub-total: cost of sampling of all items	US $3.50

26.6 Survey calendar

The preparation of a provisional calendar or timetable, indicating the sequence and estimated duration of the various component operations of the survey, is essential in survey planning and execution; such a provisional calendar should be made final as early as possible and kept constantly under review. In addition to a detailed checklist of operations, the calendar may sometimes take the form of a chart or graph.

26.7 Administrative organization

The final responsibility for survey planning should rest with the executive agency. The permanence of a survey organization, within the framework of the statistical system of a country, has definite advantages in furthering the development of specialized and experienced personnel who could undertake the continuing research needed for evolving increasingly efficient survey designs, and also for improving the quality of survey data by assessing and controlling non-sampling errors and biases. A permanent survey organization can also undertake urgent *ad hoc* surveys at short notice.

26.8 Co-ordination with other inquiries

Needed co-operation may often be obtained from different bodies by co-ordinating the survey with other compatible inquiries, using, for example, the same sampling frame, and also trying to accommodate the needs of other users, but without overloading the questionnaires.

26.9 Cartographic (mapping) work

In surveys using areal sampling units at some (or all) stages, the detailed subdivision of the universe into identifiable areal units is one of the basic important operations in a survey. Censuses of population, housing and agriculture are the most common source of such information. However, maps have to be prepared for selected areas when these are not available with the required details.

26.10 Tabulation programmes

At the very early stages of survey planning, the tabulation programme should be drawn up so that the procedures and costs involved may be investigated thoroughly and the questionnaire tested to indicate whether it is possible to gather the required information.

26.11 Methods of data collection

Data may be collected:

1. by direct investigation comprising (a) physical observation which may involve subjective methods (e.g. eye estimation of the areas under a crop in sample fields) or objective methods (e.g. physical measurements, such as the yield of a crop in the sample fields) or both, and (b) personal interview;
2. by mail inquiry;
3. by telephone;
4. from registration;
5. from transcription of records.

The objectives of the survey, the nature of the items of information, the operational feasibility and cost will often determine the method of data collection. Mail inquiries are practicable in predominantly literate cultures where respondent co-operation is ensured; the last three methods have limited applications.

Physical observations are required in inquiries involving measurements such as crop yields, housing conditions, and anthropometric studies.

The method of personal interview is widely used in economic and social surveys. Some items of information may be collected only through personal interview, which has the added advantage that the concepts and definitions can be explained to the respondents, thus reducing response errors and non-response. Direct investigation is, however, generally more expensive than other methods.

In a household inquiry, after a sample household is contacted the required data may be obtained by interviewing one or more members (the canvasser or enumeration method) or the questionnaire may be left with the household for it to be filled by the members ('householder method'), and later collected by the enumerator.

26.12 Questionnaire preparation

26.12.1 Questionnaire design

The type of questionnaire, its content, format, and the exact wording and arrangement of the questions determine to a great extent the quality of the responses. Among the many factors which should be taken into consideration in designing the questionnaire are the method of enumeration, the type of inquiry, the data to be collected, the most suitable form of the questions and their arrangement, and the processing techniques to be employed.

Although we have sometimes used the terms 'questionnaire' and 'schedule' interchangeably, a distinction might be made between the two: a schedule contains a list of items on which information is required but the exact form of the questions to be asked is not standardized but left to the judgement of the enumerators; a questionnaire contains a list of the actual questions that the enumerators would put to the respondents verbatim in a specified order. The choice between the two would be generally determined by the nature of respondent biases, the type of inquiry, the number of items, the skill and training of the enumerators, and the costs involved.

In general, a questionnaire is more suitable for inquiries where the number of items, quantitative answers and cross-tabulations are not too many, and especially in surveys on opinions and attitudes. On the other hand, a schedule is more compact and flexible: with a schedule a well-trained enumerator is generally in a better position to frame questions to elicit the required information and to meet the exigencies of the situation.

26.12.2 Details of the information to be collected

The statement of the desired information and the form in which final estimates are required would determine the details of the items of information to be collected. The problems naturally differ in different subject fields and it is not generally possible to give a formulation which will apply in all situations. A few general observations may, however, be made.

Some derived information may be obtained from the items of information collected during the survey. Examples are family composition, including family nucleus, and the socio-economic status of persons. For accurate observation, some items may require separate recording of their components or inclusion of supplementary and supporting items of information. For example, data on the total number of children borne by a woman cannot often be collected accurately unless for each marriage of a woman, account is taken separately of the children who were born alive but since died and those living, and among the second category those living in the household (or family) being enumerated and those living separately.

Some difficult or sensitive items such as income, savings, indebtedness, opinions and attitudes may better be put at the end of the schedule.

Some items of information require the inclusion of others in order to ensure proper interpretation. For example, it is necessary to ask the actual programmes which the respondents listen to in a survey of preference of radio listeners and the actual books and periodicals read in a survey of readership preferences.

The framing and arrangements of the questions or items of information should be done in such a manner as to facilitate accurate data collection and tabulation. The schedule or the questionnaire should not be too lengthy or tedious.

26.12.3 Time-reference period for data

The data collected should relate to a well-defined reference period. The time-reference period need not be identical for all items of information.

With a fixed length, the reference period can be either fixed or moving; the former is a period whose end-points are fixed, such as the week 10–16 January; the latter moves with the date of inquiry, such as the seven days preceding it.

The choice of the length and type (fixed or moving) of the reference period depends on the frequency of occurrence of the items, the manner of accounting, the memory factor, and the type of survey. In an integrated household inquiry, for example, the reference period could be a year for infrequent purchases of durable goods such as furniture, and a week or month of food consumption: those for household income and production would depend on the receipt of incomes and production cycles, such as agricultural seasons.

For most of the data in a demographic inquiry, the reference period will be the enumeration day; for economic characteristics, it may be a brief period prior to enumeration, preferably not longer than a week, but when this short period is unrepresentative of seasonal changes in the level of activity and regular periodic sample surveys are not conducted, supplementary information on 'usual' economic characteristics over a longer period may also be collected; for information on current fertility and mortality, the reference period is generally taken as twelve months preceding the enumeration, but for data on fertility history it should obviously cover the whole reproductive span of the sample women.

In continuing surveys with common ultimate stage sample units (see sections 24.3 and 26.13.1), the time-reference period for most items will be the interval since the last survey.

26.13 Survey design

26.13.1 Types of surveys

The objectives of the survey will determine the choice of the sample design and the type of survey. The following types of surveys have been distinguished by the Sub-Commission on Statistical Sampling of the United Nations Statistical Commission.

'(i) *Integrated survey.* In an integrated survey, data on several subjects (or items, or topics) are collected for the same set of sampling units for studying the relationship among items belonging to different subject fields. Such surveys are of special importance in studies on levels of living. Integrated surveys of consumption and productive enterprises are also of special importance in developing countries where the related activities are frequently undertaken in an integrated manner in the household.

'(ii) *Multi-subject survey.* When in a single survey operation several subjects, not necessarily very closely related, are simultaneously investigated for the sake of economy and convenience, the investigation may be called a multi-subject survey. The data on different subjects need not necessarily be obtained for the same set of sampling units, or even for the same type of units (e.g. households, fields, schools, etc.).

'(iii) *Continuing surveys.* The most usual example of these surveys occurs where a permanent sampling staff conducts a series of repetitive surveys which frequently include questions on the same topics in order to provide continuous series deemed of special importance to a country. Questions on the continued topics can frequently be supplemented by questions on other topics, depending upon the needs of the country.

'(iv) *Ad hoc survey.* This is a survey without any plan for repetition.

'(v) *Multi-purpose survey.* This term is sometimes used in connection with sampling organizations which conduct surveys in various fields of interest to several departments or parties, keeping in view their diverse purposes. This permits economies in the technical and other resources and their more effective use, particularly in developing countries.

'(vi) *Specialized or special-purpose survey.* This may be defined as an investigation focusing on a single set of objectives which, because of their nature or complexity, requires specialized knowledge or the use of special equipment by a technical staff with training in the subject field of enquiry.'

Multi-subject integrated surveys are generally more economical than a series of single-subject non-integrated surveys due to savings in overhead costs and in survey cost (travel to the penultimate-stage sample units such as villages, enumeration districts etc., camp setting and listing and contacting the ultimate-stage units). The economies become considerable with a permanent survey organization, when multi-stage and multi-phase sampling designs can be adopted. Further, grouping of different subjects of inquiry in the same sample of first-stage units helps in increasing the sample size at the first stage, which generally improves the efficiency of the survey estimates. In many circumstances, therefore, the efficiency attainable by separate single-subject surveys may be achieved by incurring a much smaller expenditure in an integrated multi-subject survey.

It will be noted again that only an integrated survey makes it possible to study the inter-relations of different subject fields. Further, if integration is carried out up to the penultimate-stage units (e.g. locality or a village), ecological information such as the distance from the nearest school, health and communication centres, irrigation system, existence of anti-malarial operations, etc., need be collected only once (often available from routine administrative sources) and analysed in respect of the various items of inquiry.

Recent trends have therefore been towards adopting multi-subject integrated surveys both in established survey organizations such as the U.S. Bureau of the Census and the Indian National Sample Survey and in a number of developing countries which are embarking on sampling inquiries.

A multi-subject integrated survey is generally more suitable for household inquiries. It may be impractical or inefficient for combining inquiries that require very different types of designs: there are also the problems of overloading the informant, the training of enumerators and the data processing facilities.

Often it is also found impractical or inefficient to collect some types of agricultural statistics through household sample surveys. These relate mainly to the time element involved in crop cutting (i.e. at the right time for harvesting) and the difference between the so-called 'biological yield' (i.e. the yield without any loss) and 'the economic yield' (i.e. the part of the yield available to farmers for all uses). The important point here is that to be realistic, yield surveys must be conducted under conditions of farm practices, and this is not generally possible in a multi-subject integrated inquiry.

Some specialized inquiries cannot be fitted into schemes of an integrated multi-subject survey. Thus, while generally accurate information on the demographic and socio-economic aspects of morbidity such as sex, age, days lost due to sickness, expenditure on medicines etc., could be collected through an integrated household survey, a separate diagnostic survey with a fully equipped team is required for clinical and laboratory examinations to supply information on causes of sickness: this method is currently being followed in the U.S.A. National Health Survey. Other examples are surveys of industrial establishments, fisheries surveys, etc.

26.13.2 Choice of sample design

Stratified multi-stage designs constitute the commonest type of designs used in practice. They combine the advantages (including economies) of stratification and clustering when the strata are made internally homogeneous and the first-stage units internally heterogeneous, so that a small number of first-stage units need be selected from each stratum in order to provide an efficient sample. In practice, geographically compact strata make field enumeration easier.

Usually, the primary strata are compact geographical areas. Within these, further stratification may be done on the basis of available supplementary information on the universe. The ultimate strata may be smaller homogeneous areas or groups of units having similar characteristics. Stratification could also be used at second and subsequent stages in a multi-stage design.

For a given total cost, multi-stage sampling will often give more efficient estimates than unrestricted simple random sampling, because of the usual economies of clustering and better sampling plans for the second and subsequent stages. Also, the advantages of stratification are generally considerable and will almost outweigh the slight increase in the complexity of analysis. For a highly variable universe, the very large units may be put in a separate stratum and completely enumerated: this is the general practice in surveys of establishments.

Sample selection in each stratum should be according to the most efficient method applicable and need not be uniform in all strata. For the stages for which samples can be drawn in the central offices, varying probability selection may be used, but for the ultimate and sometimes the penultimate stages, for which the samples have to be drawn by the field enumerators, simpler methods are required such as simple random or systematic selection.

The criteria of stratification and allocation of the sample size in the different strata and stages have already been noted in Chapters 12, 17, and 22. In multi-subject inquiries these evidently have to be decided on the basis of which variables are the more important: different sampling fractions for the ultimate units could be called for (e.g. for demographic *vis-à-vis* employment data), or different and appropriate samples of second-stage units within the same primary units can be selected using different criteria of size (e.g. selection of households may be based on expenditure or income criteria, whereas those for agricultural holdings may be selected on land criteria).

For computing estimates, available supplementary information may be used to obtain ratio or regression estimates in order to improve the efficiency of the estimators. The ratio and regression methods of estimation can be used in different stages and also at different levels of aggregation, but it may be desirable to have some idea of their efficiency and bias before they are used on a wide scale. Furthermore, one would need to consider the facilities available for processing when these methods are used, as they involve greater computation work.

There are considerable advantages in adopting a self-weighting design, mentioned in section 23.5, and the technique of interpenetrating sub-samples, mentioned in sections 24.6 and 25.5.4.

26.13.3 Sampling frames

The frame for sampling consists of previously available descriptions of the objects or material related to the physical field in the form of maps, lists, directories, etc., from which sampling units may be constructed and a set of sampling units selected. Generally a frame may consist either of area units or of list units covering the items being investigated in a survey. A frame becomes more valuable if it contains some supplementary information which can be used to improve sampling and estimating procedures; also, information on communications, transport, types of crops etc. may be of value in improving the design for the choice of sampling units, in programming the field work and in the formation of strata. The sampling frame should be accurate, free from omissions and duplication, adequate and up to date and the units should be identifiable without ambiguity.

Unless the sample survey is conducted simultaneously with a census – of population, housing, agriculture etc. – the initially available frame will often require amendments for up-dating. And even when the sample survey is conducted along with the census, the question of under- or over-enumeration in the census would remain to be tackled.

The problems arise particularly in the latter stages of multi-stage sampling, and often a frame may require to be constructed *ab initio.* Thus a frame of higher-stage units, such as provinces, countries, towns and villages, and enumeration areas, may be available from a census, but for the selection of households farms etc., a complete listing of such units in the selected penultimate stages will have to be made. In addition to the requirement of making the frame of ultimate units up to date and complete, such listing has the advantage that useful supplementary information may be collected with little more effort, which can be used in stratification or estimation.

Supplementary information, whether available from the frame or obtained during the listing of all the ultimate stage units in the selected penultimate stages, may also be highly useful in improving the efficiency of estimates obtained from the survey, through suitable stratification, pps or systematic selection, ratio method of estimation etc. In repeated surveys, information collected in earlier surveys may also serve to improve the frame for later surveys.

In some parts of Africa, villages are sometimes abandoned altogether and the whole population shifts elsewhere, or a village might be split into two or more parts which take on the names of the new chiefs: these make difficult the up-dating of a frame in the absence of an adequate administrative machinery.

Note: For some methods of utilizing imperfect frames, see the papers by Hansen, Hurwitz, and Jabin (1964), and Szameitat and Schaffer (1964) and the discussion at the Meeting on 'Sampling from Imperfect Frames' at the 34th Session of the International Statistical Institute, 1963 (*Proceedings*, 1964).

26.13.4 Survey periods

In countries where the economy is subject to pronounced seasonal fluctuations within a year, socio-economic inquiries might preferably be conducted for a year or multiples of a year with the work of data collection spread evenly over the survey period. This would enable not only the study of the seasonal elements but also a meaningful annual average. In a permanent survey organization where the trained field personnel are small in number, such a staggering of the survey is the only recourse.

In such cases, to minimize recall lapse and to represent all parts of the year it is necessary to adopt moving rather than fixed reference periods; for example, the reference period for the collection of information on some items of household consumption could be the seven days preceding the date of enumeration rather than a fixed calendar week.

In addition to the survey period considered above (such as one year, an agricultural season etc.), the periodicity of surveys – annual, biennial etc. – would also have to be considered.

26.14 Pilot inquiries and pre-tests

In undertaking a large-scale survey, particularly of unexplored material, it should be the general rule to conduct pre-tests and pilot and exploratory inquiries to test and improve field procedures and schedules and to train field enumerators; also to obtain information on the variability of the data and on the cost-pattern which will enable the sample design to be planned more efficiently and to finalize the tabulation programme. For example, the results of a pilot inquiry may be used to estimate the first- and second-stage components of variance relevant to a two-stage sample design which is envisaged for the main survey, and also the relevant components of cost from which it is possible to determine the optimum intensity of sampling at each of these two stages. Pilot inquiries also provide information required to determine the most effective type and size of sampling units.

Pilot inquiries required to test different methods of data collection, types of schedules and enumerators etc. may be organized in the form of interpenetrating networks of sub-samples that will enable valid comparisons to be made of the factors under study.

The pilot inquiry, which can be considered a 'dry run' for the survey proper, may be preceded by a series of small pre-tests on isolated problems of the design or enumeration procedures; the pre-test need not be based on probability samples, and even 50 to 100 households or persons covered in a pre-test could provide valuable information. One important component of pre-test in household sampling should be the practice of house-listing in some representative areas in order to train the enumerators so that the coverage errors in the survey proper are minimized.

26.15 Selection and training of enumerators

The problem of selection of enumerators depends on whether the sample survey is conducted on an *ad hoc* basis or is conducted over time as an integrated process of data collection. For the first type, part-time staff, such as teachers and housewives, can be used sometimes; but for the latter, full-time enumerators have often to be selected and incentive provided so that their performance is satisfactory and the turn-over is not high. In countries with different linguistic and ethnic groups, it is necessary to ensure that the enumerators are accepted by the population being surveyed: female enumerators may, for example, be employed to interview female respondents in some surveys. The enumerators should have nearly equal work-load, and each should cover a manageable geographical area.

It is often necessary to decide whether to employ locally-based staff confined to a few sampling units, or mobile or itinerant staff covering a wider area. The choice between the two types of staff would again depend mainly on the type of survey being undertaken.

In a small-scale inquiry, enumerators can be trained in the central office. But in a large-scale survey, this may not be practicable and the training will be organized in two phases: first the supervisors will be trained by the technical staff, and second, the supervisors in turn will go to their field offices and train the enumerators under them. It would be beneficial to organize refresher courses for both the supervisors and the enumerators, not only for the field procedures but also on the objectives and methodology of the survey. The training method normally comprises classroom teaching, provision of manuals, demonstration enumeration, field work under supervision, scrutiny of filled-in schedules, and oral and written tests. The manual for the enumerators may contain the following topics: the survey objectives and programme; preliminary work in the sample areas; how and whom to interview; detailed instructions on the questions including concepts, definitions and examples; checking completed questionnaires; preparation of summary results; and return of documents and equipment.

26.16 Supervision

Accurate field work can only result from thorough training of efficient enumerators. Nevertheless, adequate supervision must be an integral part of the field work.

In the U.S.A. Current Population Survey, the ratio of supervisor to enumerators is 1:6; in the Indian National Sample Survey and the 1969 National Survey of Family Income and Expenditure in Japan it is 1:4. The supervisor should undertake some field work, either independently or as a check on the work of the enumerators, and this information can be used to adjust the data collected by the enumerators.

It is preferable to have a copy of the filled-in forms made before they are passed to the central offices for tabulation and analysis. The enumerators should themselves carry out simple numerical calculations and consistency checks of the data. When a copy of the enumerators' filled-in schedules is retained in the supervisor's office, a percentage of both the original and the copy should be scrutinized for copying and other mistakes; and, wherever possible, gross mistakes should be referred back to.

Any correction made by the supervisor on the enumerator's schedules should not be done by erasing the enumerator's entry but should be made in such a way that both can be used for future checks and analysis.

The enumerators should also be required to keep records of the times spent on the different aspects of the field work such as journey, identification and contact of sampling units, listing, enumeration, etc., which may be useful for designing subsequent inquiries.

26.17 Processing of data

After the completion of the enumeration, the schedules should be checked as carefully as possible and those entries which are not in the form of numerical quantities should be converted into numerical form; e.g. the sexes recorded as 'male' and 'female' may be converted to 1 and 2 respectively. The first of these steps is called *editing*, and the second, *coding*.

While a precoded questionnaire can speed up a survey very greatly, it has several disadvantages, the major being the lack of checks. It is generally desirable to have the coding done in the central office.

The filled-in forms should be scrutinized by the technical staff for consistency and, as far as possible, for accuracy before they are sent on for processing. In a large-scale survey, this can be done on a sub-sample basis separately for each enumerator: if the percentage of inconsistent and inaccurate entries exceeds a certain limit for an enumerator, all the schedules filled in by him should be checked. Forms containing too many mistakes or doubtful entries may have to be rejected.

The importance of preparing a tabulation plan well in advance has been mentioned in section 26.10.

Tabulations can be manual, mechanical, or electronic. The data can be processed into final tabulations either by analysing direct from the schedules or by transferring the data to cards or tapes for mechanical or electronic processing. In small-scale sample inquiries, the main required estimates can often be obtained manually from the forms and the cross-tabulations made through mechanical or electronic processing. In large-scale inquiries too, estimates that are required quickly can often be obtained from the summaries which the enumerators may be asked to prepare; for example, for the Indian National Sample Survey on population, births and deaths in 1958–9, estimates of birth and death rates were prepared within a fortnight of the receipt of schedules: further estimates for cross-tabulations and also the computations of standard errors were made later through electronic processing.

Such preliminary results of the main items of a survey often serve important purposes. If the survey is arranged in the form of interpenetrating sub-samples, advance estimates can be obtained on earlier processing of one or more of the sub-samples (section 24.6.1).

During processing, it is advantageous to make provisions for obtaining estimates for the first-stage sample units. From a listing of these, both the final estimates and their standard errors can be computed in relation to totals and ratios: for the computation of standard errors in each stratum, two or more first-stage units should be selected with replacement, or the sampling fraction should be small when they are selected without replacement.

Checks should be introduced at every stage of processing. For the punching of cards, for example, some dummy entries might be made which would ensure the checking of the punching and the verification work.

The computations required for a large-scale survey are made on a different basis from those illustrated in this book. For each ultimate sample unit, the multiplier or the weighting factor is computed and the product of the multiplier and the value of the study variable for the sample unit is obtained. When these are summed up for all the sample units by the machine, the required estimate is obtained and a suitable procedure is adopted for obtaining its standard error.

Suppose in a household inquiry (as in Example 21.1), the sample is stratified two-stage, with villages, the first-stage units, selected with pps and with replacement, and households, the second-stage units, selected at random or systematically. For estimating a stratum total, the multiplier of each household is given by

$$\frac{\text{reciprocal of the probability of selection of the sample village}}{\text{number of sample villages in the stratum}}$$

$$\times \frac{\text{total number of households in the sample village}}{\text{number of sample households in the sample village}}$$

This multiplier will be computed for each sample household. To estimate the total consumption of food in a stratum, for example, the food consumption in a sample household will be multiplied by its multiplier, and the sum of the products for each stratum would give the estimated total food consumption in it. These stratum estimates when summed up will provide the unbiased estimate of the total food consumption in the universe. When the ultimate units in the sample penultimate units are selected with equal probability (as in this example), the multiplier will be the same for the sample ultimate units in a particular selected penultimate unit.

Of course, in a self-weighting design, for estimating the totals the multiplier will be applied only once at the overall level, and it will not be required for estimating the ratio of two totals and its variance estimator.

The advent of electronic computers in recent years has been in many ways a great help in obtaining the required estimates speedily and accurately: for a sample survey, standard errors of a host of estimates can be computed with their aid. These can also be programmed to tasks of editing with consequent improvement of consistency of the primary data and possibly of its accuracy. They make possible some forms of estimation on a scale which would have been impractical with manual and mechanical tabulations. For example, in some cases the use of the regression method of estimation on previous surveys of a series with overlapping samples can substantially improve the efficiency of the estimators. The use of computers makes possible changes in the allocation of resources between the collection of data and processing. Computers can also be used in analysis and studies, for example the interrelations of various factors by multivariate analysis (United Nations, 1964b).

However, an efficient use of computers would depend upon the availability of trained programmers, especially the local staff, continuous service facilities, and intensive utilization of computer time.

A record of time, kept on the different processing operations, would help to improve future operations.

26.18 Preparation of reports

In addition to the presentation of the numerical results, the report of a survey, either complete enumeration or a sample, should contain some discussion and interpretation of the results.

For the preparation of the report of a sample survey, the United Nations Sub-Commission on Statistical Sampling has made certain recommendations (1964b); these are summarized below.

26.18.1 General Report

This should contain a general description of the survey for the use of those who are primarily interested in the final results and broad findings rather than in the technical and statistical aspects of the sample design, execution, and analysis. The general description should include information on the following points:

1. Statement of the purposes of the survey;
2. description of the scope and coverage;
3. items of information and method of data collection;
4. date and duration – the survey and time-reference periods;
5. repetition – whether the survey is an isolated one or is one of a series of similar surveys;
6. numerical results;
7. main findings;
8. accuracy – a general indication of the accuracy achieved, distinguishing between sampling errors and non-sampling errors;
9. costs – an indication of the cost of the survey, under such headings as preliminary work, enumeration, analysis etc.;
10. assessment – the extent to which the purposes of the survey were fulfilled;
11. responsibility – the names of the organizations sponsoring and conducting the survey;
12. references – references to any available reports or papers relating to the survey.

26.18.2 Technical reports

Technical reports should be issued for surveys of particular importance and those using new techniques and procedures of special interest. In addition to covering such fundamental points as the purposes of the survey, stipulated margins of error, conditions to be fulfilled and resources available for the survey, the report should deal in detail with technical and statistical aspects of the sampling design, execution and analysis; the operations and other special aspects should also be fully covered. These aspects may be considered under the following headings:

1. *Design of the survey.* The sample design should be described in detail, sampling frames (their specification, accuracy, and adequateness), including types of sampling units, sampling fractions, particulars of stratification, sampling procedure for different stages etc.

2. *Personnel and equipment.* It is desirable to give an account of the organization of the personnel employed in planning, collecting, processing and tabulating the primary data, together with information regarding their previous training and experience. Arrangements for recruitment, training, inspection and supervision of the staff should be explained; as also should methods of checking the accuracy of the primary data at the point of collection. A brief mention of the equipment used is frequently of value to readers of the report.

3. *Statistical analysis and computational procedure.* The statistical methods followed in the estimation and in compilation of the final summary tables from the primary data should be described, the relevant formulae being reproduced

where necessary. If any elaborate processes of estimation have been used, the methods followed should be explained with reasons.

4. *Reliability and accuracy of the survey.* This should include
(i) efficiency as indicated by the standard errors deducible from the survey;
(ii) degree of agreement observed by the independent enumerators covering the same material when interpenetrating samples are used;
(iii) other non-sampling errors.

In addition, methods adopted for evaluating the accuracy of the estimates and the method of controlling and adjusting for non-sampling errors and biases should be described.

5. *Technical analysis and interpretation of survey results.* These would be of value to answer some questions or to test some hypotheses; in a demographic inquiry, for example, the fertility and mortality levels and trends can be measured and the socio-economic differentials tested.

6. *Comparisons with other sources of information.* Every reasonable effort should be made to compare the survey results with other independent sources of information. The object of this is not to throw light on the sampling error, since a well-designed survey provides adequate internal estimates of such errors, but rather to gain knowledge of biases and other non-random errors.

Disagreement between results of a sample survey and other independent sources may, of course, sometimes be due, in whole or in part, to differences in coverage, concepts and definitions or to errors in the information from other sources.

7. *Costing analysis.* The sampling method can often supply the required information with greater speed and at lower cost than a complete enumeration. For this reason, information on the costs involved in sample surveys is of particular value for the development of sample surveys within a country and is also of help to other countries.

Fairly detailed information should be given on costs of sample surveys under such headings as planning (showing separately the cost of pilot studies), travel, enumeration proper, supervision, processing, analysis and overhead costs. In addition, labour costs in man-weeks of different grades of staff, and also the time required for different operations, are worth providing, for these may suggest methods of economizing in the planning of future surveys. However, the preparation of an efficient design involves a knowledge of the various components of costs as well as of the components of variance. The concept of cost in this respect should be regarded broadly in the sense of economic cost and should therefore take account of indirect costs which may not have been charged administratively to the survey.

8. *Efficiency.* The results of a survey often provide information which enables investigations to be made on the efficiency of the sample designs, in relation to other sample designs which might have been used in the survey. The results of any such investigations should be reported. To be fully relevant, the relative costs of the different sampling plans must be taken into account when assessing the relative efficiency of different designs and intensities of sampling.

9. *Observations of technicians.* The critical observations of technicians (in statistics and the specific subject fields) in regard to the survey, or any part of it, should be given. These observations will help improve future operations.

Further Reading

Mahalanobis (1946); Murthy, Chapter 14; Murthy and Roy; United Nations, *Recommendations for the Preparation of Sample Survey Reports* (1964a); Yates, Chapters 3, 4, 5, 9, and 11.

Some references in specific subject fields are as follows:

1. *Agricultural surveys:* Food and Agriculture Organization of the United Nations, *Improvement of Agricultural Census and Survey Reports* (1969); Panse (1954; 1966); Royer; Sanderson; Sukhatme and Sukhatme (1969); Sukhatme and Sukhatme (1970, Chapter 10); Zarkovich (1963; 1965a; 1965b).
2. *Auditing and Accountancy:* Trublood and Cyert; Vance; Vance and Neter.
3. *Demographic surveys:* Blanc; Som (1973); United Nations, *Handbook of Population and Housing Census Methods, Part VI, Sampling in connexion with population and housing censuses* (1971a), and *Methodology of Demographic Sample Surveys* (1971b); United Nations Economic Commission for Africa, *Manual on Demographic Sample Surveys in Africa* (1971); U.S. Bureau of the Census (1963).
4. *Fertility and Family Planning Surveys:* Population Council; Aesádi, Klinger, and Szabady (1969).
5. *Health Surveys:* U.S. National Center for Health Statistics.
6. *Household and social surveys:* Ackoff; Gray and Corlett; Jones; Moser and Kalton; United Nations, *Handbook of Household Surveys* (1964b).
7. *Opinion and marketing research:* Blankenship; Cantril; Gallup; Parten; Payne; Stephan and McCarthy.

Appendix I List of notations and symbols

1. For *summation and product notations,* see section 1.15.
2. In general, the following *subscripts* are used:
   h for a stratum, $h = 1, 2, \ldots, L$;
   i for a first-stage unit, $i = 1, 2, \ldots, N$ for the universe, and
   $i = 1, 2, \ldots, n$ for the sample;
   j for a second-stage unit, $j = 1, 2, \ldots, M_i$ for the universe, and
   $j = 1, 2, \ldots, m_i$ for the sample;
   k for a third-stage unit, $k = 1, 2, \ldots, Q_{ij}$ for the universe, and
   $k = 1, 2, \ldots, q_{ij}$ for the sample.
3. y denotes a *study variable* (also an unbiased estimator of the total Y of the study variable for the universe);
   x denotes *another study variable*; and
   z an *ancillary variable* either for selection with probability proportional to size or for ratio or regression estimation.
4. The following *symbols* are used with additional subscripts, as required:

	Universe parameter	*Sample estimator*
Total of y	Y	y_0^* (or y)
Mean of y (per first-stage unit)	$\bar{Y}$	y_0^*/N
Proportion	P	p
Ratio	R	r
Sampling variance of estimator t	σ_t^2	s_t^2
Standard error of t	σ_t	s_t
Covariance of t and u	σ_{tu}	s_{tu}
Correlation coefficient	ρ	$\hat{\rho}$
Intraclass correlation coefficient	ρ_c	$\hat{\rho}_c$
Regression coefficient	β	$\hat{\beta}$

5. *Other notations and symbols*

f	sampling fraction (with subscripts added for strata and stages)
srs	simple random sample or simple random sampling
srswr	srs with replacement
srswor	srs without replacement
pps sampling	probability proportional to size sampling
ppswr	pps sampling with replacement
$=$	is equal to
$\neq$	is not equal to
$\approx$	is approximately equal to
$\leqslant$	is less than or equal to
$\geqslant$	is greater than or equal to
$\propto$	is proportional to
∞	infinity

SSy_i	sum of squares of deviations of y_i from the mean $\bar{y}$ (see section 1.15)
SPy_ix_i	sum of products of deviations of y_i and x_i from the respective means (see section 1.15).

6. *Greek letters used*

α (alpha)	probability point of the t or the normal distribution
β (beta)	regression coefficient
π (pye, small)	probability of selection
Π (pye, capital)	product notation
ρ (roh)	correlation coefficient
σ (sigma, small)	universe standard deviation
Σ (sigma, capital)	summation notation

Appendix II Elements of Probability and Proofs of Some Theorems in Sampling

A2.1 Introduction

In this appendix are presented first the elements of probability and then the theorems themselves required for proving some theorems in sampling.

A2.2 Elements of probability

A2.2.1 Definition of probability

We consider the classical definition of probability.

A number of events are considered equally likely if no one of them can be expected in preference to the others. Suppose there are t total possible outcomes of an experiment conducted under a given set of conditions, and that they are exhaustive (i.e. one of them must necessarily occur), mutually exclusive (i.e. no two of them can occur simultaneously), and equally likely. Let f of these t outcomes represent an event A (i.e. f are favourable to the occurrence of A), so that A happens when one of the f favourable outcomes happens and conversely.

Then the (mathematical) probability of A is

$$P(A) = f/t \tag{A2.1}$$

As $f \leqslant t$,

$$0 \leqslant P(A) \leqslant 1 \tag{A2.2}$$

$P(A) = 0$ means that the event A is impossible, and $P(A) = 1$ means that the event A is certain to occur.

Example

A six-faced die is thrown. What is the probability that (a) the number 4 turns up? (b) an even number turns up?

The six possible cases are 1, 2, 3, 4, 5, 6, and these are exhaustive and mutually exclusive; they may also be considered equally likely if the die is perfect in shape and homogeneous, and is thrown such that no one face gets any particular preference over others. That is, $t = 6$.

For (a), there is only one favourable case, namely, the figure 4, so that the probability of 4 turning up is $f/t = \frac{1}{6}$; and this holds for all the other figures.

For (b), the number of favourable cases is $f = 3$ (namely, the even numbers 2, 4, and 6), and therefore the probability that an even number will turn up is $f/t = \frac{3}{6} = \frac{1}{2}$.

A2.2.2 Theorems of total probability

Let $A_1, A_2, \ldots, A_n$ denote n events which may happen as a result of an experiment performed under a given set of conditions. Then $A_1 + A_2 + \ldots + A_n = \sum^n A_i$ will denote the occurrence of at least one of these n events. And $A_1 A_2 \ldots A_n = \prod^n A_i$ will denote the simultaneous occurrence of all these n events.

(a) *Mutually exclusive events.* If $A_1, A_2, \ldots, A_n$ are n mutually exclusive events, then the probability that at least one of these n events will occur is

$$P(A_1 + \ldots + A_n) = P(A_1) + \ldots + P(A_n) = \sum^{n} P(A_i) \qquad \text{(A2.3)}$$

If in addition the events are exhaustive, i.e. at least one of the events must occur, then

$$P(A_1 + \ldots + A_n) = \sum^{n} P(A_i) = 1 \qquad \text{(A2.4)}$$

If the non-occurrence of the event A is denoted by $\bar{A}$, then A and $\bar{A}$ being exhaustive and mutually exclusive, then $P(A) + P(\bar{A}) = 1$, so that

$$P(\bar{A}) = 1 - P(A) \qquad \text{(A2.5)}$$

For example, the probability of the figure 4 not turning up in a throw of a perfect die is $P(\bar{A}) = 1 - P(A) = \frac{5}{6}$, where A denotes the turning up of the figure 4.

(b) *Events not mutually exclusive.* If the n events $A_1, A_2, \ldots, A_n$ are not necessarily mutually exclusive, then

$$P(A_1, A_2 + \ldots + A_n) = \sum_{i}^{n} P(A_i) - \sum_{\substack{i, j \\ i<j}} P(A_iA_j) + \sum_{\substack{i, j, k \\ i<j<k}} P(A_iA_jA_k) - \ldots - (-1)^n \, P(A_1A_2 \ldots A_n) \qquad \text{(A2.6)}$$

A2.2.3 Theorems of compound probability

The conditional probability that A_2 will occur when it is known that A_1 has occurred is denoted by $P(A_2|A_1)$.

If $A_1, A_2, \ldots, A_n$ are n events that may result from an experiment, then the probability of the simultaneous occurrence of all these n events is

$$P(A_1A_2 \ldots A_n) = P(A_1)P(A_2|A_1)\,P(A_3|A_1A_2) \ldots P(A_n|A_1A_2 \ldots A_{n-1}) \qquad \text{(A2.7)}$$

provided that each of the factors on the right-hand side is non-zero.

When each one of n events $A_1, A_2, \ldots, A_n$ is independent of the others, the events are said to be *mutually independent,* and in this case

$$P(A_1A_2 \ldots A_n) = P(A_1)\,P(A_2) \ldots P(A_n) = \prod^{n} P(A_i) \qquad \text{(A2.8)}$$

i.e. the probability of the simultaneous occurrence of the n events is the product their unconditional probabilities.

For example, when two perfect dice are tossed, the probability of 4 being thrown up by both is, since the throws are independent, $\frac{1}{6} \times \frac{1}{6} = \frac{1}{36}$.

A2.2.4 Mathematical expectation

(a) *Definition.* A variable that may take any of the values of a specified set with a specific relative probability for the different units is called a *random variable* or a *stochastic variable.* For example, with a perfect die the number occurring in its tossing is a random variable which takes the values 1, 2, . . ., 6 each with the same associated probability of $\frac{1}{6}$.

Let a random variable Y_i take the values $Y_1, Y_2, \ldots, Y_N$ with respective probabilities $p_1, p_2, \ldots, p_N$, the set $Y_1, Y_2, \ldots, Y_N$ containing all the possible values of Y_i. These values need not all be different, but they are assumed to arise from an exhaustive set of mutually exclusive events $\left(\sum^{N} p_i = 1\right)$.

The mathematical expectation of Y_i is defined as

$$E(Y_i) = Y_1 p_1 + Y_2 p_2 + \ldots + Y_N p_N = \sum^{N} Y_i p_i \qquad \text{(A2.9)}$$

where the summation is taken over all the possible values of Y_i.

For example, the mathematical expectation of the number being thrown up by the tossing of a perfect die is

$$1 \times \tfrac{1}{6} + 2 \times \tfrac{1}{6} + \ldots + 6 \times \tfrac{1}{6} = 3.5$$

(b) *Some properties of expected values*

(i) The expected value of a constant is the constant itself. For, if $Y_1 = Y_2 = \ldots = Y_N = a$, and the probability is p_i, then the expected value of $Y_i = a$ is, from equation (A2.9),

$$E(Y_i) = \sum^{N} a p_i = a \sum^{N} p_i = a \qquad \text{(A2.10)}$$

since $\sum^{N} p_i = 1$, as the events are exhaustive.

(ii) $E(aY_i) = a \,.\, E(Y_i)$ (A2.11)

(iii) $E(a + Y_i) = a + E(Y_i)$ (A2.12)

(iv) The mathematical expectation of the sum of a number of random variables is the sum of their expectations.

$$E(X_i + Y_i + Z_i + \ldots) = E(X_i) + E(Y_i) + E(Z_i) + \ldots \qquad \text{(A2.13)}$$

(v) A number of random variables is said to be *independent* if the probability of any one of them to take a given value does not depend on the values taken by the other variables. Thus two variables X_i and Y_j which take respective values $X_1, X_2, \ldots, X_N$ and $Y_1, Y_2, \ldots, Y_N$ will be called independent if

$$P(X_i = X_i;\; Y_j = Y_j) = P(X_i = X_i)\, P(Y_j = Y_j)$$

for all i and j.

(vi) The mathematical expectation of the product of a number of independent random variables is the product of their expectations.

If $X_i, Y_i, Z_i, \ldots$ are the independent random variables, then

$$E(X_i Y_i Z_i \ldots) = E(X_i)\, E(Y_i)\, E(Z_i) \ldots \qquad \text{(A2.14)}$$

A2.2.5 Variance and covariance

(a) *Definition of variance.* The variance of a random variable Y_i is defined as

$$V(Y_i) = E(Y_i - EY_i)^2 \tag{A2.15}$$

$$= \sum^{N} (Y_i - EY_i)^2 p_i$$

$$= E(Y_i^2) - (EY_i)^2 \tag{A2.16}$$

from the definition of expected values, given in equation (A2.9), EY_i denoting for brevity $E(Y_i)$; $Y_1, Y_2, \ldots, Y_N$ are all the possible values of Y_i and $p_1, p_2, \ldots, p_N$ are the probabilities associated with the possible values.

(b) *Some properties of the variance.* The variance of a constant is zero, i.e.

$$V(a) = 0 \tag{A2.17}$$

Also

$$V(aY_i) = a^2 V(Y_i) \tag{A2.18}$$

and

$$V(a + Y_i) = V(Y_i) \tag{A2.19}$$

(c) *Covariance.* The covariance between two random variables X_i and Y_j is defined as

$$\text{Cov}(X_i, Y_j) = E[(X_i - EX_i)(Y_j - EY_j)] \tag{A2.20}$$

$$= 0 \quad \text{if } X_i \text{ and } Y_j \text{ are independent} \tag{A2.21}$$

(d) *Variance of a linear combination of random variables.* By definition,

$$V(X_i + Y_j) = E[(X_i + Y_j) - E(X_i + Y_j)]^2$$

$$= E[(X_i - EX_i) + (Y_j - EY_j)]^2$$

$$= E(X_i - EX_i)^2 + E(Y_j - EY_j)^2 + 2E[(X_i - EX_i)(Y_j - EY_j)]$$

$$= V(X_i) + V(Y_j) + 2\,\text{Cov}(X_i, Y_j) \tag{A2.22}$$

This result can be generalized to the case of a number of random variables. Thus for three variables X_i, Y_i, Z_i,

$$V(X_i + Y_i + Z_i) = V(X_i) + V(Z_i) + V(Z_i) + 2\,\text{Cov}(X_i, Y_i) + 2\,\text{Cov}(X_i, Z_i) + 2\,\text{Cov}(Y_i, Z_i) \tag{A2.23}$$

When the variables are independent, the covariance terms are zero and

$$V(X_i + Y_i + Z_i + \ldots) = V(X_i) + V(Y_i) + V(Z_i) + \ldots \tag{A2.24}$$

Corollary

$$V(X_i - Y_j) = V(X_i) + V(Y_j) - 2\,\text{Cov}(X_i, Y_j) \tag{A2.25}$$

$$= V(X_i) + V(Y_j) \quad \text{when } X_i \text{ and } Y_j \text{ are independent} \tag{A2.26}$$

A2.3 Proofs of some theorems in sampling

A2.3.1 Number of possible simple random samples

(a) There are ${}^N C_n$ or

$$\binom{N}{n} = \frac{N!}{n!\,(N-n)!}$$

ways of selecting n units out of a total of N units without replacement and disregarding the order of selection of the n units, where $n! = 1 \times 2 \times \ldots \times n$ is the product of the first n natural numbers and is known as factorial n. By convention $0! = 1$.

Thus if $N = 6$, $n = 2$, then in sampling without replacement and disregarding the order, the number of possible samples is (see section 2.4.1)

$$\binom{6}{2} = \frac{6!}{3!\,2!} = 15$$

(b) There are N^n ways of selecting n units out of a total of N units with replacement and considering the order of selection.

Thus if $N = 6$, $n = 2$, then the number of such possible samples is $6^2 = 36$ (see section 2.5.1).

A2.3.2 Expectation and sampling variance of the sample mean

A simple random sample of size n is drawn from the universe of size N. Let Y_i ($i = 1, 2, \ldots, N$) be the value of the variable for the ith universe unit. The universe mean is by definition

$$\overline{Y} = \frac{1}{N}(Y_1 + Y_2 + \ldots + Y_N) = \frac{1}{N}\sum_{i=1}^{N} Y_i \qquad \text{(A2.27)}$$

and the universe variance per unit

$$\sigma^2 = \frac{1}{N}\sum^{N} (Y_i - \overline{Y})^2 \qquad \text{(A2.28)}$$

Let us denote by y_i ($i = 1, 2, \ldots, n$) the value of the variable for the ith sample unit (i.e. the unit selected at the ith draw). The sample mean is, by definition,

$$\overline{y} = \frac{1}{n}(y_1 + y_2 + \ldots + y_n) = \frac{1}{n}\sum^{n} y_i \qquad \text{(A2.29)}$$

We shall obtain the expectation and variance of the sample mean separately for sampling with and without replacement.

(a) *Sampling with replacement.* To prove:

$$E(\overline{y}) = \overline{Y} \qquad \text{(A2.30)}$$

$$\text{Var}(\overline{y}) = \sigma_{\overline{y}}^2 = \frac{\sigma^2}{n} \qquad \text{(A2.31)}$$

Proof.

$$E(\overline{y}) = E\left(\frac{1}{n}\sum^{n} y_i\right) = \frac{1}{n}E\left(\sum^{n} y_i\right) = \frac{1}{n}\sum^{n} E(y_i)$$

from section A2.2.4.

Now in sampling with replacement, as the successive draws are independent and the universe remains the same throughout the sampling process, y_i ($i = 1, 2, \ldots, n$) can assume any one of the values $Y_1, Y_2, \ldots, Y_n$ with the same probability $1/N$, so that

$$E(y_i) = \sum^{N} Y_i \,.\, P(y_i = Y_i)$$

$$= \sum^{N} Y_i \frac{1}{N} = \overline{Y}$$

Hence

$$E(\overline{y}) = \frac{1}{n} n\overline{Y} = \overline{Y}$$

For the second equation, by definition the variance of $\overline{y}$ is

$$\mathrm{Var}(\overline{y}) = \sigma_{\overline{y}}^2 = E(\overline{y} - E\overline{y})^2$$

$$= E\left[\frac{1}{n} \sum^{n} (y_i - Ey_i)\right]^2$$

$$= \frac{1}{n^2} E\left[\sum^{n} (y_i - Ey_i)\right]^2$$

$$= \frac{1}{n^2} \sum_{i=1}^{n} E(y_i - Ey_i)^2 + \frac{1}{n^2} \sum_{i \neq j}^{n} E(y_i - Ey_i)\, E(y_j - Ey_j)$$

$$= \frac{1}{n^2} \sum^{n} \mathrm{Var}(y_i) + \frac{1}{n^2} \sum_{i \neq j}^{n} \mathrm{Cov}(y_i, y_j) \qquad \text{(A2.32)}$$

Now

$$\mathrm{Var}(y_i) = E(y_i - \overline{Y})^2$$

$$= \sum^{N} (Y_i - \overline{Y})^2 P(y_i = Y_i)$$

$$= \frac{1}{N} \sum^{N} (Y_i - \overline{Y})^2 = \sigma^2 \qquad \text{(A2.33)}$$

$$\mathrm{Cov}(y_i, y_j) = E(y_i - \overline{Y})(y_j - \overline{Y})$$

$$= \sum_{i,j}^{N} (Y_i - \overline{Y})(Y_j - \overline{Y}) P(y_i = Y_i; y_j = Y_j) \qquad \text{(A2.34)}$$

In sampling with replacement y_i and y_j are independent, i.e. y_i can take any of the values $Y_1, Y_2, \ldots, Y_N$ irrespective of the value y_j takes, so that

$$P(y_i = Y_i;\; y_j = Y_j) = P(y_i = Y_i)\, P(y_j = Y_j)$$

$$= \frac{1}{N} \cdot \frac{1}{N} = \frac{1}{N^2}$$

Thus

$$\mathrm{Cov}(y_i, y_j) = \frac{1}{N^2} \sum_{i,j}^{N} (Y_i - \overline{Y})(Y_j - \overline{Y})$$

$$= \frac{1}{N^2} \sum_{i}^{N} (Y_i - \overline{Y}) \sum_{j}^{N} (Y_j - \overline{Y}) = 0 \qquad \text{(A2.35)}$$

as $\sum^{N} (Y_i - \overline{Y}) = \sum^{N} (y_j - \overline{Y}) = 0$, being the sum of the deviations from the mean.

From equation (A2.32), using equations (A2.33) and (A2.35), we obtain

$$\sigma_{\overline{y}}^2 = \frac{1}{n^2} n\sigma^2 = \frac{\sigma^2}{n}$$

Note: A shorter proof would be as follows.

$$\text{Var}(\bar{y}) = \text{Var}\left[\frac{1}{n}\sum^{n} y_i\right] = \frac{1}{n^2}\,\text{Var}\left(\sum^{n} y_i\right) = \frac{1}{n^2}\sum^{n}\text{Var}(y_i)$$

since the y_is are independent (from equation A2.24). Using equation (A2.33), the final result is obtained.

(b) *Sampling without replacement.* To prove:

$$E(\bar{y}) = \bar{Y} \tag{A2.36}$$

$$\text{Var}(\bar{y}) = \sigma_{\bar{y}}^2 = \frac{\sigma^2}{n}\,\frac{N-n}{N-1}$$

$$= \frac{S^2}{n}\,(1-f) \tag{A2.37}$$

where

$$S^2 = \frac{1}{N-1}\sum^{N}(Y_i - \bar{Y})^2$$

$$= \frac{N}{N-1}\,\sigma^2 \tag{A2.38}$$

and $f = n/N$ is the sampling fraction.

Proof. As before

$$E(\bar{y}) = \frac{1}{n}\sum^{n} E(y_i)$$

In sampling without replacement also, y_i can take any one of the values $Y_1, Y_2, \ldots, Y_N$ with the same probability $1/N$ (see note 1). Therefore $E(\bar{y}) = \bar{Y}$.

To prove (A2.37), we proceed up to (A2.34) as for srs with replacement. In srs without replacement, however,

$$P(y_i = Y_i;\ y_j = Y_j)$$
$$= P(y_i = Y_i)\,P(y_j = Y_j \mid y_i = Y_i)$$
$$= \frac{1}{N}\cdot\frac{1}{N-1} \quad \text{if } i \neq j$$

since y_j can take any value excepting Y_i (the value which is already known to have been assumed by y_i) with probability $1/(N-1)$. So

$$\text{Cov}(y_i, y_j) = \frac{1}{N(N-1)}\sum_{i \neq j}^{N}(Y_i - \bar{Y})(Y_j - \bar{Y})$$

$$= \frac{1}{N(N-1)}\sum_{i}^{N}(Y_i - \bar{Y})\left[\sum_{j}^{N}(Y_j - \bar{Y}) - (Y_i - \bar{Y})\right]$$

$$= \frac{1}{N(N-1)}\left[\sum_{i}^{N}(Y_i - \bar{Y})\sum_{j}^{N}(Y_j - \bar{Y}) - \sum_{i}^{N}(Y_i - Y)^2\right]$$

$$= -\frac{1}{N(N-1)}\sum^{N}(Y_i - \bar{Y})^2 \qquad \text{as } \sum^{N}(Y_j - \bar{Y}) = 0$$

$$= -\frac{\sigma^2}{N-1} \qquad \text{by definition} \tag{A2.39}$$

As there are $n(n-1)$ covariance terms in (y_i, y_j) for $i \neq j$,

$$\sum_{i \neq j}^{n} \text{Cov}(y_i, y_j) = -\frac{n(n-1)\,\sigma^2}{N-1}$$

From equations (A2.32), (A2.33), and (A2.39),

$$\begin{aligned}\sigma_{\bar{y}}^2 &= \frac{n\sigma^2}{n^2} - \frac{n(n-1)}{n^2(N-1)}\,\sigma^2 \\ &= \frac{\sigma^2}{n}\left(1 - \frac{n-1}{N-1}\right) \\ &= \frac{\sigma^2}{n}\,\frac{N-n}{N-1} = \frac{S^2}{n}(1-f)\end{aligned}$$

Notes

1. In srs with replacement, the probability that the ith universe unit is selected on any draw is clearly $1/N$, since the units are returned after selection so that the universe remains the same throughout the sampling process.

   In srs without replacement, the probability that a specific universe unit is selected on the first draw is $1/N$. The probability that it is selected on the second draw is given by the probability that it is not selected in the first draw multiplied by the conditional probability that it is selected on the second draw, i.e.

$$\frac{N-1}{N} \cdot \frac{1}{N-1} = \frac{1}{N}$$

   Similarly, the probability that it is selected on the third draw is

$$\frac{N-1}{N} \cdot \frac{N-2}{N-1} \cdot \frac{1}{N-2} = \frac{1}{N}$$

   and so on. Hence the result.

2. A specific universe unit may be included in the sample of n units at any of the n draws; the events that the unit is included in the first draw, second draw, . . ., nth draw are also mutually exclusive. Therefore the probability that the specific unit is included in the sample is, from A2.2.2(a), the sum of the probabilities of these n mutually exclusive events, i.e. n/N, since each of the n mutually exclusive events has the same probability $1/N$.

3. In section 2.3, a simple random sample of n units out of the universe of N units has been defined as the process in which each combination of n units (numbering ${}^N C_n$ in sampling without replacement and N^n in sampling with replacement) has the same chance of selection as every other combination: a characteristic of simple random sampling is that the probability that a universe unit will be selected at any given draw is the same as that on the first draw, namely $1/N$ (see note 1). The above proofs have been derived from this characteristic and the theorems on probability. Alternative proofs which are derived from the definition of simple random sampling, if required, will be found in the advanced theoretical text-books.

Corollaries

$$E(N\bar{y}) = \sum^{N} Y_i \qquad \text{(A2.40)}$$

in srs with and without replacement.

$$\text{Var}(N\bar{y}) = N^2\,\frac{\sigma^2}{n} \qquad \text{(A2.41)}$$

in srs with replacement.

$$\mathrm{Var}(N\bar{y}) = N^2 \frac{\sigma^2}{n} \frac{N-n}{N-1} \tag{A2.42}$$

in srs without replacement.

Proof. Equation (A2.40) follows from equations (A2.30) and (A2.36) on multiplying both sides by N.

Equations (A2.41), (A2.42) follow from (A2.31) and (A2.37) on multiplying both sides by N^2.

A2.3.3 Unbiased estimators of universe variances

Defining the per unit variance of a sample as

$$s^2 = \frac{1}{n-1} \sum^n (y_i - \bar{y})^2 \tag{A2.43}$$

to prove that

$$E(s^2) = \sigma^2 \tag{A2.44}$$

in srs with replacement,

$$E(s^2) = \frac{N}{N-1} \sigma^2 = S^2 \tag{A2.45}$$

in srs without replacement.

Proof. The sum of squares $\sum^n (y_i - \bar{y})^2$ may be written as

$$\begin{aligned}\sum^n (y_i - \bar{y})^2 &= \sum^n [(y_i - \bar{Y}) - (\bar{y} - \bar{Y})]^2 \\ &= \sum^n (y_i - \bar{Y})^2 + n(\bar{y} - \bar{Y})^2 - 2(\bar{y} - \bar{Y}) \sum^n (y_i - \bar{Y}) \\ &= \sum^n (y_i - \bar{Y})^2 + n(\bar{y} - \bar{Y})^2 - 2n(\bar{y} - \bar{Y})^2 \\ &= \sum^n (y_i - \bar{Y})^2 - n(\bar{y} - \bar{Y})^2\end{aligned}$$

as $(\bar{y} - \bar{Y})$ is a constant, and $\sum^n (y_i - \bar{Y}) = \sum^n y_i - n\bar{Y} = n\bar{y} - n\bar{Y} = n(\bar{y} - \bar{Y})$.

Now

$$\begin{aligned}E\left[\sum^n (y_i - \bar{Y})^2\right] &= \sum^n E(y_i - \bar{Y})^2 \\ &= \sum^n \mathrm{Var}(y_i) \qquad \text{by definition} \\ &= n\sigma^2 \end{aligned} \tag{A2.46}$$

from equation (A2.33).

In *sampling with replacement*, from equation (A2.31)

$$\begin{aligned}E[n(\bar{y} - \bar{Y})^2] &= nE(\bar{y} - \bar{Y})^2 \\ &= n\,\mathrm{Var}(\bar{y}) \\ &= n\sigma^2/n = \sigma^2\end{aligned} \tag{A2.47}$$

Therefore, in sampling with replacement, taking the expectation of both sides of equation (A2.43) and using equations (A2.46) and (A2.47), we obtain

$$E(s^2) = \frac{1}{n-1}(n\sigma^2 - \sigma^2) = \sigma^2$$

In *sampling without replacement,* from equation (A2.37),

$$\begin{aligned} E[n(\bar{y} - \bar{Y})^2] &= n\,E(\bar{y} - \bar{Y})^2 = n\,\text{Var}(\bar{y}) \\ &= n\frac{\sigma^2}{n}\frac{N-n}{N-1} = \sigma^2\frac{N-n}{N-1} \end{aligned} \tag{A2.48}$$

Therefore, in sampling without replacement, taking the expectation of both sides of equation (A2.43) and using equations (A2.46) and (A2.48), we obtain

$$\begin{aligned} E(s^2) &= \frac{1}{n-1}\left[n\sigma^2 - \sigma^2\frac{N-n}{N-1}\right] \\ &= \frac{\sigma^2}{n-1}\frac{N(n-1)}{N-1} \\ &= \frac{N\sigma^2}{N-1} = S^2 \end{aligned}$$

by definition of S^2 (equation A2.38).

A2.3.4 Unbiased estimators of $\sigma_{\bar{y}}^2$

$$E\left(\frac{s^2}{n}\right) = \sigma_{\bar{y}}^2 \tag{A2.49}$$

in srs with replacement

$$E\left[\frac{s^2}{n}(1-f)\right] = \sigma_{\bar{y}}^2 \tag{A2.50}$$

in srs without replacement where s^2 is defined in equation (A2.43).

Proof. For srs with replacement, we have seen that

$$\sigma_{\bar{y}}^2 = \sigma^2/n$$

and

$$E(s^2) = \sigma^2$$

Therefore

$$\begin{aligned} E\left(\frac{s^2}{n}\right) &= \frac{1}{n}E(s^2) \\ &= \sigma^2/n = \sigma_{\bar{y}}^2 \end{aligned}$$

For srs without replacement, we have seen that

$$\sigma_{\bar{y}}^2 = \frac{S^2}{n}(1-f)$$

and $E(s^2) = S^2$. Therefore,

$$E\left[\frac{s^2}{n}(1-f)\right] = \frac{1-f}{n}E(s^2) = \frac{(1-f)S^2}{n} = \sigma_{\bar{y}}^2$$

Corollaries. In srs with replacement

$$E\left(N^2\frac{s^2}{n}\right) = \text{Var}(N\bar{y}) \tag{A2.51}$$

In srs without replacement,

$$E\left[N^2 \frac{s^2}{n}(1-f)\right] = \text{Var}(N\bar{y}) \qquad \text{(A2.52)}$$

The results follow immediately from equations (A2.49) and (A2.50) multiplying both sides by N^2.

A2.3.5 Estimation of universe proportion P

The universe proportion P has been defined in equation (2.67) as

$$P = N'/N \qquad \text{(A2.53)}$$

where N' is the number of universe units possessing a certain attribute, and N is the total number of units in the universe. The results proved above are applicable here also by defining for the universe, $Y_i = 1$ if the ith unit ($i = 1, 2, \ldots, N$) has the attribute, and $Y_i = 0$, otherwise; and similarly for the sample. For the universe $\sum^{N} Y_i = N'$; $\bar{Y} = Y/N = N'/N = P$.

1. *To prove that the sample proportion p is an unbiased estimator of P.*
   The sample proportion p has been defined as

$$p = n'/n = \sum^{n} y_i/n = \bar{y} \qquad \text{(A2.54)}$$

where n' is the number of sample units possessing the attribute, and n the total number of sample units; $y_i = 1$, if the ith sample unit has the attribute and $y_i = 0$, otherwise.

Taking the expectation of both sides of equation (A2.54), we obtain

$$E(p) = E(\bar{y}) = \bar{Y} = P$$

from equation (2.67).

2. *Sampling variance of p.* We have seen in equation (2.68) that the universe variance of Y_i is

$$\sigma^2 = P(1-P) \qquad \text{(A2.55)}$$

An unbiased estimator of σ^2 has been defined in equation (2.73), namely

$$s^2 = \frac{1}{n-1}\sum^{n}(y_i - y)^2$$

$$= \frac{1}{n-1}\left(\sum^{n} y_i^2 - ny^2\right) = \frac{np(1-p)}{n-1} \qquad \text{(A2.56)}$$

since $y_i = 1$ or 0; $\sum^{n} y_i^2 = n' = np$; and $\bar{y}^2 = p^2$.

The sampling variance of p is

$$\sigma_p^2 = \sigma_{\bar{y}}^2 = \sigma^2/n$$

$$= \frac{P(1-P)}{n} \quad \text{in srswr} \qquad \text{(A2.57)}$$

$$= \frac{P(1-P)}{n} \cdot \frac{N-n}{N-1} \quad \text{in srswor} \qquad \text{(A2.58)}$$

3. *Unbiased estimators of* σ_p^2. The unbiased estimator of σ_p^2 is

$$s_p^2 = \frac{p(1-p)}{n-1} \quad \text{in srs with replacement} \tag{A2.59}$$

$$= (1-f)\frac{p(1-p)}{n-1} \quad \text{in srs without replacement} \tag{A2.60}$$

Proof: Since

$$s^2 = \frac{np(1-p)}{n-1}$$

$$E(s_p^2) = E\left(\frac{s^2}{n}\right) = \frac{\sigma^2}{n}$$

$$= \frac{P(1-P)}{n} = \sigma_p^2 \qquad \text{in srswr}$$

and

$$E(s_p^2) = E\left[(1-f)\frac{s^2}{n}\right]$$

$$= \frac{\sigma^2}{n}\frac{N-n}{N-1}$$

$$= \frac{P(1-P)}{n} \cdot \frac{N-n}{N-1} = \sigma_p^2 \qquad \text{in srswor}$$

A2.3.6 Variance of the ratio of random variables

(a) If y and x are two random variables with expectations Y and X, variances σ_y^2 and σ_x^2, and covariance $\sigma_{yx} = \rho\sigma_y\sigma_x$ (where ρ is the correlation coefficient), the sampling variance of the ratio of two random variables $r = y/x$ is given approximately by

$$\sigma_r^2 = \frac{1}{X^2}(\sigma_y^2 + R^2\sigma_x^2 - 2R\sigma_{yx}) \tag{A2.61}$$

This is proved below for a simple random sample.

(b) In simple random sampling, if variables y_i and x_i are measured on each unit of a simple random sample of size n out of N universe units, the estimator $r = \bar{y}/\bar{x} = y/x$ is biased for the ratio $R = Y/X$ (where Y and X are the respective universe totals of the variables and $y = N\bar{y}$ and $x = N\bar{x}$ are the corresponding unbiased sample estimators). We shall derive the expected value, the bias and the sampling variance and the variance estimator of the estimator r.

(i) *Expected value of* r. Let $e = (y - Y)/Y$ and $e' = (x - X)/X$. Then, since $E(y) = Y$, and $E(x) = X$, $E(e) = 0 = E(e')$; $\text{Var}(e) = \text{Var}(y)/Y^2$, $\text{Var}(e') = \text{Var}(x)/X^2$, and $\text{Cov}(e, e') = \text{Cov}(y, x)/YX$. As $y = Y(1+e)$, and $x = X(1+e')$, we have

$$r = y/x$$

$$= Y(1+e)/X(1+e')$$

$$= R(1+e)(1+e')^{-1} \tag{A2.62}$$

If we assume that $|e'| < 1$, i.e. x lies between 0 and $2X$, which is likely to happen when the sample size is large, the term $(1+e')^{-1}$ in expression (A2.62) may be expanded by Taylor's theorem. With this assumption and taking the expectation of both sides of (A2.62), we get

$$E(r) = R\,E(1 + e - e' + e'^2 - ee' + \ldots) \tag{A2.63}$$

If we further assume that terms involving second and higher powers of e and e' are negligible, then we have, to this order of approximation, $E(r) = R$ since $E(e) = 0 = E(e')$.

(ii) *Bias of r.* If we assume that only terms involving third and higher powers of e and e' are negligible, we have

$$\begin{aligned} E(r) &= R[1 + 0 - 0 + \mathrm{Var}(e') - \mathrm{Cov}(e, e')] \\ &= R\left[1 + \frac{\mathrm{Var}(x)}{X^2} - \frac{\mathrm{Cov}(y, x)}{YX}\right] \end{aligned} \tag{A2.64}$$

for $\mathrm{Var}(e') = E(e'^2)$, and $\mathrm{Cov}(e, e') = E(e, e')$, as $E(e) = 0 = E(e')$. Therefore, the bias of r is approximately

$$\begin{aligned} E(r) - R &= R\left[\frac{\mathrm{Var}(x)}{X^2} - \frac{\mathrm{Cov}(y, x)}{YX}\right] \\ &= R(\mathrm{CV}_x^2 - \rho\, \mathrm{CV}_y\, \mathrm{CV}_x) \\ &= \frac{1}{X^2}[R\, \mathrm{Var}(x) - \mathrm{Cov}(y, x)] \end{aligned} \tag{A2.65}$$

where ρ is the correlation coefficient between the two variables, and CV_y and CV_x are the coefficients of variation of y and x respectively.

To the order of approximation assumed, the bias will be zero, if

$$R = \mathrm{Cov}(y, x)/\mathrm{Var}(x)$$

and this will occur when the regression line of y on x is a straight line passing through the origin (0, 0).

The bias of the estimator r, given in expression (A2.65), can be estimated by replacing X, R, $\mathrm{Var}(x)$ and $\mathrm{Cov}(y, x)$ by their respective sample estimators. The estimator of the bias is itself subject to bias. If the distributions are not skewed, the coefficients of variation will not be very large, and $|\rho| \leqslant 1$, in which case the bias, given by the above expression, will be small. An upper bound of the magnitude of the relative bias $\left|\frac{E(r) - R}{R}\right|$ is given by $\mathrm{CV}_y\, \mathrm{CV}_x$.

(iii) *Sampling variance of r.* As the estimator r is biased, its mean square error is, by definition,

$$\begin{aligned} \mathrm{MSE}(r) &= E(r - R)^2 \\ &= R^2 E[(e - e')^2 (1 + e')^{-2}] \end{aligned}$$

using equation (A2.62) and simplifying. If we assume as before that $|e'| < 1$ and that terms involving second and higher powers of e and e' are negligible, then r is considered to be unbiased to this order of approximation, and the variance and the mean square error become the same. In this case,

$$\begin{aligned} \mathrm{MSE}(r) &= \mathrm{Var}(r) \\ &= R^2 [E(e - e')^2] \\ &= R^2\left[\frac{\mathrm{Var}(y)}{Y^2} + \frac{\mathrm{Var}(x)}{X^2} - \frac{2\,\mathrm{Cov}(y, x)}{YX}\right] \\ &= R^2(\mathrm{CV}_y^2 + \mathrm{CV}_x^2 - 2\rho\, \mathrm{CV}_y\, \mathrm{CV}_x) \\ &= \frac{1}{X^2}[\mathrm{Var}(y) + R^2\, \mathrm{Var}(x) - 2R\, \mathrm{Cov}(y, x)] \end{aligned} \tag{A2.66}$$

The variance estimator of r is obtained on replacing X, R, Var(y), Var(x) and Cov(y, x) by their respective sample estimators. Thus, the approximate variance estimator is given by equation (2.49) for simple random sampling without replacement. The variance estimators are generally biased.

Note: The bias in the ratio estimator $y_R = Xy/x$ is X times that in r (equation A2.65). Compare equations (2.36) and (3.12).

A2.3.7 Varying probability sampling

With the notations used in section 5.3, to prove that in varying probability sampling with replacement,

$$E\left(y_0^* = \frac{1}{n}\sum^n y_i^* = \frac{1}{n}\sum^n \frac{y_i}{\pi_i}\right) = Y \qquad \text{(A2.67)}$$

$$\text{Var}(y_0^*) = \sigma^2_{y_0^*} = \frac{1}{n}\sum^N \left(\frac{Y_i}{\pi_i} - Y\right)^2 \pi_i \qquad \text{(A2.68)}$$

$$E\left[s^2_{y_0^*} = \sum^n (y_i^* - y_0^*)^2/n(n-1)\right] = \sigma^2_{y_0^*} \qquad \text{(A2.69)}$$

Proof. In sampling with replacement at each draw

$$E(y_i^*) = E\left(\frac{y_i}{\pi_i}\right) = \sum^N \frac{Y_i}{\pi_i}\pi_i = Y$$

so that
$$E(y_0^*) = \frac{1}{n}\sum^n E(y_i^*) = \frac{1}{n}nY = Y$$

To prove (A2.68),

$$\text{Var}(y_0^*) = \sigma^2_{y_0^*} = \frac{1}{n^2}\text{Var}\left(\sum^n y_i^*\right) = \frac{1}{n^2}\sum^n \text{Var}(y_i^*)$$

as the y_i^*s are independent, since sampling is with replacement from (A2.24). Now

$$\begin{aligned}\text{Var}(y_i^*) &= \sigma^2_{y_i^*}\\ &= E(y_i^* - Ey_i^*)^2\\ &= E(y_i^* - Y)^2\\ &= \sum^N (Y_i/\pi_i - Y)^2\pi_i\\ &= \sum^N Y_i^2/\pi_i - Y^2 \end{aligned} \qquad \text{(A2.70)}$$

since $\sum^N \pi_i = 1$. Thus,

$$\begin{aligned}\sigma^2_{y_0^*} &= \frac{1}{n^2}\sum^n \text{Var}(y_i^*)\\ &= \frac{1}{n^2}n\sigma^2_{y_i^*}\\ &= \sigma^2_{y_i^*}/n\\ &= \frac{1}{n}\sum^N\left(\frac{Y_i}{\pi_i} - Y\right)^2\pi_i\\ &= \frac{1}{n}\left(\sum^N \frac{Y_i^2}{\pi_i} - Y^2\right)\end{aligned}$$

To prove (A2.69), we define the per unit variance of a varying probability sample as

$$s^2_{y_i^*} = \frac{1}{n-1} \sum^n (y_i^* - y_0^*)^2 \tag{A2.71}$$

The sum of squares $\sum^n (y_i^* - y_0^*)^2$ may be written as (see section A2.3.3)

$$\sum^n (y_i^* - y_0^*)^2 = \sum^n (y_i^* - Y)^2 - n(y_0^* - Y)^2$$

Now,

$$E\left[\sum^n (y_i^* - Y)^2\right] = \sum^n E(y_i^* - Y)^2$$

$$= \sum^n \text{Var}(y_i^*) = n\sigma^2_{y_i^*} \tag{A2.72}$$

from equation (A2.70). Also,

$$E[n(y_0^* - Y)^2] = n\,E(y_0^* - Y)^2$$

$$= n\,\text{Var}(y_0^*)$$

$$= n\sigma^2_{y_i^*}/n = \sigma^2_{y_i^*} \tag{A2.73}$$

from equation (A2.68).

Therefore, taking the expectation of both sides of equation (A2.71) and using equations (A2.72) and (A2.73), we obtain

$$E(s^2_{y_i^*}) = \frac{1}{n-1}(n\sigma^2_{y_i^*} - \sigma^2_{y_i^*}) = \sigma^2_{y_i^*} \tag{A2.74}$$

i.e. $s^2_{y_i^*}$ is an unbiased estimator of $\sigma^2_{y_i^*}$. Therefore $s^2_{y_0^*} = s^2_{y_i^*}/n$ is an unbiased estimator of $\sigma^2_{y_0^*} = \sigma^2_{y_i^*}/n$.

A2.3.8 Cluster sampling

With the notations of section 6.2.1, to prove that in simple random sample of clusters

(a)
$$\sigma^2 = \sigma_w^2 + \sigma_b^2 \tag{6.5}$$

or,
Total variance (per unit) = Within-cluster variance + Between-cluster variance

(b)
$$\sigma_b^2 = \frac{\sigma^2}{M_0}[1 + (M_0 - 1)\rho_c] \tag{6.9}$$

Proof. (a) By definition $\sigma^2 = \dfrac{1}{NM_0} \displaystyle\sum_i^N \sum_j^{M_0} (Y_{ij} - \bar{Y})^2$ so that (6.5)

$$NM_0\,\sigma^2 = \sum_i \sum_j (Y_{ij} - \bar{Y})^2 = \sum_i \sum_j (Y_{ij} - \bar{Y}_i + \bar{Y}_i - \bar{Y})^2$$

$$= \sum_i \left[\sum_j (Y_{ij} - \bar{Y}_i)^2 + M_0(\bar{Y}_i - \bar{Y})^2 + 2(\bar{Y}_i - \bar{Y})\sum_j (Y_{ij} - \bar{Y}_i)\right]$$

$$= \sum_i \sum_j (Y_{ij} - \bar{Y}_i)^2 + M_0 \sum_i (\bar{Y}_i - \bar{Y})^2$$

as the third term with the square brackets is zero.

Thus $\sigma^2 = \frac{1}{NM_0} \sum_i \sum_j (Y_{ij} - \bar{Y}_i)^2 + \frac{1}{N} \sum_j (\bar{Y}_i - \bar{Y})^2 = \sigma_w^2 + \sigma_b^2$

by definitions of σ_w^2 (equation 6.6) and σ_b^2 (equation 6.7).

(b) By definition, $\sigma_b^2 = \sum_i (\bar{Y}_i - \bar{Y})^2/N$ Now, (6.7)

$$(\bar{Y}_i - \bar{Y}) = \left(\frac{1}{M_0} \sum_j Y_{ij} - \bar{Y}\right) = \frac{1}{M_0}\left(\sum_j Y_{ij} - M_0 \bar{Y}\right) = \frac{1}{M_0}\left(\sum_j Y_{ij} - \sum_j \bar{Y}\right)$$

$$= \frac{1}{M_0} \sum_j (Y_{ij} - \bar{Y})$$

So, $$\sigma_b^2 = \frac{1}{NM_0^2} \sum_i \left[\sum_j (Y_{ij} - \bar{Y})\right]$$

$$= \frac{1}{NM_0^2} \left[\sum_i \sum_j (Y_{ij} - \bar{Y})^2 + \sum_i \sum_{j \neq j'} (Y_{ij} - \bar{Y})(Y_{ij'} - \bar{Y})\right]$$

$$= \frac{1}{NM_0^2} [NM_0 \sigma^2 + NM_0 (M_0 - 1) \sigma^2 \rho_0]$$

by definitions of σ^2 (equation 6.5) and ρ_c (equation 6.8). This, when simplified, gives equation (6.9).

A2.3.9 Stratified simple random sampling

With the notations used in section 10.2, to prove that

(a) $$E\left(y = \sum^L y_{h0}^*\right) = \sum^L Y_h = Y \qquad \text{(A2.75)}$$

(b) $$\sigma_y^2 = \sum^L \sigma_{y_{h0}^*}^2 \qquad \text{(A2.76)}$$

where $$\sigma_{y_{h0}^*}^2 = N_h^2 \frac{\sigma_h^2}{n_h} \quad \text{in srswr} \qquad \text{(A2.77)}$$

$$= N_h^2 \frac{\sigma_h^2}{n_h} \frac{N_h - n_h}{N_h - 1}$$

$$= N_h^2 \frac{S_h^2}{n_h} (1 - f_h) \quad \text{in srswor} \qquad \text{(A2.78)}$$

and $$f_h = n_h/N_h$$

(c) $$E\left(s_y^2 = \sum^L s_{y_{h0}^*}^2\right) = \sum^L \sigma_{y_{h0}^*}^2 = \sigma_y^2 \qquad \text{(A2.79)}$$

Proof. (a) From equation (A2.40), we know that y_{h0}^* is an unbiased estimator of Y_h in every stratum, so that

$$E(y) = E\left(\sum^L y_{h0}^*\right) = \sum^L E(y_{h0}^*) = \sum^L Y_h = Y$$

(b) For each stratum, from equation (A2.41) in sampling with replacement, we have

$$\sigma_{y_{h0}^*}^2 = N_h^2 \frac{\sigma_h^2}{n_h}$$

Then,
$$\begin{aligned}\sigma_y^2 &= \mathrm{Var}(y)\\ &= \mathrm{Var}\left(\sum^L y^*_{ho}\right)\\ &= \sum^L \mathrm{Var}(y^*_{ho}) + \sum^L_{h\neq h'} \mathrm{Cov}(y^*_{ho}, y^*_{h'o})\\ &= \sum^L \mathrm{Var}(y^*_{ho})\\ &= \sum^L \sigma^2_{y^*_{ho}}\end{aligned}$$

as the covariance terms vanish because samples are drawn independently in the different strata.

The proof for sampling without replacement follows similarly, noting that for each stratum from equation (A2.42)

$$\sigma^2_{y^*_{ho}} = N_h^2 \frac{S_h^2}{n_h}(1 - f_h)$$

Corollary. The sampling variance of the unbiased estimator of the overall mean is

$$\mathrm{Var}(y/N) = \sigma_y^2/N^2 \qquad \text{(A2.80)}$$

σ_y^2 being defined by equations (A2.76)–(A2.78)

(c) For each stratum, from equation (A2.51) in srs with replacement,

$$\begin{aligned}s^2_{y^*_{ho}} &= \frac{1}{n_h(n_h-1)} \sum^{n_h} (y^*_{hi} - y^*_{ho})^2\\ &= \frac{N_h^2}{n_h(n_h-1)} \sum^{n_h} (y_{hi} - \bar{y}_h)^2 \qquad \text{(A2.81)}\end{aligned}$$

is an unbiased estimator of $\sigma^2_{y^*_{ho}} = N_h^2\sigma_h^2/n_h$. Hence $s_y^2 = \sum^L s^2_{y^*_{ho}}$ is an unbiased estimator of $\sigma_y^2 = \sum^L \sigma^2_{y^*_{ho}}$.

Similarly in srs without replacement, from equation (A2.52), $[N_h^2 s_h^2/n_h(n_h - 1)](1 - f_h)$ is an unbiased estimator of $\sigma^2_{y^*_{ho}} = N_h^2\ S_h^2/n\ (1 - f_h)$. Hence the result (A2.79) also for srs without replacement.

Corollary. An unbiased variance estimator of the variance of the overall mean σ_y^2/N^2 is

$$s_y^2/N^2 \qquad \text{(A2.82)}$$

A2.3.10 Optimum allocation in stratified single-stage sampling

To prove the result given in section 12.3.2, we use the well-known Cauchy-Schwarz inequality, which states that if a_h and b_h are two sets of positive numbers, then

$$\left(\sum a_h^2\right)\left(\sum b_h^2\right) \geqslant \left(\sum a_h b_h\right)^2 \qquad \text{(A2.83)}$$

there being equality only if b_h is proportional to a_h, i.e. if b_h/a_h is a constant for all h.

Note: The theorems for stratified varying probability sampling can be proved on similar lines as for stratified srs.

Now, the optimum allocation is obtained on minimizing the variance function V, given by expression (12.1), for a given total cost C, obtained using expression (12.5), or *vice versa*; this is equivalent to minimizing the product VC. Taking only those parts of V and C that depend on n_h, we have to minimize

$$\left(\sum N_h^2 V_h/n_h\right)\left(\sum n_h c_h\right) \tag{A2.84}$$

From the Cauchy-Schwarz inequality (A2.83), by putting $a_h = N_h\sqrt{(V_h/n_h)}$ and $b_h = \sqrt{(n_h c_h)}$, we see that

$$\left(\sum N_h^2 V_h/n_h\right)\left(\sum n_h c_h\right) \geqslant \left[\sum N_h \sqrt{(V_h c_h)}\right]^2 \tag{A2.85}$$

The minimum value of expression (A2.84), i.e. equality in (A2.85), occurs only when

$$\frac{b_h}{a_h} = \frac{n_h \sqrt{c_h}}{N_h \sqrt{V_h}} = \text{constant} = k, \text{ say} \tag{A2.86}$$

or

$$n_h = kN_h \sqrt{(V_h/c_h)} \tag{A2.87}$$

i.e. n_h should be proportional to $N_h \sqrt{(V_h/c_h)}$.

Summing both sides of (A2.87), we get

$$n = \sum n_h = k \sum N_h \sqrt{(V_h/c_h)} \tag{A2.88}$$

which gives the value of k; substituting this in (A2.87), we get

$$n_h = \frac{nN_h \sqrt{(V_h/c_h)}}{\sum N_h \sqrt{(V_h/c_h)}} \tag{A2.89}$$

The above expression for n_h presumes the value of n, which would depend on whether the total cost or the variance is specified. If the total cost is fixed at C', then we substitute the n_h values obtained from equation (A2.89) in the cost function (12.5), and solve for n, which gives

$$n = (C' - c_0)\frac{\sum N_h \sqrt{(V_h/c_h)}}{\sum N_h \sqrt{(V_h c_h)}} \tag{A2.90}$$

Substituting this value of n in equation (A2.89), we get equation (12.6).

Similarly, if the variance is fixed at V' then we substitute the optimum n_h in the variance function (12.1), whence

$$n = \left[\sum N_h \sqrt{(V_h c_h)}\right]\left[\sum N_h \sqrt{(V_h/c_h)}\right]\Big/ V' \tag{A2.91}$$

substituting this value of n in equation (A2.89), we get equation (12.7).

Notes

1. See notes 2 and 3 to section 12.3.2.
2. The Neyman allocation follows from the above formulation as indicated in section 12.3.3. A direct proof may also be derived noting that $c_h = \bar{c}$, a constant, and $n = (C' - c_0)/\bar{c}$.
3. The proofs for stratified srs without replacement (note 3 to section 12.3.2, note to section 12.6.1 and note 4 to section 12.6.2) can be derived similarly.

Table A3.1 Random Numbers

03 47 43 73 86	36 96 47 36 61	46 98 63 71 62	33 26 16 80 45	60 11 14 10 95
97 74 24 67 62	42 81 14 57 20	42 53 32 37 32	27 07 36 07 51	24 51 79 89 73
16 76 62 27 66	56 50 26 71 07	32 90 79 78 53	13 55 38 58 59	88 97 54 14 10
12 56 85 99 26	96 96 68 27 31	05 03 72 93 15	57 12 10 14 21	88 26 49 81 76
55 59 56 35 64	38 54 82 46 22	31 62 43 09 90	06 18 44 32 53	23 83 01 30 30
16 22 77 94 39	49 54 43 54 82	17 37 93 23 78	87 35 20 96 43	84 26 34 91 64
84 42 17 53 31	57 24 55 06 88	77 04 74 47 67	21 76 33 50 25	83 92 12 06 76
63 01 63 78 59	16 95 55 67 19	98 10 50 71 75	12 86 73 58 07	44 39 52 38 79
33 21 12 34 29	78 64 56 07 82	52 42 07 44 38	15 51 00 13 42	99 66 02 79 54
57 60 86 32 44	09 47 27 96 54	49 17 46 09 62	90 52 84 77 27	08 02 73 43 28
18 18 07 92 46	44 17 16 58 09	79 83 86 19 62	06 76 50 03 10	55 23 64 05 05
26 62 38 97 75	84 16 07 44 99	83 11 46 32 24	20 14 85 88 45	10 93 72 88 71
23 42 40 64 74	82 97 77 77 81	07 45 32 14 08	32 98 94 07 72	93 85 79 10 75
52 36 28 19 95	50 92 26 11 97	00 56 76 31 38	80 22 02 53 53	86 60 42 04 53
37 85 94 35 12	83 39 50 08 30	42 34 07 96 88	54 42 06 87 98	35 85 29 48 39
70 29 17 12 13	40 33 20 38 26	13 89 51 03 74	17 76 37 13 04	07 74 21 19 30
56 62 18 37 35	96 83 50 87 75	97 12 25 93 47	70 33 24 03 54	97 77 46 44 80
99 49 57 22 77	88 42 95 45 72	16 64 36 16 00	04 43 18 66 79	94 77 24 21 90
16 08 15 04 72	33 27 14 34 09	45 59 34 68 49	12 72 07 34 45	99 27 72 95 14
31 16 93 32 43	50 27 89 87 19	20 15 37 00 49	52 85 66 60 44	38 68 88 11 80
68 34 30 13 70	55 74 30 77 40	44 22 78 84 26	04 33 46 09 52	68 07 97 06 57
74 57 25 65 76	59 29 97 68 60	71 91 38 67 54	13 58 18 24 76	15 54 55 95 52
27 42 37 86 53	48 55 90 65 72	96 57 69 36 10	96 46 92 42 45	97 60 49 04 91
00 39 68 29 61	66 37 32 20 30	77 84 57 03 29	10 45 65 04 26	11 04 96 67 24
29 94 98 94 24	68 49 69 10 82	53 75 91 93 30	34 25 20 57 27	40 48 73 51 92
16 90 82 66 59	83 62 64 11 12	67 19 00 71 74	60 47 21 29 68	02 02 37 03 31
11 27 94 75 06	06 09 19 74 66	02 94 37 34 02	76 70 90 30 86	38 45 94 30 38
35 24 10 16 20	33 32 51 26 38	79 78 45 04 91	16 92 53 56 16	02 75 50 95 98
38 23 16 86 38	42 38 97 01 50	87 75 66 81 41	40 01 74 91 62	48 51 84 08 32
31 96 25 91 47	96 44 33 49 13	34 86 82 53 91	00 52 43 48 85	27 55 26 89 62
56 67 40 67 14	64 05 71 95 86	11 05 65 09 68	76 83 20 37 90	57 16 00 11 66
14 90 84 45 11	75 73 88 05 90	52 27 41 14 86	22 98 12 22 08	07 52 74 95 80
68 05 51 18 00	33 96 02 75 19	07 60 62 93 55	59 33 82 43 90	49 37 38 44 59
20 46 78 73 90	97 51 40 14 02	04 02 33 31 08	39 54 16 49 36	47 95 93 13 30
64 19 58 97 79	15 06 15 93 20	01 90 10 75 06	40 78 78 89 62	02 67 74 17 33
05 26 93 70 60	22 35 85 15 13	92 03 51 59 77	59 56 78 06 83	52 91 05 70 74
07 97 10 88 23	09 98 42 99 64	61 71 62 99 15	06 51 29 16 93	58 05 77 09 51
68 71 86 85 85	54 87 66 47 54	73 32 08 11 12	44 95 92 63 16	29 56 24 29 48
26 99 61 65 53	58 37 78 80 70	42 10 50 67 42	32 17 55 85 74	94 44 67 16 94
14 65 52 68 75	87 59 36 22 41	26 78 63 06 55	13 08 27 01 50	15 29 39 39 43
17 53 77 58 71	71 41 61 50 72	12 41 94 96 26	44 95 27 36 99	02 96 74 30 83
90 26 59 21 19	23 52 23 33 12	96 93 02 18 39	07 02 18 36 07	25 99 32 70 23
41 23 52 55 99	31 04 49 69 96	10 47 48 45 88	13 41 43 89 20	97 17 14 49 17
60 20 50 81 69	31 99 73 68 68	35 81 33 03 76	24 30 12 48 60	18 99 10 72 34
91 25 38 05 90	94 58 28 41 36	45 37 59 03 09	90 35 57 29 12	82 62 54 65 60
34 50 57 74 37	98 80 33 00 91	09 77 93 19 82	74 94 80 04 04	45 07 31 66 49
85 22 04 39 43	73 81 53 94 79	33 62 46 86 28	08 31 54 46 31	53 94 13 38 47
09 79 13 77 48	73 82 97 22 21	05 03 27 24 83	72 89 44 05 60	35 80 39 94 88
88 75 80 18 14	22 95 75 42 49	39 32 82 22 49	02 48 07 70 37	16 04 61 67 87
90 96 23 70 00	39 00 03 06 90	55 85 78 38 36	94 37 30 69 32	90 89 00 76 33

53 74 23 99 07	61 32 28 69 84	94 62 67 86 24	98 33 41 19 95	47 53 53 38 09
63 38 06 86 54	99 00 65 26 94	02 82 90 23 07	79 62 67 80 60	75 91 12 81 19
35 30 58 21 46	06 72 17 10 94	25 21 31 75 96	49 28 24 00 49	55 65 79 78 07
63 43 36 82 69	65 51 18 37 88	61 38 44 12 45	32 92 85 88 65	54 34 81 85 35
98 25 37 55 26	01 91 82 81 46	74 71 12 94 97	24 02 71 37 07	03 92 18 66 75
02 63 21 17 69	71 50 80 89 56	38 15 70 11 48	43 40 45 86 98	00 83 26 91 03
64 55 22 21 82	48 22 28 06 00	61 54 13 43 91	82 78 12 23 29	06 66 24 12 27
85 07 26 13 89	01 10 07 82 04	59 63 69 36 03	69 11 15 83 80	13 29 54 19 28
58 54 16 24 15	51 54 44 82 00	62 61 65 04 69	38 18 65 18 97	85 72 13 49 21
34 85 27 84 87	61 48 64 56 26	90 18 48 13 26	37 70 15 42 57	65 65 80 39 07
03 92 18 27 46	57 99 16 96 56	30 33 72 85 22	84 64 38 56 98	99 01 30 98 64
62 95 30 27 59	37 75 41 66 48	86 97 80 61 45	21 53 04 01 63	45 76 08 64 27
08 45 93 15 22	60 21 75 46 91	98 77 27 85 42	28 88 61 08 84	69 62 03 42 73
07 08 55 18 40	45 44 75 13 90	24 94 96 61 02	57 55 66 83 15	73 42 37 11 61
01 85 89 95 66	51 10 19 34 88	15 84 97 19 75	12 76 39 43 78	64 63 91 08 25
72 84 71 14 35	19 11 58 49 26	50 11 17 17 76	86 31 57 20 18	95 60 78 46 75
88 78 28 16 84	13 52 53 94 53	75 45 69 30 96	73 89 65 70 31	99 17 43 48 76
45 17 75 65 57	28 40 19 72 12	25 12 74 75 67	60 40 60 81 19	24 62 01 61 16
96 76 28 12 54	22 01 11 94 25	71 96 16 16 88	68 64 36 74 45	19 59 50 88 92
43 31 67 72 30	24 02 94 08 63	38 32 36 66 02	69 36 38 25 39	48 03 45 15 22
50 44 66 44 21	66 06 58 05 62	68 15 54 35 02	42 35 48 96 32	14 52 41 52 48
22 66 22 15 86	26 63 75 41 99	58 42 36 72 24	58 37 52 18 51	03 37 18 39 11
96 24 40 14 51	23 22 30 88 57	95 67 47 29 83	94 69 40 06 07	18 16 36 78 86
31 73 91 61 19	60 20 72 93 48	98 57 07 23 69	65 95 39 69 58	56 80 30 19 44
78 60 73 99 84	43 89 94 36 45	56 69 47 07 41	90 22 91 07 12	78 35 34 08 72
84 37 90 61 56	70 10 23 98 05	85 11 34 76 60	76 48 45 34 60	01 64 18 39 96
36 67 10 08 23	98 93 35 08 86	99 29 76 29 81	33 34 91 58 93	63 14 52 32 52
07 28 59 07 48	89 64 58 89 75	83 85 62 27 89	30 14 78 56 27	86 63 50 80 02
10 15 83 87 60	79 24 31 66 56	21 48 24 06 93	91 98 94 05 49	01 47 59 38 00
55 19 68 97 65	03 73 52 16 56	00 53 55 90 27	33 42 29 38 87	22 13 88 83 34
53 81 29 13 39	35 01 20 71 34	62 33 74 82 14	53 73 19 09 03	56 54 29 56 93
51 86 32 68 92	33 98 74 66 99	40 14 71 94 58	45 94 19 38 81	14 44 99 81 07
35 91 70 29 13	80 03 54 07 27	96 94 78 32 66	50 95 52 74 33	13 80 55 62 54
37 71 67 95 13	20 02 44 95 94	64 85 04 05 72	01 32 90 76 14	53 89 74 60 41
93 66 13 83 27	92 79 64 64 72	28 54 96 53 84	48 14 52 98 94	56 07 93 89 30
02 96 08 45 65	13 05 00 41 84	93 07 54 72 59	21 45 57 09 77	19 48 56 27 44
49 83 43 48 35	82 88 33 69 96	72 36 04 19 76	47 45 15 18 60	82 11 08 95 97
84 60 71 62 46	40 80 81 30 37	34 39 23 05 38	25 15 35 71 30	88 12 57 21 77
18 17 30 88 71	44 91 14 88 47	89 23 30 63 15	56 34 20 47 89	99 82 93 24 98
79 69 10 61 78	71 32 76 95 62	87 00 22 58 40	92 54 01 75 25	43 11 71 99 31
75 93 36 57 83	56 20 14 82 11	74 21 97 90 65	96 42 68 63 86	74 54 13 26 94
38 30 92 29 03	06 28 81 39 38	62 25 06 84 63	61 29 08 93 67	04 32 92 08 00
51 29 50 10 34	31 57 75 95 80	51 97 02 74 77	76 15 48 49 44	18 55 63 77 09
21 31 38 86 24	37 79 81 53 74	73 24 16 10 33	52 83 90 94 96	70 47 14 54 36
29 01 23 87 88	58 02 39 37 67	42 10 14 20 92	16 55 23 42 45	54 76 09 11 06
95 33 95 22 00	18 74 72 00 18	38 79 58 69 32	81 76 80 26 92	82 80 84 25 39
90 84 60 79 80	24 36 59 87 38	82 07 53 89 35	96 35 23 79 18	05 98 90 07 35
46 40 62 98 82	54 97 20 56 95	15 74 80 08 32	16 46 70 50 80	67 72 16 42 79
20 31 89 03 43	38 46 82 68 72	32 14 82 99 70	80 60 47 18 97	63 49 30 21 30
71 59 73 05 50	08 22 23 71 77	91 01 93 20 49	82 96 59 26 91	66 39 67 08 60

Reprinted by permission from a larger table in *Statistical Tables for Biological, Agricultural and Medical Research*, by R. A. Fisher and F. Yates. Oliver and Boyd, Edinburgh (6th edition, 1963).

Table A3.2 Selected values of the t-distribution

Degrees of freedom	Probability (α)					
	0.5	0.2	0.1	0.05	0.01	0.001
1	1.000	3.078	6.314	12.706	63.657	636.619
2	0.816	1.886	2.920	4.303	9.925	31.598
3	0.765	1.638	2.353	3.182	5.841	12.924
4	0.741	1.533	2.132	2.776	4.604	8.610
5	0.727	1.476	2.015	2.571	4.032	6.869
6	0.718	1.440	1.943	2.447	3.707	5.959
7	0.711	1.415	1.895	2.365	3.499	5.408
8	0.706	1.397	1.860	2.306	3.355	5.041
9	0.703	1.383	1.833	2.262	3.250	4.781
10	0.700	1.372	1.812	2.228	3.169	4.587
11	0.697	1.363	1.796	2.201	3.106	4.437
12	0.695	1.356	1.782	2.179	3.055	4.318
13	0.694	1.350	1.771	2.160	3.012	4.221
14	0.692	1.345	1.761	2.145	2.977	4.140
15	0.691	1.341	1.753	2.131	2.947	4.073
16	0.690	1.337	1.746	2.120	2.921	4.015
17	0.689	1.333	1.740	2.110	2.898	3.965
18	0.688	1.330	1.734	2.101	2.878	3.922
19	0.688	1.328	1.729	2.093	2.861	3.883
20	0.687	1.325	1.725	2.086	2.845	3.850
25	0.684	1.316	1.708	2.060	2.787	3.725
30	0.683	1.310	1.697	2.042	2.750	3.646
40	0.681	1.303	1.684	2.021	2.704	3.551
60	0.679	1.296	1.671	2.000	2.660	3.460
120	0.677	1.289	1.658	1.980	2.617	3.373
∞ (normal)	0.674	1.282	1.645	1.960	2.576	3.291

Reprinted by permission from a larger table in *Statistical Tables for Biological, Agricultural and Medical Research* by R. A. Fisher and F. Yates. Oliver and Boyd, Edinburgh (6th edition, 1963).

Table A3.3 Values of Coefficient of Variation (CV) per unit in the universe for different values of the universe proportion P

P	CV (%)	P	CV (%)	P	CV (%)
0.01	995	0.25	173	0.65	73
0.02	700	0.30	153	0.70	65
0.03	569	0.35	136	0.75	58
0.04	490	0.40	127	0.80	50
0.05	436	0.45	111	0.85	41
0.10	300	0.50	100	0.90	33
0.15	238	0.55	90	0.95	23
0.20	200	0.60	82	0.99	10

$CV = \sqrt{[(1-P)/P]}$

Table A3.4 Sample size (n) required to ensure desired coefficient of variation of sample estimator (e) in sampling with replacement

Desired CV of sample esti-mator (e)(%)	*Value of universe CV per unit (%)*												
	5	*10*	*20*	*30*	*40*	*50*	*60*	*70*	*80*	*90*	*100*	*150*	*200*
25	1	1	1	1	3	4	6	8	10	13	16	36	64
20	1	1	1	2	4	6	9	12	16	20	25	56	100
15	1	1	2	4	7	11	16	22	28	36	44	100	178
10	1	1	4	9	16	25	36	49	64	81	100	225	400
5	1	4	16	36	64	100	144	196	256	324	400	900	1600
4	2	5	25	56	100	156	225	306	400	506	625	1406	2500
3	3	11	44	108	178	278	400	544	711	900	1111	2500	4444
2.5	4	16	64	144	256	400	576	784	1024	1296	1600	3600	6400
2	5	25	100	225	400	625	900	1225	1600	2025	2500	5625	10 000
1	25	100	400	900	1600	2500	3600	4900	6400	8100	10 000	22 500	40 000

n = (Universe CV per unit/Desired CV of sample estimator)^2

Appendix IV Hypothetical universe of 600 households in 30 villages in 3 zones

Table A4.1 *

Zone	*Village*	*Area (km^2)*	*Previous census population*	*Number of households*	*Size of the households*	*Total popula-tion*
I	1	8.7	69	17	7 5 5 4 6 2 3 5 5 6 5 4 4 4 5 3 3	76
	2	10.6	82	18	6 5 4 5 4 5 6 5 3 5 4 4 5 3 3 5 6 4	82
	3	15.0	110	26	6 6 3 5 3 4 5 5 4 4 4 3 7 5 4 6 2 5 5 6 1 5 5 4 6 3	116
	4	6.2	80	18	6 3 6 3 6 3 4 5 4 4 4 5 6 3 5 1 3 5	76
	5	9.6	92	24	5 4 6 5 4 5 6 5 4 4 7 6 6 5 4 4 5 6 3 4 3 3 5 3	112
	6	7.3	65	17	3 4 4 6 5 7 3 5 4 5 4 6 4 5 3 3 6	77
	7	4.5	72	20	6 4 4 5 4 5 6 4 3 5 4 6 5 5 2 4 5 4 3 4	88
	8	10.6	108	24	5 3 3 7 4 4 6 6 4 5 3 7 6 4 5 6 3 5 1 3 5 6 4 4	109
	9	5.4	106	24	5 3 5 3 4 6 5 4 6 5 6 3 6 5 6 6 3 5 4 4 5 4 6 2	111
	10	3.5	80	22	4 4 5 5 4 4 5 4 3 5 5 3 4 5 4 3 4 3 5 4 4 1	88
II	1	5.8	72	15	8 4 4 6 5 5 5 7 3 5 6 6 6 4 4	78
	2	11.4	102	22	9 6 5 5 4 3 5 5 3 10 4 4 6 5 4 7 4 5 3 6 4 5	112
	3	5.8	73	17	6 5 5 3 6 5 3 7 3 5 5 6 6 5 3 4 3	80
	4	7.8	84	19	6 4 6 4 5 4 4 5 7 6 4 6 4 5 6 5 3 4 5	93
	5	6.5	98	20	8 4 3 5 6 3 7 5 5 5 6 6 7 3 6 6 7 2 5 6	105
	6	9.0	84	19	4 3 5 6 5 6 5 5 5 3 5 4 5 5 3 9 4 5 6	93
	7	7.3	85	19	7 5 5 3 5 4 4 5 8 4 4 6 5 7 4 5 5 5 4	95
	8	7.0	102	23	4 5 5 4 6 5 5 3 4 4 8 5 4 4 5 6 4 5 4 6 7 5 6	114

* This table is continued on page 342.

Table A4.1 contd.

Zone	*Village*	*Area (km²)*	*Previous census population*	*Number of households*	*Size of the households*	*Total population*
	9	10.5	122	25	8 4 4 5 5 4 7 5 6 4 5 5 8 5 3 6 3 4 6 5 4 7 3 5 6	127
	10	11.1	102	23	7 4 5 6 5 4 6 4 4 7 5 2 5 4 4 9 6 5 6 2 4 4 5	113
	11	6.3	86	18	4 5 4 4 7 5 5 7 5 9 6 5 7 4 3 5 4 5	94
III	1	10.0	78	15	8 6 7 4 6 5 5 2 7 7 7 5 6 4 4	83
	2	14.2	112	21	9 4 5 8 6 3 6 5 5 5 6 2 8 5 8 4 5 7 5 7 8	121
	3	8.2	97	18	8 8 4 8 3 6 6 6 4 6 6 8 7 6 4 5 6 4	105
	4	12.5	117	21	7 5 6 7 8 4 5 6 6 6 6 5 7 5 9 8 6 6 6 4 7	129
	5	6.5	106	20	7 3 5 6 6 7 5 6 5 5 4 10 7 6 5 8 6 4 6 3	114
	6	10.0	115	21	7 4 6 7 10 4 5 5 6 7 6 3 8 7 6 4 6 5 5 3 7	121
	7	7.0	110	21	7 6 4 8 1 6 6 7 4 6 6 7 6 7 3 6 6 7 2 5 6	116
	8	12.5	104	17	3 8 5 7 6 10 7 8 4 8 7 6 6 5 7 6 6	109
	9	10.2	103	16	6 5 7 6 4 5 8 4 7 8 7 6 10 7 6 4 5	105

Appendix V Case study: Indian National Sample Survey, 1964-5

A5.1 Introduction

As a case study, the main features of the planning, execution and analysis of the Indian National Sample Survey (NSS) for the period July 1964–June 1965 are given in this chapter. These are based on *Technical Paper on Sample Design, The National Sample Survey, Nineteenth Round, July 1964–June 1965* by A. S. Roy and A. Bhattacharyya.

The National Sample Survey was started in 1950 with the object of obtaining comprehensive and continuing information on economic, social, demographic, and agricultural characteristics through sample surveys on a country-wide basis. The information collected is utilized for planning, research and other purposes by the Central and State Governments, the Planning Commission and other interested organizations. The NSS is a continuing, multi-subject, integrated survey and is conducted in the form of successive 'rounds': each round covers several topics of current interest in a specific survey period. The scope, period, sample design and programme of each round are fixed by taking into account the requirements of its users and the resources available for that period. Since 1958–9, the survey period has been made one complete year coinciding approximately with the agricultural year.

A5.2 Objectives of the survey

The survey for 1964–5 was designed to provide information on the following topics: population, current and historical fertility and current mortality; employment and unemployment; indebtedness of rural labour households; land utilization, acreage and production of the major cereal crops; rural retail prices; and integrated socio-economic activities of households. The survey covered the whole of India, but excluded a few specific areas, the latter accounting for less than 0.5 per cent of the total estimated population. Hotel residents and persons in boarding houses were included, but inmates of hospitals, nursing homes, jails etc. were not covered by the survey: persons without fixed abode were included only for the demographic inquiry. The results were required for the nineteen state and union territories and separately for the rural and the urban sectors, but the crop survey was designed to provide estimates for all the major cereal crops taken together for rural India as a whole. The total cost of the field survey was about Rs. 8 million (£1 Sterling = Rs. 18.00, and $1 US = Rs. 7.50 at the present official exchange rates) and an equivalent sum is estimated to have been spent on the tabulation, analysis and preparation of reports.

A5.3 Budget and cost control

From the early planning stages, preliminary budgets were prepared on the basis of the experiences of a few of the preceding rounds and were examined on the basis of the time records of the staff engaged in field enumeration and processing.

A5.4 Administrative organization

A Programme Advisory Committee consisting of the representatives of the Planning Commission, Central Ministries, State Governments, the Central Statistical Organization, the NSS Directorate and the Indian Statistical Institute advised the Department of Statistics under Cabinet Secretariat of the Government of India on the overall planning, subject coverage, methodology, tabulation programme, fixation of priorities, and other related matters of the NSS. The final decisions of these points were taken by the Department of Statistics.

The Directorate of the National Sample Survey under the Department of Statistics is responsible for the field work and the Indian Statistical Institute for the technical work, including the planning of the survey, processing, analysis, and the preparation of reports and studies. Field work is done by full time and mostly permanent enumerators and supervisory staff, a large number of whom have Bachelor of Arts, Science, or Commerce degrees. The technical work is undertaken by full time, qualified personnel. Thus the benefits of a permanent survey organization are realized.

Fourteen out of the fifteen states participated in the survey on a matching basis. They surveyed an additional matching sample, selected in an identical manner, used the same concepts and definitions, followed the same survey procedure, but processed and analysed their own data. The sample surveyed by the Central Government agency is known as the 'central sample' and that surveyed by the State Government agencies the 'state sample'. The central and state samples were 'linked' members, described later, in a system of interpenetrating sub-samples. The data of the central and the state samples are sometimes pooled together to provide more efficient estimates than either sample.

A5.5 Co-ordination with other bodies

The interests of government departments and other institutions were ascertained and accommodated to the extent possible, as advised by the Programme Advisory Committee. Close co-ordination with the Office of the Registrar General, revenue departments etc. was maintained for obtaining and updating the frames for the strata and the first-stage units (generally villages in the rural sector and urban blocks in the urban sector, the maps for which were obtained from the two agencies mentioned).

A5.6 Questionnaire preparation

For reasons mentioned in section 26.12.1, schedules rather than questionnaires were used. Information was collected with required details; the time reference periods for the data depended on the items. The list of schedules canvassed is given later.

A5.7 Methods of data collection and supervision

For the socio-economic inquiries, the data were collected by interviewing a sample of households, and the data on crop acreage and yield rates were obtained by direct physical observation of the acreage and the actual harvesting of crops (known as crop-cutting experiments). Information was also obtained for a sub-sample of households on areas under different crops by the interview method with a view to comparing its utility with that of the detailed crop schedule based on physical observation: this comparison has not yet been made.

There was, on an average, one supervisor for every four enumerators. The supervisors also undertook field checks of the enumerators' work.

A5.8 Survey design

A5.8.1 Type of survey

The survey was a multi-subject inquiry on the topics mentioned in section A5.2; for a sub-sample of households, the integrated survey approach was taken for consumption and productive enterprises (section 26.13.1).

A5.8.2 Choice of sample design

The sample design was stratified multi-stage and the design was similar but completely independent for the rural and the urban areas. In the rural sector, groups of villages formed the strata, 353 in number; in the urban sector, towns and cities in a state were generally grouped into 2 strata according to their 1961 population, there being a total of 37 urban strata from the whole of India.

For the household inquiries, the sampling plan was stratified two-stage; within each stratum, the first-stage units were villages in the rural sector, and urban blocks in the urban sector, and households formed the second-stage units. The first-stage units were selected circular systematically with probability proportional to size, the measure of size being related to population, and the second-stage units were selected linear systematically with equal probability. The sample design was self-weighting for the household inquiries, at the state level for the rural sector and at the stratum level for the urban sector. For crop-area surveys, the sampling plan was stratified two-stage, and for yield-rate surveys it was four-stage, the successive stages being villages, clusters of plots, crop plots and circular cuts. The same sample villages were used for both the households and the crop inquiries.

The first-stage units were selected in the form of four independent, interpenetrating sub-samples. The first-stage units were selected at the Indian Statistical Institute and the households, clusters of plots etc., selected in the field by the enumerators.

Experiences of the past years provided information on the allocation of the all-India sample sizes (first-stage units) to the different strata, and the allocation of the second-stage units for the different types of inquiries. The total sample size is in particular the product of the total number of enumerators and the number of villages and urban blocks that can be surveyed by an enumerator during the whole survey period. With the work load for an enumerator being more or less fixed, the demand for a big sample in order to provide estimates at the state and lower levels had to be met by increasing the number of enumerators. The total number of enumerators in the nineteenth round was 752 for the central sample (with a reserve of 10 per cent), of whom 706 were to survey both the rural and the urban areas and 46 in the urban areas only. The saving arising from the multi-subject nature of the survey was also utilized for having a large number of sample fsu's. The total number of first-stage units for the central sample was set at 8472 villages and 4572 urban blocks.

For the reasons mentioned in section 26.13.4, the survey period was the agricultural year, divided into six equal periods of two months each, called sub-rounds, and one-sixth of the sample surveyed in each sub-round. This ensured firstly the employment of a smaller number of skilled and well-trained enumerators, and secondly the representation of all the four seasons so that the seasonal fluctuations were taken into account.

For the household inquiries, sampling was the most intensive for the population schedule, with overall sampling fractions of 0.2 per cent in the rural sector and 0.4 per cent in the urban for the central sample, and least intensive for the integrated schedules 17 and 17(suppl.), with overall sampling fractions of 0.01 per cent in the rural sector and 0.02 per cent in the urban sector.

The time required for enumerating the different schedules was obtained on the basis of the records in the previous rounds and for the schedules which were canvassed for the first time on the basis of a try-out. In large-scale surveys, journey time accounts for an appreciable portion of the total time and depends on the size of the area an enumerator covers, the number of sample first-stage units and the general transport facilities. In the present survey, an enumerator's area of operation in the rural sector was the stratum in which he was posted. Most of these strata were less than 4000 square miles in area and this was considered to be a manageable size for an enumerator. When a stratum exceeded 4000 square miles in area, that stratum was sub-divided into two or more parts known as investigation zones and the sample villages to be surveyed by an enumerator were selected from one of these zones. An enumerator was allotted twelve villages in his stratum and a similar procedure was followed for the urban blocks, which achieved some savings in the journey time, but only marginal savings could be achieved in the enumeration time. For the socio-economic inquiries, the preparation of sampling frames for selecting households was simplified and a schedule was canvassed in a sub-sample of households selected for another schedule. For crop surveys, a sample of clusters of plots instead of a direct sample of individual plots was chosen: this was because the time taken for identifying the plots is comparatively large in relation to the time taken for recording the land utilization of such plots. Also, the same sample of clusters of plots was surveyed in all the crop seasons instead of surveying fresh clusters during each season. Finally, when the sample village or the urban block was big, the enumerators were allowed to confine the survey only to a part of a sample village or the urban block so that their individual work load remained within limits. For household inquiries, each sample village and urban-block was visited and surveyed only once during the whole round. The crop survey was conducted in each of the four seasons – autumn, winter, spring and summer. The crop-area survey was conducted in all the twenty-four sample villages of the rural sector while the crop-yield survey was taken up in one-fourth of the villages only. The price inquiry was conducted in a fixed set of sample villages which has been continuing since the sixteenth round July 1960–June 1961. These sample villages were 491 in number. The time standards for enumerating different schedules and the average work load of an enumerator within a sample village or an urban block are shown in Table A5.1.

A5.8.3 Sample design in the rural sector

Some details of the sample design in the rural sector are given in this section. The sampling frame used for the first-stage units, namely villages, was the Primary

Table A5.1 Average time requirements for different schedules and journey between sample villages/urban blocks, and the average work-load within a sample village/ urban block: Indian National Sample Survey, 19th Round, July 1964 – June 1965

Schedule no.	*Description*	*Sector*	*Time standard*	*Sample size*
0.1	General schedule (for listing)	Rural	2.5 days/village	120 hh/village
0.2	General schedule (for listing)	Urban	3.5 days/block	160 hh/village
3.01	Rural retail price (monthly)	Rural	1.5 days/village	419 villages
4.1	Investigator's time record for socio-economic enquiry	Rural and urban	–	–
4.2	Investigator's time record for crop surveys and price enquiry	Rural	–	–
5.0	Land utilization survey	Rural	30 plots/day	60 plots per crop-cutting village, and 20 plots per non-crop-cutting village
5.1	Crop cutting experiments	Rural	2 plots/day	6 cuts per crop-cutting village
5.2	Driage experiment	Rural	–	6 cuts per crop-cutting village
10	Urban labour force	Urban	8 hh/day	6.7 hh/block
10.1	Employment, unemployment and indebtedness of rural labour households	Rural	8 hh/day	8 hh/village
12	Population, births and deaths	Rural and urban	25 hh/day	20 hh/village and block
16	Integrated household schedule (detailed)	Rural and urban	1 hh/day	2 hh/village and block
17	Integrated household schedule (abridged)	Rural and urban	2 hh/day	1 hh/village and block
17 (Suppl.)	Integrated household survey (abridged) – land utilization	Rural	1 hh/day	1 hh/village and block
–	Field progress and reports	Rural and urban	–	–
–	Journey to a village/block	Rural and urban	1 day per village/block	–

Note

1. Schedules 5.0, 5.1, 5.2, and 17 (Suppl.) were taken up in each of the four crop seasons.
2. For Schedule 5.1, the sample size may go up to 12 cuts, depending upon the nature of sowing of pure and mixed crops.
3. 'hh' denotes household(s).

Census Abstract (PCA) of the 1961 population census. The rural PCA provided a complete list of all the 1961 census villages with their identification particulars, as well as supplementary information, for example on population, area, number of houses and households, number of literates, number of workers in different occupational groups such as agriculture, industry etc. The abstract was available in the form of booklets (manuscript copy), each booklet giving the list of villages for one *tehsil/taluk/thana* of a state.

The villages were selected with probability proportional to size, the size of a village being related to its population. Each village was assigned a size in such a way that the size became a simple indicator of the village population and also an integer. The strict probability proportional to population sampling was not favoured because it would have involved heavy computational work at the selection stage, would have been unnecessary for those items which are not strongly related to population, and would have made it extremely difficult to achieve a self-weighting design. The average population of villages having populations between 0–499, 500–999, 1000–1999, 2000–2999, and so on were computed for each of these population classes separately for each state and union territory. The size 1 was assigned to villages with population less than 500. The size of a village in any other population size class was taken as the ratio of the average population on the class to the average population of villages in the first class, rounded off to a suitable integer. The total size of a *tehsil* was obtained by cumulating the sizes of all the villages contained in that *tehsil.* Similarly, the total size of a region was obtained by cumulating the sizes of all tehsils contained in that region (there were forty-eight regions, formed by grouping contiguous districts within a state mainly on the basis of information on topography, crop pattern, and population density: the ultimate rural strata were formed within these regions).

The all-India central sample of 8472 villages was allocated to the different states (and urban territories) on a joint consideration of the rural populations, area under food crops, and the enumerator strength. This allocation was modified to ensure a minimum sample size of 360 villages in each state, which was made a multiple of 24 in order to facilitate having four sub-samples of six villages each per stratum. The allocation to any rural stratum was always 24 sample villages in order to achieve uniform work load for the enumerators all over India.

The total number of sample villages allocated to a state (or union territory) was divided by 24 to get the total number of strata to be formed in the state (or union territory). The number was allocated to the regions in that state in proportion to the region sizes. The proportional allocations were adjusted to obtain an integral number of strata in all the regions. After deciding the number of strata, the next step was to determine the stratum sizes for a state. The strata were made of equal size within a state: the first consideration was that it would justify having equal allocations of sample size per stratum, and the second that it would considerably help to implement the self-weighting design. With this idea in view, the total size of the state, that is the sum of the sizes of all villages in that state, was divided by the proposed number of strata (say L) to yield the average stratum size (say Z') for that state. This average size was rounded off and made a multiple of 12 in order to attain an integral interval at the selection stage as 12 villages (central and state samples taken together) were selected systematically in each sub-sample of a rural stratum. The formation of strata had the following requirements to satisfy:

1. Each stratum should be a geographically compact area.
2. The stratum boundaries should not cut across the regions since estimates were required to be obtained separately for each region; (region-wise estimates can be built up even if the regions contain some part-strata, but such estimates are liable to large sampling errors due to the 'randomness' of sample size on which they are based.)
3. The stratum boundaries should not cut across the state blocks which were compact groups of districts in a big state. Each state block was an administrative unit of the NSS Directorate and there would have arisen some administrative difficulties of travelling allowance, inspection, supervision, etc. if the

enumerator under the charge of one state block had to survey sample villages falling in another state block.
4. Different parts of a stratum should be similar with respect to population densities and altitude.
5. There should be good transport and communication facilities among different parts of a stratum so that an enumerator posted in that stratum could travel throughout it without much difficulty.
6. The original size (say Z'') of a stratum (i.e. sum of the sizes of villages in it) should not differ by more than 10 per cent from the planned average figure Z, because otherwise the large adjustment to be done in order to equalize this Z'' to Z would lower the sampling efficiency of the pps design.

In all, 353 compact strata were formed by grouping contiguous *tehsils* such that the first, second, and the sixth conditions were completely satisfied and conditions three, four and five satisfied to the extent possible. Finally, the original stratum size Z'' was made equal to Z by slightly increasing or decreasing the sizes of the villages belonging to the stratum.

In each stratum, four independent sub-samples of 12 villages each were selected circular systematically with probability proportional to the size, the size being defined earlier. Out of these 12 villages in a sub-sample, the six villages with odd orders of selection constituted the central sample and the rest, with even orders of selection, the state sample. The linking of central and state samples ensured a better spread of samples over the whole stratum and consequently a better estimate when the central and the state sample data are pooled together.

The interval for systematic selection for either the central or the state sample was obtained as $I = Z/6$; the interval I, an integer, was the same for all the strata in a state since stratum sizes were equalized during the formation of the strata. Four independent random starts were used for selecting the four sub-samples of the 12 villages each. The procedure of pps circular systematic selection has been explained in section 5.4.3.

Because of pps selection, larger villages occurred more frequently in the sample. This resulted in too heavy a work load for some enumerators. The planned work load involved the listing of 120 households per sample village on an average, i.e. a total of 1440 households per year per enumerator. To reduce the work load of household listing in large villages, the hamlets were grouped in such a manner as to contain approximately the same population and the survey confined to one hamlet group, selected at random with equal probability. The number of hamlet groups to be formed in a village was specified by the Indian Statistical Institute, but the formation and selection of the hamlet groups were done by the enumerators on reaching the sample villages.

For selection of households for the socio-economic inquiries, the households were not stratified according to the available information for this would have complicated the computation of the estimates and would have rendered difficult the task of making the design self-weighting. The same objectives of stratification were to a large extent achieved by arranging the households in the manner described in section 4.3.3. From the sampling frame thus prepared, a sample of 22 households on an average was selected linear systematically with the interval and a random start prescribed for this purpose. For schedule 16, a sub-sample of 2 households on an average was selected systematically with interval 11 and a specified random start for this schedule; for schedule 17, from the sample households selected for schedule 16, a sub-sample of 1 household on an average was selected linear systematically with interval 2 and a random start 1 or 2, and the next household in the frame was surveyed for schedule 17 (the actual sample household for schedule 17 thus did not belong to the total of 22 households referred to earlier); in the remaining 20 households, schedule 12 was canvassed.

The following methods were used for selecting the sample of clusters or plots in the sample villages.

1. Suppose the sample village was cadastrally surveyed and a village map showing the location and boundaries of the plots was available without much trouble. Then clusters of size 10 (i.e. of 10 plots each) were formed by combining plots with major survey numbers 1–10, 11–20, 21–30 etc., and similarly clusters of size 5 were groups of major survey numbers, 1–5, 6–10, 11–15, 16–20, etc. Then 4 clusters of 5 plots each were selected in villages meant for land utilization only and 6 clusters of 10 plots each were selected in sample villages meant for both acreage and yield-rate surveys. The actual procedure of selection was as follows:

   Suppose N is the highest survey number in a village/hamlet-group: n the number of clusters to be selected; c the cluster size = 5 or 10; and N' is N increased as little as possible to become a multiple of cluster size c.

   Then n plots, known as 'basic plots', were selected circular systematically with interval I and random start R where $I = (N'/n)$ and $I \leqslant R \leqslant N$. Next clusters were formed around the selected basic plots, e.g. if the basic plot with survey number 472 was selected, the corresponding sample cluster was taken as a group of plots with survey numbers 471–480 and 471–475 for $c = 10$ and 5 respectively.
2. Suppose the village map was not available but a list of all plots in the village was available. In this case, all the plots in the village were first given a continuous sampling serial number. The selection of basic plots and formation of clusters around them were exactly the same as in method 1 with the only difference that survey numbers were replaced by sampling serial numbers.
3. In case neither the village map nor the list of plots was available, plots were selected indirectly through a sample of households. A sample of 4 or 6 households was selected systematically depending on whether the village was meant for just acreage survey or both acreage and yield-rate surveys. All the plots possessed by these sample households within a 5 mile radius of the sample village were taken up for survey.

Self-weighting sample. The sample was made self-weighting with respect to the unbiased estimator of the state total of a study variable of the household inquiries in the following manner.

The following notations are used for a state

h: subscript for a stratum;
i: subscript for a village;
j: subscript for a household;
L: total number of strata;
n_h: number of sample villages in any sub-sample of the hth stratum ($n_h = 6$);
Z_h: total size of the hth stratum ($Z_h = Z_0$, as the strata were of equal size);
Z_{hi}: size of the ith sample village in the hth stratum;

D_{hi}: number of hamlet-groups formed in the hith sample village ($D_{hi} = 1$, if the whole village was surveyed);
I_{hi}: interval for selecting the combined sample in the hith sample village;
m_{hi}: number of households in the combined sample in the hith sample village;
y_{hij}: value of the study variable for the jth sample household in the hith sample village ($j = 1, 2, \ldots, m_{hi}$);
Y: total value of the study variable for the state;

From estimating equations (21.15), (21.17) and (21.18), an unbiased estimator of Y, based on any sub-sample, is given by

$$y = \sum_{h=1}^{L} \frac{1}{6} \sum_{i=1}^{6} \frac{Z_0}{Z_{hi}} D_{hi} I_{hi} \sum_{j=1}^{m_{hi}} y_{hij}$$

$$= \sum_{h}^{L} \sum_{i}^{6} \sum_{j}^{m_{hi}} w_{hij} y_{hij} \qquad \text{(A5.1)}$$

where

$$w_{hij} = \frac{Z_0}{6Z_{hi}} D_{hi} I_{hi}$$

is the multiplier for y_{hij}. The object is to make w_{hij} a constant, w_0, within a state. Now $Z_h = \sum_i Z_{hi} = Z_0$ is the same for all the strata within a state, and the values of Z_{hi} and D_{hi} are already determined by the village population. So the only item that can be properly chosen so as to equalize all the w_{hij} values is I_{hi}; i.e.

$$I_{hi} = \frac{6w_0}{Z_0} \cdot \frac{Z_{hi}}{D_{hi}} \qquad \text{(A5.2)}$$

which determines the interval for selecting the combined sample in a sample village, when the value of the multiplier w_0 is fixed.

The unbiased estimator y then becomes

$$y = w_0 \sum_{h}^{L} \sum_{i}^{6} \sum_{j}^{m_{hi}} y_{hij} \qquad \text{(A5.3)}$$

The next problem is to find an appropriate w_0 that will ensure the desired sample size in terms of households. From equation (A5.3), putting $y_{hij} = 1$, the estimated total number of households in the state is

$$m = w_0 \sum^{L} \sum^{6} m_{hi}$$

$$= w_0 \times \text{number of sample households per sub-sample in the state} \qquad \text{(A5.3)},$$

To be more precise about the value of w_0, we can replace y by its true value Y, and write

$$w_0 = \frac{\text{total number of households in the state}}{\text{number of sample households per sub-sample in the state}}$$

$$= \text{reciprocal of the overall sampling fraction} \qquad \text{(A5.5)}$$

Now, the number of sample households per sub-sample in the state = nm_0, where n is the number of sample villages (in any sub-sample) for the state and m_0 the average number of households planned to be selected per sample village. In this particular case, $n = 6L$ and $m_0 = 22$. The true total number of households at the mid-point of the survey period, i.e. on 1 January 1965, is of course unknown; it is replaced by y' obtained by first projecting the rural population of the state from the 1961 Census (and the values of the previous NSS rounds) and then dividing that by the average household size (obtained also from the previous NSS rounds). Thus w_0 finally works out as

$$w_0 = y'/nm_0 = y'/132L \tag{A5.6}$$

First, w_0 is to be obtained from equation (A5.6) for each state, and then the values of I_{hi} are determined from equation (A5.2) for a sample village, noting that the factor $6w_0/Z_0$ is the same for all the sample villages in a state.

When, as in most cases, I_{hi} as computed was a fraction, it was rounded off in the randomized manner, as described in note 3 to section 12.3. The rounded off (integer) values of I_{hi} were given to the enumerators.

The intervals obtained by equation (A5.2) ensured that the total sample size (number of households) in the state would be near the desired value $nm_0 = 132L$, but they do not ensure anything for the individual sample villages. Some enumerators were allotted too many big villages, and as each enumerator had to survey 12 sample villages, the total sample size for them was much larger than the planned figure of 12 x 22 households. In such cases, to provide relief to the enumerators, they were allowed to survey a smaller number of sample households than was strictly required for a self-weighting design: the shortfall was made good at the scrutiny stage by repeating some of the filled-in schedules.

Once the design was made self-weighting for the combined sample, it became automatically so for individual household inquiries, for constant fractions of the combined sample households were surveyed for each of these inquiries in all the sample villages.

Estimating procedures. With the same notation as for the *rural sector*, the unbiased estimators for the state totals of study variables were as follows.

Schedule 0.1 (complete enumeration of the households in the sample village or the selected hamlet-group):

$$y = \sum_{h=1}^{L} \frac{1}{6} \sum_{i=1}^{6} \frac{Z_0}{Z_{hi}} D_{hi} I_h y_{hi}$$

$$= I_0 \sum_{h}^{L} \sum_{i}^{6} D_{hi} \frac{y_{hi}}{Z_{hi}} \tag{A5.7}$$

where $I_h = Z_h/n_h$ and $Z_0 = Z_0/6$.

Schedules 10.1, 12, 16, and 17:

$$y = w_0 \sum_{h}^{L} \sum_{i}^{6} \sum_{j}^{m_{hi}} y_{hij} \tag{A5.8}$$

where w_0 is the constant multiplier for the corresponding schedule.

Information on fertility history was collected for one-fifth of the sample households for Schedule 12; the constant multiplier for such information was therefore $5w_0$, where w_0 was the constant multiplier for Schedule 12.

Schedule 3.01: An estimator of the average price P of a commodity was

$$p = \sum_{i=1}^{n'} y_i/n' \tag{A5.9}$$

where n' is the number of reporting villages in the state and the sub-sample considered.

In the *urban sector,* as there were 2 strata in each state (i.e. $L = 2$), the unbiased estimators took the following form:

Schedule 0.2:

$$y = \sum_{h=1}^{2} I_h \sum_{i=1}^{n_h} D_{hi} \frac{y_{hi}}{Z_{hi}} \tag{A5.10}$$

where $I_h = Z_h/n_h$.

Schedule 10, 12, 16, and 17:

$$y = \sum_{h=1}^{2} w_h \sum_{i=1}^{n_h} \sum_{j=1}^{m_{hi}} y_{hij} \tag{A5.11}$$

where w_h is the constant multiplier for a schedule in the hth stratum.

For *crop-surveys,* an unbiased estimator of the acreage under a particular cereal crop in a season from a sub-sample in a state was

$$a = \sum^{L} a_h \tag{A5.12}$$

where

$$a_h = I \sum_{i=1}^{L} \frac{D_{hi}}{Z_{hi}} \frac{M'_{hi}}{m_{hi}} \sum_{j=1}^{m_{hj}} a_{hij} \tag{A5.13}$$

where M'_{hi} is the adjusted highest survey number/sampling serial number (method 1 or 2) or total number of households (method 3) in the ith sample village of the hth stratum, m_{hi} is the number of sample plots or households planned for survey, and a_{hij} is the area under the specified cereal crop for the jth sample plot or in all plots of the jth sample household in the ith sample village.

An estimator of the yield rate (Schedule 5.1) for a cereal crop in a season from sub-sample 1 or 2 is given by

$$x = \sum^{L} a_h\bar{y}_h \Big/ \sum^{L} a_h \tag{A5.14}$$

where a_h is given by equation (A5.13), and $\bar{y}_h$ is the simple average of the yield rates over the crop-cuts taken for the crop in the hth stratum; the summation Σ extends over all strata reporting crop-cutting experiments. The yield rates were obtained on the basis of the sample cuts of radius 4′; the yields of all the plants inside the circle together with half the yield of the border plants of the 4′ circle constituted the yield of a cut.

While calculating the production estimates, the estimate of yield rate for sub-sample 1 was multiplied by the corresponding area estimate, pooled over sub-samples 1 and 3, and the estimate of yield rate for sub-sample 2 was multiplied by the corresponding area estimate, pooled over sub-samples 2 and 4. Thus two independent estimates y_1 and y_2 of total production were obtained. For obtaining the dry weights, these two production estimates were multiplied by the driage factors based on sub-samples 1 and 2 respectively. The driage factor was obtained on the basis of the sample cuts of radius 2′3″, including the full border. If d and g were the weights of dry crop and green crop respectively, the driage factor was given by $\Sigma d/\Sigma g$ where Σ denotes summation over all cuts taken on that crop in all strata. For paddy crop, the production figure related to 'clean rice' and for this the figure for dry paddy was multiplied by 0.662.

Substitution of casualties. For socio-economic inquiries, all casualty households, villages and blocks were substituted. Similarly for crop survey, all casualty plots/clusters/households/villages were substituted by the previously surveyed corresponding units in the same sub-sample and stratum.

Combined estimators. If y_l ($l = 1, 2, 3, 4$) be the lth sub-sample estimator of the state total of a study variable, the combined estimator was

$$y_0 = \tfrac{1}{4}(y_1 + y_2 + y_3 + y_4) = \tfrac{1}{4}\sum_{h=1}^{L}(y_{h1} + y_{h2} + y_{h3} + y_{h4}) \qquad \text{(A5.15)}$$

where y_{hl} is the lth sub-sample estimator of the hth stratum total.

The ratio $R = Y/X$ was estimated by $r_l = y_l/x_l$ from the lth sub-sample and the combined estimator of R was

$$r = y_0/x_0 = (y_1 + y_2 + y_3 + y_4)/(x_1 + x_2 + x_3 + x_4) \qquad \text{(A5.16)}$$

The combined estimator of the area under a crop was similarly the mean of the four sub-sample estimates. The combined production estimator y_0 for a season is the mean of the two production estimators y_1 and y_2. Whenever season-wise total crop estimates were obtained, the estimate for the year was the sum of the season-wise estimates.

Variance estimators (see also section 25.5.4). Two unbiased variance estimator of y are

$$s_y^2 = \tfrac{1}{12}\sum_{h=1}^{L}\sum_{l=1}^{4}(y_{hl} - \bar{y}_h)^2 \qquad \text{(A5.17)}$$

and

$$s_y'^2 = \tfrac{1}{12}\sum_{l}^{4}(y_l - y_0)^2 \qquad \text{(A5.18)}$$

where

$$\bar{y}_h = \tfrac{1}{4}\sum_{l}^{4} y_{hl}$$

$s_y'^2$ is easier to compute but less efficient than s_y^2. Unbiased variance estimators of the acreage estimator a in equation (A5.12) are obtained similarly.

Two variance estimators of r are

$$s_r^2 = \frac{1}{12}\frac{1}{x^2}\sum_{h}^{L}\sum_{l}^{4}\left[(y_{hl} - \bar{y}_h)^2 + r^2(x_{hl} - \bar{x}_h)^2 - 2r(y_{hl} - \bar{y}_h)(x_{hl} - \bar{x}_h)\right] \qquad \text{(A5.19)}$$

and

$$s_r'^2 = \tfrac{1}{12}\sum_{l}^{4}(r_l - r)^2 \qquad \text{(A5.20)}$$

the latter being easier to compute but less efficient than the former.

A variance estimator of the combined production estimator y_0 is

$$s^2_{y_0} = \frac{1}{4}(y_1 - y_2)^2 \tag{A5.21}$$

A5.9 Pilot inquiries and pre-tests

The previous 18 rounds, covering the period 1950–64, acted as pilot inquiries, the results of which were used to design the sample for the 19th round effectively, including information on the variability and the cost. In addition, special tryouts were organized for new schedules introduced in this survey.

A5.10 Selection and training of enumerators

The type of enumerators has been mentioned in paragraph A5.4. The training of enumerators was done in two phases, the supervisors being trained by the technical staff and in their turn training the enumerators (section 26.15).

A5.11 Processing of data

The schedules were first scrutinized by the field supervisors and then by the technical staff of the Indian Statistical Institute. Queries of a technical nature from the field were answered by the technical staff.

A5.12 Some results

Some results of the survey obtained from the central sample are given in Table A5.2. The coefficients of variation of the estimates are seen to be reasonably small.

Table A5.2 Some estimates from the Indian National Sample Survey, 18th round, July 1964 – June 1965

Item		*Estimate*	*CV (%)*	*No. of villages/ urban blocks*	*No. of house-holds*
1. Rural India:	(a) Birth rate	37.0/1000	0.75	8472	169 440
	(b) Death rate	13.0/1000	1.49	8472	169 440
2. Urban India:	(a) Birth rate	31.9/1000	1.13	4572	91 440
	(b) Death rate	8.0/1000	1.69	4572	91 440
3. Total monthly consumer expenditure in the urban sector in some States:					
	(a) Andhra Pradesh	Rs. 208 million	6.20	384	768
	(b) Gujerat	Rs. 163 million	2.40	192	384
	(c) Rajasthan	Rs. 111 million	6.68	216	432
	(d) West Bengal	Rs. 329 million	6.28	432	864

The edited information was put into punch cards and the final tabulation done by mechanical processing. Electronic processing is also resorted to for special computations and analysis.

Further reading

Full details of the survey methods of the Indian National Sample Survey, 1964–5, are given by Roy and Bhattacharya; Murthy gives in Chapter 15 the survey methods of the Indian National Sample Survey, 1958–9, and in Chapter 16 the Family Living Survey in urban areas of India, 1958–9, the latter from Nanjamma (1963).

The following are references to some readily available case studies.

Agricultural surveys: (a) conducted by the Indian Statistical Institute; Mahalanobis (1944, 1946, 1968); (b) in India: Panse and Sukhatme, and Sukhatme and Panse; (c) Survey of Fertilizer Practice in England and Wales: Yates, sections 4.5, 4.23, 5.10, 5.19, 6.19, 7.17, 7.23, 8.3, 8.6, 8.9, 8.17, 8.19, 10.7, and 11.10.

Auditing and accounting: Trublood and Cyert.

Demographic surveys: (a) in Greece, Deming (1966), Chapter 12; (b) in India, Som *et al.* (1961), and Office of the Registrar General, *Sample Registration of Births Deaths in India; Rural 1965–68*; (c) in Pakistan: Pakistan Institute of Development Economics, *Report of the Population Growth Estimation Experiment.* (d) in Trinidad and Tobago: Harewood; (e) in the U.S.: the U.S. Bureau of the Census Technical paper No.7 (summaries in Hansen *et al.*, Vol. I, Chapter 12, and in Kish, section 10.4).

Family planning surveys: Acsádi, Klinger and Szabady (1969, 1970).

Health Surveys: U.S. National Center for Health Statistics.

Social Surveys in Great Britain: Gray and Corlett; Moser and Kalton.

Marketing research survey in Great Britain, 1952: Moser and Kalton, Chapter 8.

Employment and unemployment survey of Greece, 1962: Raj, Appendix I.

Surveys of retail stores and annual survey of manufactures in the U.S.A., Hansen *et al.*, Vol. I, Chapter 12.

Inventory surveys: Deming (1960), Chapters 6, 7, and 8; and Deming (1966), Chapter 11.

Slonim in Chapter XVII gives a simple account of a large number of sample surveys.

For development of sampling in Great Britain, Moser (1949, 1955), and Moser and Kalton; in India, Mahalanobis (1946); and in Sweden, Dalenius.

References

This list comprises publications referred to under 'Further Reading' at the end of each chapter and others cited in the text. A bibliography of the literature on sampling, if required, will be found in Yates's *Sampling Methods for Censuses and Surveys* and Murthy's *Sampling Theory and Methods.*

ACKOFF, R. L. (1953). *Design of Social Research.* University of Chicago Press, Illinois.

ACSÁDI, G., KLINGER, A., and SZABADY, E. (1969). *Survey Techniques in Fertility and Family Planning Research: Experience in Hungary.* Central Statistical Office, Demographic Research Institute, Budapest.

ACSÁDI, G., KLINGER, A., and SZABADY, E. (1970). *Family Planning in Hungary, Main Results of the 1966 Fertility and Family Planning (TCS) Study.* Central Statistical Office, Demographic Research Institute, Budapest.

BAILEY, N. T. J. (1951). 'On estimating the size of mobile population from recapture data.' *Biometrika,* 38, 293–306.

BLANKENSHIP, A. (1945). *How to Conduct Consumer and Opinion Research* (2nd edn.). Harpers, New York.

BEHMOIRAS, J. P. (1965). *La situation démographique au Tchad; Résultats provisoires de l'enquête démographique 1964.* République du Tchad, Commissariat général au plan, Société d'Etudes pour le Développement Economique et Sociale, Paris.

BENJAMIN, B. (1955). 'Quality of response in census taking.' *Population Studies,* 8, 288–93.

BENJAMIN, B. (1968). *Health and Vital Statistics.* Allen and Unwin, London.

BLANC, R. (1962). *Manuel de recherches démographiques en pays sous-développés.* Institut National de la Statistique et des Etudes Economiques, Paris. (Reprinted and translated at the United Nations Economic Commission for Africa, *Manual of Demographic Research in Under-developed countries,* E/CN.14/ASPP/L.14.)

BROOKES, B. C., and DICK, W. F. L. (1969). *Introduction to Statistical Method* (2nd edn., cased and paperback). Heinemann, London.

CANTRIL, H. (1944). *Gauging Public Opinion.* University Press, Princeton.

CHAKRAVARTI, I. M., LAHA, R. G., and ROY, J. (1967). *Handbook of Methods of Applied Statistics, Vol. II, Planning of Surveys and Experiments.* Wiley, New York and London.

CHANDRA SEKAR, C. and DEMING, W. E. (1949). 'On a method of estimating birth and death rates and the extent of registration.' *Journal of the American Statistical Association,* 44, 101–15.

CLAUSEN, J. A. and FORD, R. N. (1947). 'Controlling bias in mail questionnaire.' *Journal of the American Statistical Association,* 42, 497–511.

COCHRAN, W. G. (1963). *Sampling Techniques* (2nd edn.). Wiley, New York and London.

CONWAY, F. (1967). *Sampling – An Introduction for Social Scientists.* Allen and Unwin, London.

DALENIUS, T. (1957). *Sampling in Sweden: Contributions to the Method and Theories on Sample Survey Practice.* Almqvist and Wiksell, Stockholm.

DAS, N. C. (1969). *Special Study on Morbidity.* The National Sample Survey No. 119. The Cabinet Secretariat, Government of India, New Delhi.

DAS GUPTA, A. (1958). Determination of fertility level and trend in defective registration areas. *Bulletin de l'Institut international de statistique,* 36, 127–36.

DEMING, W. E. (1960). *Sample Design in Business Research.* Wiley, New York and London.

DEMING, W. E. (1966). *Some Theory of Sampling.* Dover Publications, New York, and Constable, London. (First published in 1950 by Wiley, New York).

DESABIE, M. J. (1959 & 1962). *Théorie et Pratique des Sondages,* Tomes I et II. Institut National de la Statistique et des Etudes Economiques, Paris.

EL-BADRY, M. A. (1956). 'A sampling procedure for mailed questionnaire.' *Journal of the American Statistical Association,* 51, 209–27.

ERICSON, W. A. (1969). 'Subjective Bayesian models in sampling finite populations: stratification.' In *New Developments in Survey Sampling* edited by N. L. Johnson and H. Smith, Jr. Wiley, New York and London.

FELLER, W. (1957). *An Introduction to Probability Theory and its Applications* (2nd edn.). Wiley, New York.

FELLIGI, I. P., (1964). 'Response variance and its estimation.' *Journal of the American Statistical Association,* 59, 1016–41.

FISHER, R. A. and YATES, F. (1963). *Statistical Tables for Use in Bibliogical, Agricultural and Medical Research* (6th edn.). Oliver and Boyd, Edinburgh.

FOOD AND AGRICULTURAL ORGANIZATION OF THE UNITED NATIONS (1969). *Improvement of Agricultural Census and Survey Reports.* African Commission on Agricultural Statistics, Fourth Session, Algiers (AGS: AF/4/69/4). Rome.

FRANKEL, L. R. (1969). The role of accuracy and precision of response in sample surveys. In *New Developments in Survey Sampling* edited by W. L. Johnson and H. Smith, Jr. Wiley, New York and London.

GALLUP, G. (1948). *A Guide to Public Opinion Polls.* University Press, Princeton.

GINI, C. and GALVANI, L. (1929). Di una applicazione de metodos rappresentativo all' ultimo censimento italiano della popolazione. *Annali di Statistica,* 6, 1–107.

GLASS, D. V. and GREBENIK, E. (1954). *The Trend and Pattern of Fertility in Great Britain: A Report on the Family Census of 1946, Part I (Report).* H.M.S.O., London.

GRAY, P. G. and CORLETT, T. (1950). 'Sampling for the Social Survey.' *Journal of the Royal Statistical Society,* A. 113, 150–206.

HÁJEK, J. and DUPAČ, V. (1967). *Probability in Science and Engineering.* Academia, Prague and Academic Press, New York.

HAMMERSLEY, J. M. (1953). 'Capture-recapture analysis.' *Biometrika,* 40, 265–78.

HANSEN, M. H., HURWITZ, W. N. and BERSHAD, M. A. (1961). 'Measurement errors in censuses and surveys.' *Bulletin of the International Statistical Institute,* 38, 359–74.

HANSEN, M. H., HURWITZ, W. N. and JABINE, T. B. (1964). 'The use of imperfect lists for probability sampling at the U.S. Bureau of the Census.' *Bulletin of the International Statistical Institute,* 40, 497–517.

HANSEN, M. H., HURWITZ, W. N. and MADOW, W. G. (1953). *Sampling Survey Methods and Theory, Vol. I, Methods and Applications, Vol. II, Theory.* Wiley, New York, Chapman and Hall, London.

HANSEN, M. H., HURWITZ, W. N., and PRITZKER, L. (1964). 'The estimation and interpretation of gross differences and the simple response variance.' In *Contribution to Statistics* edited by C. R. Rao. Pergamon Press, Oxford and Statistical Publishing Society, Calcutta, pp. 111–36.

HANSEN, M. H. and PRITZKER, L. (1956). 'The Post-Enumeration Survey of the 1950 Census of Population: some results, evaluation, and implications.' Paper presented at the Annual Meeting of the Population Association of America, Ann Arbor, Michigan, 19 May 1956. Mimeographed.

HANSEN, M. H. and TEPPING, B. J. (1969). 'Progress and problems in survey methods and theory illustrated by the United States Bureau of the Census.' In *New Developments in Survey Sampling* edited by N. L. Johnson and H. Smith, Jr. Wiley, New York and London.

HAREWOOD, J. (1968). *Continuous Sample Survey of Population, General Reports, Rounds 1–8.* Continuous Sample Survey of Population, Publication no. 11. Central Statistical Office, Trinidad and Tobago, Port of Spain.

HENDRICKS, W. A. (1949). 'Adjustment for bias by non-response in mailed surveys.' *Agricultural Economics Research,* 1, 52–6.

HENDRICKS, W. A. (1956). *The Mathematical Theory of Sampling.* The Scarecrow Press, New Brunswick and Bailey Bros. & Swinfen, London.

HERBERGER, L. (1971). 'Organization and functioning of the Micro-census in the Federal Republic of Germany.' *Population Data and Use of Computers with Special Reference to Population Research,* German Foundation for Developing Countries, Berlin and Federal Statistical Office, Wiesbaden. Mimeographed.

HESS, I., RIEDEL, D. C. and FITZPATRICK, T. B. (1961). *Probability Sampling of Hospitals and Patients,* University of Michigan, Ann Arbor.

INTERNATIONAL STATISTICAL INSTITUTE (1964). *Proceedings of the 34th Session, Bulletin of the International Statistical Institute,* 40, bk. 1.

INTERNATIONAL UNION FOR THE SCIENTIFIC STUDY OF POPULATION (1959). *Problems in African Demography: A Colloquium.* Paris.

JOHNSON, P. O. (1949). *Statistical Methods in Research.* Prentice-Hall, Englewood, N.J.

JONES, D. C. (1949). *Social Surveys.* Hutchinson, London.

KENDALL, M. G. and BUCKLAND, W. R. (1970). *A Dictionary of Statistical Terms* (3rd edn.). Oliver and Boyd, Edinburgh and Hafner, New York.

KENDALL, M. G., and SMITH, B. BABINGTON. (1954). *Tables of Random Sampling Numbers.* Cambridge University Press, Cambridge and New York.

KENDALL, M. G., and STUART, A. (1968). *The Advanced Theory of Statistics,* Vol. 3 (2nd edn.). Griffin, London and Hafner, New York.

KISER, C. V. (1934). 'Pitfalls in sampling for population study.' *Journal of the American Statistical Association,* 29, 250–56.

KISH, L. (1967). *Survey Sampling.* Wiley, New York and London.

KISH, L. (1971). 'Special aspects of demographic samples.' *International Population Conference, London 1969, Tome 1,* International Union for the Scientific Study of Population, Liège.

KROTKI, K. J. (1966). 'The problem of estimating vital rates in Pakistan.' *World Population Conference, 1965, Vol. III.* United Nations, Sales No.: 66.XIII.7.

LAHIRI, D. B. (1958). Observations on the use of interpenetrating samples in India. *Bulletin de l'Institut international de statistique,* 36, 144–52.

LAURIAT, P. and CHINTAKANANDA, A. (1965). 'Technique to measure population growth: Survey of Population Change in Thailand.' *United Nations World Population Conference, Belgrade* (B6/V/E/507).

LESLIE, P. H. (1952). 'The estimation of population parameters from data obtained by means of the capture-recapture method, II. The estimation of total numbers.' *Biometrika,* 39, 363–88.

LINCOLN, F. C. (1930). 'Calculating waterfowl abundance on the basis of banding returns.' *Circ. U.S. Department of Agriculture, No. 118.*

LINDLEY, D. V. (1965). *Introduction to Probability and Statistics from a Bayesian Viewpoint, Pts. 1 and 2.* Cambridge University Press, Cambridge.

MACURA, M. and BALABAN, V. (1961). 'Yugoslav experience in evaluation of population censuses and sampling.' *Bulletin of the International Statistical Institute,* 38, 375–99.

MAHALANOBIS, P. C. (1940). 'A sample survey of acreage under jute in Bengal.' *Sankhyā,* 4, 511–30.

MAHALANOBIS, P. C. (1944). 'On large-scale sample surveys.' *Philosophical Transactions of Royal Society,* B, 231, 329–451.

MAHALANOBIS, P. C. (1946). 'Recent experiments in statistical sampling in the Indian Statistical Institute.' *Journal of the Royal Statistical Society,* A, 108, 326–78. Reprinted in *Sankhyā,* 20, 1–68 (1958) and by Asia Publishing House, London and Statistical Publishing Society, Calcutta (1963).

MAHALANOBIS, P. C. (1960). 'A method of fractile graphical analysis.' *Econometrica,* 28, 325–51. Reprinted in *Sankhyā,* A, 23, 325–58.

MAHALANOBIS, P. C. (1968). *Sample Census of Area Under Jute in Bengal, 1940.* Statistical Publishing Society, Calcutta.

MOSER, C. A. (1949). 'The use of sampling in Great Britain.' *Journal of the American Statistical Association,* 44, 231–59.

MOSER, C. A. (1955). 'Recent developments in the sampling of human populations in Great Britain.' *Journal of the American Statistical Association,* 50, 1195–1214.

MOSER, C. A. and KALTON, G. (1971). *Survey Methods in Social Investigation* (2nd edn., cased and paperback). Heinemann, London.

MURTHY, M. N. (1967). *Sampling Theory and Methods.* Statistical Publishing Society, Calcutta.

MURTHY, M. N. and ROY, A. S. (1970). 'A problem in integration of surveys – a case study.' *Journal of the American Statistical Association,* 65, 123–35.

NADOT, R. (1966). *Afrique Noire, Madagascar, Comores – démographie comparée, 3 – Fecondité: Niveau.* Institut National de la Statistique et des Etudes Economiques, Paris.

NANJAMMA, C. (1963). 'Technical paper on sample designs of working class and middle class family living surveys.' *Sankhyā,* B, 25, 359–418.

NARAIN, R. D. (1954). 'The general theory of sampling on successive occasions.' *Bulletin of the International Statistical Institute,* 34, 87–9.

NETER, J. and WAKSBERG, J. (1965). *Response Errors in Collection of Expenditures Data by Household Interviews: An Experimental Study.* Bureau of the Census, Technical Paper No. 11. U.S. Department of Commerce, Washington.

NISSELSON, H. and WOOLSEY, T. D. (1959). 'Some problems of the household interview design for the National Health Survey.' *Journal of the American Statistical Association,* 54, 88–101.

OFFICE OF THE REGISTRAR GENERAL, INDIA (1970). *Sample Registration of Births Deaths in India: Rural 1965–68.* Ministry of Home Affairs, New Delhi.

PAKISTAN INSTITUTE OF DEVELOPMENT ECONOMICS (1968). *Report of the Population Growth Estimation Experiment 1961–1963.* Karachi.

PANSE, V. G. (1954). *Estimation of Crop Yields.* Food and Agriculture Organization of the United Nations, Rome.

PANSE, V. G. (1966). *Some Problems of Agricultural Census Taking.* Food and Agricultural Organization of the United Nations, Rome.

PARTEN, M. B. (1950). *Surveys, Polls and Samples: Practical Procedures.* Harpers, New York.

PAYNE, S. L. (1951). *The Art of Asking Questions.* University Press, Princeton and Oxford University Press, London.

POPULATION COUNCIL (1970). *A Manual for Surveys of Fertility and Family Planning: Knowledge, Attitudes, and Practice,* New York.

RAJ, D. (1968). *Sampling Theory.* McGraw-Hill, New York and London.

RAND CORPORATION. (1955). *A Million Random Digits.* Free Press, Glencoe.

RAO, C. R. (1973). *Advanced Statistical Methods in Biometric Research* (2nd edn.). Hafner, New York.

ROY, A. S. and BHATTACHARYYA, A. (1968). *Technical Paper on Sample Design, Nineteenth Round, July 1964–June 1965.* The National Sample Survey No. 125. The Cabinet Secretariat, Government of India, New Delhi.

ROYER, J. (1958). *Handbook on Agricultural Sampling Surveys in Africa, I. Principles and Examples.* Food and Agricultural Organization of the United Nations, Commission for Technical Co-operation in Africa South of Sahara, and Government of France.

SAMPFORD, M. R. (1962). *An Introduction to Sampling Theory with Applications to Agriculture.* Oliver and Boyd, Edinburgh.

SANDERSON, F. H. (1954). *Methods of Crop Forecasting.* Harvard University Press, Cambridge, Massachusetts.

SCHMITT, S. A. (1969). *Measuring Uncertainty: An Elementary Introduction.* Addison-Wesley, Reading, Massachusetts.

SCOTT, C. (1967). Sampling for demographic and morbidity surveys in Africa. *Review of the International Statistical Institute,* 35, 154–71.

SCOTT, C. (1968). 'Vital rate surveys in Tropical Africa: some new data affecting sample design.' In *The Population of Tropical Africa* edited by J. C. Caldwell and C. Okonjo. Longmans, London.

SLONIM, M. J. (1960). *Sampling.* (Third Paperback printing 1967). Simon and Schulter, New York.

SNEDECOR, G. W. and COCHRAN, W. G. (1967). *Statistical Methods* (6th edn.). The Iowa State University Press, Ames.

SOM, R. K. (1958–9). 'On sampling design in opinion and marketing research.' *Public Opinion Quarterly,* 32, 564–66.

SOM, R. K. (1959). 'Self-weighting sample design with an equal number of ultimate stage units in each of the selected penultimate stage units.' *Calcutta Statistical Association Bulletin,* 9, 59–66.

SOM, R. K. (1965). 'Use of Interpenetrating samples in demographic studies.' *Sankhyā,* B, 27, 329–42.

SOM, R. K. (1973). *Recall Lapse in Demographic Enquiries.* Asia Publishing House, Bombay, London, and New York.

SOM, R. K., DE, A. K., DAS, N. C., PILLAI, B. T., MUKHERJEE, H., SARMA, S. M. U. (1961). *Preliminary Estimates of Birth and Death Rates and of the Rate of Growth of Population, Fourteenth Round, July 1958–July 1959.* The National Sample Survey No. 48. The Cabinet Secretariat, Government of India, New Delhi.

STEPHAN, F. F. and McCARTHY, P. J. (1958). *Sampling Opinions.* Wiley, New York and Chapman and Hall, London.

STUART, A. (1962). *Basic Ideas of Scientific Sampling.* Griffin, London.

SUKHATME, P. V. and SUKHATME, B. V. (1969). 'On some methodological aspects of sample surveys of agriculture in developing countries.' In *New Developments in Survey Sampling* edited by N. L. Johnson and H. Smith, Jr. Wiley, New York and London.

SUKHATME, P. V. and SUKHATME, B. V. (1970). *Sampling Theory of Surveys with Applications* (2nd edn.). Food and Agriculture Organization of the United Nations, Rome and Asia Publishing House, Bombay.

SUKHATME, P. V. and SETH, G. R. (1952). 'Non-sampling errors in surveys.' *Journal of the Indian Society of Agricultural Statistics,* 4, 5–41.

SZAMEITAT, K. and SCHAFFER, K.-A. (1964). 'Imperfect frames in statistics and the consequences for their use in sampling.' *Bulletin of the International Statistical Institute,* 40, 517–38.

THIONET, P. (1953). *Applications des méthodes de sondage.* Institut National de la Statistique et des Etudes Economiques, Paris.

THIONET, P. (1958). *La Théorie des sondages.* Institut National de la Statistique et des Etudes Economiques, Paris.

TIPPET, L. H. C. (1952). *The Methods of Statistics* (4th edn.). Williams and Moorgate, London.
TRUBLOOD, R. M., and CYERT, R. M. (1957). *Sampling Techniques in Accountancy.* Prentice-Hall, Englewood.
UNITED NATIONS (1961). *The Mysore Population Study*, Sales No.: 61.XIII.3, New York.
UNITED NATIONS (1964a). *Recommendations for the Preparation of Sample Survey Reports (Provisional Issue).* Sales No.: 64.XVII.7, New York.
UNITED NATIONS (1964b). *Handbook of Household Surveys.* Sales No. 64.XVII.13, New York.
UNITED NATIONS (1970). *A Short Manual on Sampling, Vol. II, Computer Programmes for Sample Designs.* Sales No.: E.71.XVII.4, New York.
UNITED NATIONS (1971a). *Handbook of Population and Housing Census Methods, Part VI, Sampling in connexion with population and housing censuses.* Sales No.: E.70.XVII.9, New York.
UNITED NATIONS (1971b). *Methodology of Demographic Sample Surveys.* Sales No.: E.71.XVII.11. New York.
UNITED NATIONS. (1972). *A Short Manual on Sampling, Vol. I. Elements of Sample Survey Theory.* Sales No.: E. 72. XVII. 5, New York.
UNITED NATIONS ECONOMIC COMMISSION FOR AFRICA (1971). *Manual on Demographic Sample Surveys in Africa* (Draft), Addis Ababa. Mimeographed.
U.S. BUREAU OF THE CENSUS (1960). *The Accuracy of Census Statistics with and without Sampling.* Technical Paper No. 2. U.S. Department of Commerce, Washington.
U.S. BUREAU OF THE CENSUS (1963). *The Current Population Survey – A Report on Methodology.* Technical Paper No. 7. U.S. Department of Commerce, Washington.
U.S. BUREAU OF THE CENSUS (1968). *Sampling Lectures.* U.S. Department of Commerce, Washington.
U.S. NATIONAL CENTER FOR HEALTH STATISTICS (1963). *Origin, Progress, and Operation of the U.S. National Health Survey.* Vital and Health Statistics Series 1, No. 1, U.S. Department of Health, Education, and Welfare, Washington.
U.S. NATIONAL CENTER FOR HEALTH STATISTICS. *Vital and Health Statistics, Series 1. Programs and Collection Procedures, and Series 2, Data Evaluation and Methods Research.* U.S. Department of Health Education, and Welfare, Washington.
VANCE, L. L. (1950). *Scientific Method for Auditing: Applications of Statistical Sampling Theory to Auditing Procedure.* University of California Press, Berkely and Cambridge University Press, London.
VANCE, L. L. and NETER, J. (1950). *Statistical Sampling for Auditors and Accountants.* Wiley, New York.
WEBER, A. A. (1967). *Les Méthodes de Sondage.* Bureau régional de l'Europe, Organisation mondiale de la Santé, Copenhagen. Mimeographed.
YAMANE, T. (1967). *Elementary Sampling Theory.* Prentice-Hall, Englewood Cliffs, N.J., and London.
YATES, F. (1960). *Sampling Methods for Censuses and Surveys* (3rd edn.). Griffin, London and Hafner, New York.
YERUSHALMY, J. and NEYMAN, J. (1947). *Public Health Reports,* 67.
YULE, G. U. and KENDALL, M. G. (1950). *An Introduction to the Theory of Statistics.* (14th edn.). Griffin, London.
ZARKOVICH, S. S. (1962). Agricultural statistics and multisubject household surveys. *Monthly Bulletin of Agricultural Economics and Statistics,* 11, 1–5.
ZARKOVICH, S. S. (1963). *Sampling methods and Censuses, vol. II. Quality of Statistical Data* (Draft, mimeographed). Printed in 1966. Food and Agriculture Organization of the United Nations.
ZARKOVICH, S. S. (1965a). *Sampling Methods and Censuses.* Food and Agriculture Organization of the United Nations.
ZARKOVICH, S. S., ed. (1965b). *Estimation of Areas in Agricultural Statistics.* Food and Agriculture Organization of the United Nations, Rome.

Answers to Exercises

Chapter 2 (page 45)

1. Following Example 2.1, $y = 2.91 \pm 0.245$* acres per plot; $y_0^* = 291.0 \pm 24.5$ acres; CV for both, 8.42%.

2. Following Example 2.3, (a) 940 597 ± 19 128; the 95% probability limits are 903 106 and 978 088; (b) 12.49 ± 0.254 cattle per farm; the 95% probability limits are 11.99 and 12.99.

5. Using the methods of section 2.13, the number of households possessing radios is 40 ± 13; the number of persons in these households is 175 ± 60.

7. For (a) using the methods of section 2.9 and considering the sample as having been drawn with replacement, (i) 25 902 ± 1797; (ii) 6093 ± 650; (iii) 0.2331 ± 0.01872. Here $n = 43$, $N = 325$. For (b) use equation (2.76) of section 2.13, which gives the standard error of the preparation absent as 0.00 7224. Here n has to be taken as 3427.

8. Considering the sample as with replacement and following Example 2.2, (a) 2460 ± 175; (b) \$47 880 ± 1215; (c) \$18 143 ± 1164; (d) 3.73 ± 0.2657; (e) \$27.49 ± \$1.76; (f) \$7.38 ± \$0.534; (g) \$19.46 ± \$1.476; (h) 0.3789 ± 0.0758.

Chapter 3 (page 51)

1. $y_R^* = 962\,096 \pm 14\,218$, CV 1.48%; $y_0^* = 943\,609 \pm 19\,188$, CV 2.03%.

2. 961 348 ± 12 012, CV 1.25%.

3. 959 620 ± 14 081, CV 1.47%.

4. 957 579 ± 11 349, CV 1.19%.

Chapter 4 (page 58)

1. Following the method of Example 4.1, the estimated total area under wheat is 9855 ± 238 acres, CV 2.4%.

Chapter 5 (page 73)

1 Using the methods of section 5.3, $y_0^* = 351\,664$ acres, CV 3.84%.

2. Using the methods of section 5.5, $\hat{p} = 0.2045 \pm$ ± 0.011 085, $y_0^* = 1178 \pm 63.84$. For details of computations, see the reference.

3. For (a), use the methods of section 5.3, and for (b), section 5.7. (a) 19 943 ± 1242 acres; CV 6.23% (b) 19 453 ± 946 acres, CV 4.86%.

Chapter 6 (page 80)

1. Following the method of note 4 in section 6.4 and taking M_0 = number of persons per *kraal* = 3427/43 = 79.7, then the estimated intraclass correlation coefficient is 0.085.

2. Relative cost efficiencies 100, 94, 86, 92; optimum size is 1 ft row.

Chapter 7 (page 91)

1. Using equation (7.8), and noting that the permissible margin of error $d = 0.1\ P$, where P is the universe proportion, the number of wells is 44 for $P = 0.9$ to 400 for $P = 0.5$. For details, see the references.

2. See exercise 7, Chapter 1, using $P = 0.5$, $n = 400$ persons.

3. In Example 2.2, the CV of the estimator obtained from 20 sample households was 0.0863. Using equation (7.22), the required sample size is 60 households.

4. Assuming normality and using Table 7.2, Table 5, the estimated s.d. is 18, whereas the true s.d. is 16.1.

Chapter 10 (page 120)

1. As students are often asked to analyse the data of a stratified srs in the above form, the required computations are given in a table. The estimated standard errors of the stratum means will be obtained on dividing column (14) by N_h.

2. For (a) use equations (10.5(d)) and (10.15); $y = 1\,353\,572$ households, CV 9.12%. For (b) use equations (10.55) and (10.15); gain 407%.

3. From equation (10.51), average household size = = 3022/598.8 = 4.95; from equation (10.52), its standard error is 0.1306 and CV 2.64%. Note that in this case, the use of separate ratio estimates for the estimate of a ratio has not led to an improvement.

4. For convenience and checking, a table is provided with additional columns (4)–(7); y_{hi} is the wheat acreage of the ith farm ($i = 1, 2, \ldots n_h$) in the hth stratum ($h = 1, 2, \ldots, 6$), r_h is the number and p_h the proportion of farms growing wheat in the hth stratum.

* The sign ± after an estimate indicates the estimated standard error of the estimate.

Ch. 10 Ans. Table 1

Stratum	$\bar{y}_h = \sum_{i=1}^{n_h} \frac{y_{hi}}{n_h}$	y^*_{ho} $N_h\bar{y}_h$	$\left(\sum_{i=1}^{n_h} y_{hi}\right)^2$	$\frac{\left(\sum_{i=1}^{n_h} y_{hi}\right)^2}{n_h}$	SSy_{hi}	$n_h(n_h-1)$	$\frac{SSy_{hi}}{n_h(n_h-1)}$	$s^2_{y^*_{ho}} = N_h^2$.col.(12)	$s^2_{y^*_{ho}}$	CV of y^*_{ho} (%)
(1)	(6)	(7)	(8)	(9)	(10)	(11)	(12)	(13)	(14)	(15)
I	4.05	2572	383 161	2504	3075	23 256	0.132210	53 310	231	9.0
II	10.31	5878	2 024 929	14 673	9580	18 906	0.506696	164 626	406	6.9
III	15.29	7263	3 090 564	26 874	7208	13 110	0.549773	124 043	352	4.8
IV	23.16	7017	2 859 481	39 171	12 248	5256	2.330295	213 942	462	6.6
V	28.71	2565	363 609	17 315	990	420	2.357833	18 676	137	5.3
All strata	12.21★ (y/N)	25 293 (y)						574 597 (s_y^2)	758★ s_y	3.0★ CV_y (%)

★ Not additive.

Ch. 10 Ans. Table 2

Stratum (acres)	*No. of farms in stratum* N_h	*No. of farms sampled* n_h	$\sum_{i=1}^{n_h} y_{hi}$	$\sum_{i=1}^{n_h} y_{hi}^2$	r_h	$p_h = r_h/n_h$
1–5	435	25	0	0	0	0
6–20	519	26	3	9	1	0.0385
21–50	357	16	6	36	1	0.0625
51–150	519	17	159	3969	8	0.4706
151–300	400	26	762	38 510	20	0.7692
300	266	15	1371	164 737	15	1
All strata	2496 (N)	125 (n)	2301	207 231	45 (h)	0.3600 (p)★

★ Not additive.

For (i), using the formulae of sections 2.9 and 2.12, the estimated total area of wheat = (2496 x 2301/125) = 45 946 ± 8142 acres, CV 17.7%. Estimated number of farms growing wheat = $N_p = N\sqrt{[p(1-p)/(n-1)]}$ = 109, CV 12.1%.

For (ii), using the formulae of section 10.7 (1), $y = \Sigma N_h \bar{y}_h$ = 41 106 ± 4444 acres, CV 10.8%; using the results of section 10.3, the estimated number of farms growing wheat = $\Sigma N_h p_h$ = 860, with estimated standard error $\sqrt{[\Sigma N_h^2 p_h(1-p_h)/(n_h-1)]}$ = 79, CV 10.2%. Note the gain in efficiency in stratification after sampling.

Chapter 11 (page 134)

1. Following the method of section 11.2, y = 38 016 ± 3028 acres.

Chapter 12 (page 142)

1. Use equations (12.6) and (12.27); the results are given in Table 12.6.
2. Using the Neyman allocation (equation (12.10)), with $V_h = P_h(1-P_h)$ and n = 2000, the allocations are 1222, 167 and 611 persons.

Chapter 15 (page 178)

1. (a) Using the methods of section 15.2.6 (b–i), note 2, the average weight of a tablet is 1.08 ± 0.035 g; (b) using the methods of section 15.4, note 3, the proportion of sub-standard tablets is 0.058 ± 0.020; (c) using the method of section 15.2.5, note 2, the ratio is 0.170 ± 0.0020.
2. (a) Using equations (15.13) and (15.14), the estimated total number of cattle in the area is 28 820 ± 3427, with CV 11.89%. (b) (i) Following the method of section 15.2.6 (b–ii), the average number of cattle per farm is, from equations (15.45) and (15.46), 12.4595 ± 1.0940, with CV 8.40%; (ii) following the methods of section 15.2.6 (b–iii), the average is 13.0508 ± 1.195, with CV 9.16%; and

(iii) following the methods of section 15.2.6 (b–i), the average is 13.9090 ± 1.654, with CV 11.89%.

3. Following the methods of section 15.2.3, the estimated total number of beetles is 24 594 ± 2706. For details of computation, see the reference in the exercise.

Chapter 16 (page 191)

1. Following the method of section 16.3.1, the estimated total number of cattle is 28 421 ± 2899, with CV 10.20%.

2. Following section 16.3.2, the estimated total number of cattle is 24 188, with CV 9.79%. For details of computation, see the reference.

Chapter 17 (page 200)

1. Using equation (17.11), $m_0 = 2.9$ or 3. From the cost function, $n = (C - c_0)/(c_1 + m_0 c_2) = 336$.

2. Following the same method as for Example 17.2, estimated variance of the total income = 3180 + 653 049 = 3 833 529.

3. (a) Following the same method as for Example 17.1, $\hat{V}_1 = 0.0788$, $\hat{V}_2 = 2.1374$, and the unbiased variance estimate of the mean = 0.000788 + + 0.0021374 = 0.0029254.
(b) From equation (17.3), expected variance = 0.0843766.

Chapter 20 (page 241)

1. Although the sample design is stratified three-stage srs with villages, fields and plots selected at different stages, we can use the notations of a stratified two-stage design as only one plot is selected from each field. Denote by y_{hij} the yield of the jth sample field in the ith sample village of the hth stratum ($h = 1, 2, \ldots, 10; i = 1, 2; j = 1, 2$). Then if $a_0 = \frac{1}{160}$ acre is the area of plot selected in each sample field, the unbiased estimator of the total yield of the hth stratum is $y^*_{h0} = \frac{1}{2}(y^*_{h1} + y^*_{h2})$ with an unbiased variance estimator $\frac{1}{4}(y^*_{h1} - y^*_{h2})^2$, where $y^*_{hi} = \frac{1}{2} A_h(y_{hi1} + y_{hi2})/a_0$, and A_h is the total area of the hth stratum. The estimated total yield is 27 716 ± 2187 metric tonnes (1 metric tonne = 2204.6 lb).

2. Following the same methods as for Example 20.1, the estimated average number of adults per household in the three zones combined is 2.72 ± 0.097.

3. See sections 20.2.10 (b) and 10.5. The ratio estimate of the average household size is y_{RS}/x_{RS} = 3017.67/598.78 = 5.04 ± 0.0306, with the estimated CV of 0.61%, a very considerable improvement over the CV of the ratio of the unbiased estimators, 3.22% in Example 10.2.

Chapter 21 (page 255)

1. Following the method of Example 21.1, the results are: (a) total expenditure on cereals, Rs. 18 370 605 ± 2 230 232; (b) per capita expenditure on cereals, Rs. 9.79 ± 1.498; (c) average household size, 5.73 ± 0.442.

2. Following Example 21.1, the estimated average number of adults per household is 2.67 ± 0.078.

Chapter 24 (page 277)

1. Using equations (24.26) and (24.29), $N^* = 1087 \pm 159$.

Index of Names and Organizations

Index of Subjects